Carbon-Based Materials AND COMPOSITES FOR WATER AND WASTEWATER TREATMENT

Carbon-Based Materials AND COMPOSITES FOR WATER AND WASTEWATER TREATMENT

Edited by

ABU HASSAN NORDIN
Universiti Teknologi MARA (UiTM), Malaysia

AHMAD ILYAS RUSHDAN
Universiti Teknologi Malaysia (UTM), Malaysia

LUQMAN CHUAH ABDULLAH
Universiti Putra Malaysia (UPM), Malaysia

ELSEVIER

Elsevier
Radarweg 29, PO Box 211, 1000 AE Amsterdam, Netherlands
125 London Wall, London EC2Y 5AS, United Kingdom
50 Hampshire Street, 5th Floor, Cambridge, MA 02139, United States

ISBN: 978-0-443-33681-2

For information on all Elsevier publications visit our website at https://www.elsevier.com/books-and-journals

Publisher: Matthew Deans
Acquisitions Editor: Glyn Jones
Editorial Project Manager: Prachi Mittal
Production Project Manager: Maria Shalini
Cover Designer: Mark Rogers

Typeset by TNQ Tech

Contents

Contributors

Luqman Chuah Abdullah
Department of Chemical and Environmental Engineering, Faculty of Engineering, Universiti Putra Malaysia, Serdang, Malaysia

Norfarhana Abdul Samad
Centre of Chemical Technology, Department of Petrochemical Engineering, Politeknik Tun Syed Nasir Syed Ismail, Pagoh Education Hub, Johor, Malaysia

Kenit Acharya
Department of Chemistry, Institute of Science, Humanities and Liberal Studies, Indus University, Ahmedabad, Gujarat, India

Augustine Agi
Faculty of Chemical and Process Engineering Technology, College of Engineering Technology, Universiti Malaysia Pahang, Gambang, Pahang, Malaysia

Muhammad Abbas Ahmad Zaini
Centre of Lipid Engineering and Applied Research (CLEAR), Ibnu Sina Institute for Scientific and Industrial Research, Universiti Teknologi Malaysia, Skudai, Johor, Malaysia; Department of Chemical Engineering, Faculty of Chemical and Energy Engineering, Universiti Teknologi Malaysia, Skudai, Johor, Malaysia

Muhammad Syabbatul Al Hamdi Mustaffa AlBakri
Bioresource Technology Division, School of Industrial Technology, Universiti Sains Malaysia, Gelugor, Penang, Malaysia; Green Biopolymer, Coatings & Packaging Cluster, School of Industrial Technology, Universiti Sains Malaysia, Gelugor, Penang, Malaysia

Ikechukwu Ogadimma Alisi
Department of Industrial Chemistry, Faculty of Physical Sciences, Federal University Dutsin-Ma (FUDMA), Dutsin-Ma, Katsina State, Nigeria

Muhammad Aliyu
Department of Chemical Technology, Faculty of Science, Chulalongkorn University, Bangkok, Thailand; Center of Excellence in Catalysis for Bioenergy and Renewable Chemicals (CBRC), Faculty of Science, Chulalongkorn University, Bangkok, Thailand

Emin Bayraktar
ISAE-Supmeca, School of Mechanical and Manufacturing Engineering, Saint-Ouen, France

I. Budiman
Research Center for Biomass and Bioproducts, National Research and Innovation Agency (BRIN), Cibinong, Indonesia

Vishwajit Chavda
Department of Chemistry, Institute of Science, Humanities and Liberal Studies, Indus University, Ahmedabad, Gujarat, India; Department of Environmental Engineering, Kyungpook National University, Daegu, South Korea

Faten Ermala Che Othman
Department of Chemical and Biomolecular Engineering, National University of Singapore, Singapore

N. Chitraningrum
Research Center for Biomass and Bioproducts, National Research and Innovation Agency (BRIN), Cibinong, Indonesia

S.F. Mohd Din
Faculty of Chemical and Energy Engineering, Universiti Teknologi Malaysia (UTM), Skudai, Johor, Malaysia

Anees Ameera Fauzi
Research Center for Biomass and Bioproducts, National Research and Innovation Agency (BRIN), Cibinong, Indonesia; Faculty of Chemical and Energy Engineeing, Universiti Teknologi Malaysia, Johor, Malaysia

Mansi Gandhi
Department of Chemistry, Gujarat Technological University, Ahmedabad, Gujarat, India

Hafeez Adesina Gazali
Department of Industrial Chemistry, Faculty of Physical Sciences, Federal University Dutsin-Ma (FUDMA), Dutsin-Ma, Katsina State, Nigeria

Asif Hafeez
Department of Materials, National Textile University, Faisalabad, Pakistan

Nurul Sahida Binti Hassan
Centre of Hydrogen Energy, Institute of Future Energy, Johor, Malaysia

Balkis Hazmi
Institute of Nanoscience and Nanotechnology (ION2), Universiti Putra Malaysia, Serdang, Selangor, Malaysia

Y.M.N.S. Ismail
Faculty of Chemical and Energy Engineering, Universiti Teknologi Malaysia (UTM), Skudai, Johor, Malaysia; Department of Environment Pulau Pinang, Aras Bawah, Zon B, Wisma Persekutuan, Seberang Perai Utara, Pulau Pinang, Malaysia

Adikwu Gowon Jacob
Department of Industrial Chemistry, Faculty of Physical Sciences, Federal University Dutsin-Ma (FUDMA), Dutsin-Ma, Katsina State, Nigeria

Pallavi Jain
Department of Chemistry, SRM Institute of Science and Technology, Delhi NCR Campus, Modinagar, Uttar Pradesh, India

Rawesh Kumar
Chemistry Department, Patna University, Patna, Bihar, India

C.A. Luqman
Laboratory of Biocomposite, Institute of Tropical Forestry and Forest Products (INTROP), Universiti Putra Malaysia, Serdang, Selangor, Malaysia; Advanced Engineering Materials and Composites Research Centre (AEMC), Department of Mechanical and Manufacturing Engineering, Faculty of Engineering, Universiti Putra Malaysia, UPM Serdang, Selangor Darul Ehsan, Malaysia

Mohamad Haafiz Mohamad Kassim
Bioresource Technology Division, School of Industrial Technology, Universiti Sains Malaysia, Gelugor, Penang, Malaysia; Green Biopolymer, Coatings & Packaging Cluster, School of Industrial Technology, Universiti Sains Malaysia, Gelugor, Penang, Malaysia; Research Collaboration Center for Nanocellulose, BRIN - Andalas University, Padang, Indonesia

Mohamad Faisal Mohamad Zauqi
Bioresource Technology Division, School of Industrial Technology, Universiti Sains Malaysia, Gelugor, Penang, Malaysia; Green Biopolymer, Coatings & Packaging Cluster, School of Industrial Technology, Universiti Sains Malaysia, Gelugor, Penang, Malaysia

N.B. Mohamed
Faculty of Chemical and Energy Engineering, Universiti Teknologi Malaysia (UTM), Skudai, Johor, Malaysia

Nurul Fazita Mohammed Rawi
Bioresource Technology Division, School of Industrial Technology, Universiti Sains Malaysia, Gelugor, Penang, Malaysia; Green Biopolymer, Coatings & Packaging Cluster, School of Industrial Technology, Universiti Sains Malaysia, Gelugor, Penang, Malaysia; Research Collaboration Center for Nanocellulose, BRIN - Andalas University, Padang, Indonesia

Azrul Nurfaiz Mohd Faizal
Institute of Ecosystem Science Borneo (IEB), Universiti Putra Malaysia Sarawak, Bintulu, Sarawak, Malaysia; Centre of Lipid Engineering and Applied Research (CLEAR), Ibnu Sina Institute for Scientific and Industrial Research, Universiti Teknologi Malaysia, Skudai, Johor, Malaysia

Muhammad Asyraf Muhammad Rizal
Faculty of Mechanical Engineering, Universiti Teknologi Malaysia, Johor Bahru, Johor, Malaysia; Centre for Advanced Composite Materials (CACM), Universiti Teknologi Malaysia, Johor Bahru, Johor, Malaysia

M.H. Mohamed Noor
Faculty of Chemical and Energy Engineering, Universiti Teknologi Malaysia (UTM), Skudai, Johor, Malaysia

Abu Hassan Nordin
Faculty of Applied Sciences, Universiti Teknologi MARA (UiTM), Arau Campus, Perlis, Malaysia

Mohd Nurazzi Norizan
Bioresource Technology Division, School of Industrial Technology, Universiti Sains Malaysia, Gelugor, Penang, Malaysia; Green Biopolymer, Coatings & Packaging Cluster, School of Industrial Technology, Universiti Sains Malaysia, Gelugor, Penang, Malaysia; Research Collaboration Center for Nanocellulose, BRIN - Andalas University, Padang, Indonesia

Siti Muhamad Nur Husna
Faculty of Pharmacy and Biomedical Sciences, MAHSA University, Jenjarom, Selangor, Malaysia; Department of Primary Care Medicine, Faculty of Medicine, Universiti Malaya, Wilayah Persekutuan Kuala Lumpur, Malaysia

Abiodun Busuyi Ogbesejana
Department of Industrial Chemistry, Faculty of Physical Sciences, Federal University Dutsin-Ma (FUDMA), Dutsin-Ma, Katsina State, Nigeria

Nicky Rahmana Putra
Faculty of Engineering Technology and Science, Higher College of Technology, Ruwais, United Arab Emirates

Sapna Raghav
Department of Chemistry, Shri Jagdishprasad Jhabarmal Tibbrewala (JJT) University, Jhunjhunu, Rajasthan, India

Hilmi Abdul Rahman
Chemical Management Unit, University Laboratory Management Centre, Universiti Teknologi Malaysia, Skudai, Johor, Malaysia

Yuvrajsinh B. Rajput
Department of Chemistry, Institute of Science, Humanities and Liberal Studies, Indus University, Ahmedabad, Gujarat, India

Umer Rashid
Department of Chemical Technology, Faculty of Science, Chulalongkorn University, Bangkok, Thailand; Center of Excellence in Catalysis for Bioenergy and Renewable Chemicals (CBRC), Faculty of Science, Chulalongkorn University, Bangkok, Thailand; Faculty of Applied Sciences, UCSI University, Cheras, Kuala Lumpur, Malaysia

Ahmad Ilyas Rushdan
Department of Chemical Engineering, Faculty of Chemical and Energy Engineering, Universiti Teknologi Malaysia (UTM), Skudai, Johor, Malaysia; Centre for Advanced Composite Materials (CACM), Universiti Teknologi Malaysia, Johor Bahru, Johor, Malaysia

Khubab Shaker
Department of Materials, National Textile University, Faisalabad, Pakistan

Novia Amalia Sholeha
College of Vocational Studies, Bogor Agricultural University (IPB University), Bogor, Indonesia

Prashant Singh
Department of Chemistry, Atma Ram Sanatan Dharma College, University of Delhi, New Delhi, India

Uduma Anya Uduma
Department of Industrial Chemistry, Faculty of Physical Sciences, Federal University Dutsin-Ma (FUDMA), Dutsin-Ma, Katsina State, Nigeria

Abdul Wafi
Research Center for Food Technology and Processing, National Research and Innovation Agency (BRIN), Playen, Gunungkidul, Yogyakarta, Indonesia

Sandeep Yadav
Department of Chemistry, SRM Institute of Science and Technology, Delhi NCR Campus, Modinagar, Uttar Pradesh, India

Ahmad Syahmi Zaini
Centre of Lipid Engineering and Applied Research (CLEAR), Ibnu Sina Institute for Scientific and Industrial Research, Universiti Teknologi Malaysia, Skudai, Johor, Malaysia

CHAPTER 1

Emerging trends of carbon-based materials and composites in water and wastewater treatment applications: A bibliometric analysis

Abu Hassan Nordin[1], Ahmad Ilyas Rushdan[2] and Luqman Chuah Abdullah[3]
[1]Faculty of Applied Sciences, Universiti Teknologi MARA (UiTM), Arau Campus, Perlis, Malaysia; [2]Faculty of Chemical and Energy Engineering, Universiti Teknologi Malaysia (UTM), Skudai, Johor, Malaysia; [3]Department of Chemical and Environmental Engineering, Faculty of Engineering, Universiti Putra Malaysia, Serdang, Malaysia

1. Introduction

The rapid expansion of industrial activities, urbanization, and population growth has intensified pressure on global water resources, leading to the continuous release of complex contaminants into aquatic environments. Conventional wastewater treatment systems, originally designed to remove bulk organic matter and nutrients, are increasingly challenged by the presence of emerging contaminants, including pharmaceuticals, personal care products, endocrine-disrupting compounds, pesticides, and industrial additives [1,2]. These pollutants are often detected at trace concentrations but exhibit high persistence, bioaccumulation potential, and ecotoxicological risks, raising serious concerns regarding water security, ecosystem integrity, and public health.

In response to these challenges, advanced water and wastewater treatment technologies have been actively explored, among which carbon-based materials and composites have attracted substantial scientific and technological interest. Materials such as activated carbon, biochar, graphene-based structures, carbon nanotubes, and hybrid carbon composites offer distinctive advantages, including high specific surface area,

Carbon-Based Materials and Composites for Water and Wastewater Treatment
ISBN 978-0-443-33681-2
https://doi.org/10.1016/B978-0-443-33681-2.00011-9

tunable porosity, diverse surface functionalities, chemical stability, and strong affinity toward a wide range of contaminants [2,3]. Recent advances in material synthesis and functionalization have further enhanced the performance of carbon-based systems, enabling selective adsorption, catalytic degradation, and integration with hybrid treatment processes [4–6]. As a result, carbon-based materials have evolved from conventional adsorbents into multifunctional platforms for water and wastewater treatment applications.

Despite the rapid growth of research in this area, the expanding body of literature remains fragmented across multiple disciplines, material classes, and application domains. Existing review studies predominantly focus on specific materials or treatment mechanisms, often overlooking broader research patterns, knowledge structures, and developmental trajectories within the field. Consequently, there is a lack of systematic understanding regarding how research on carbon-based materials and composites has evolved over time, which materials and applications dominate the field, and where emerging research frontiers and knowledge gaps are located.

Bibliometric analysis provides a powerful and objective approach to address these limitations by quantitatively mapping publication trends, citation dynamics, influential journals, leading authors, institutional collaborations, and thematic evolution within a research domain. By applying bibliometric tools, it becomes possible to identify research hotspots, assess scientific impact, and uncover emerging directions that may guide future innovation and policy-relevant research.

In this context, the present study conducts a comprehensive bibliometric analysis of research on carbon-based materials and composites for water and wastewater treatment applications. Specifically, this work aims to (i) analyze publication and citation trends over time, (ii) identify the most influential journals, and countries, (iii) examine collaborative networks and geographical research distribution, and (iv) elucidate emerging themes and research frontiers within the field. By providing a structured and data-driven overview of the research landscape, this study offers valuable insights for researchers, engineers, and policymakers seeking to advance the development and application of carbon-based technologies for sustainable water and wastewater treatment.

2. Methodology

The Web of Science (WoS) database represents a comprehensive citation indexing platform widely used for academic and scientific investigations. It curates more than 20,000 rigorously selected journals across a broad range of fields, including engineering, natural sciences, social sciences, and the humanities. The establishment of WoS has substantially enhanced researchers' ability to examine the evolution of scholarly topics at multiple analytical scales, namely the macro level (research disciplines), meso level (journals and institutions), and micro level (individual authors) [7]. In general, publications appearing in WoS-indexed journals are systematically cataloged, allowing scholars to track knowledge development across temporal, geographical, institutional, and disciplinary dimensions [8]. On December 30, 2025, an advanced search was made in WoS using the following search string: "TS = ((Carbon) OR (("emerging contaminant*" OR "emerging pollutant") AND (removal OR remediation OR treatment)))", with a time span from 2010 to 2025. To maximize retrieval coverage, wildcard symbols were appended to selected keywords so that all variants sharing common word stems were captured in the search output [9]. The initial search returned 839 records; following a rigorous relevance screening process, 704 publications, including journal articles, conference proceedings, and book chapters were retained for in-depth analysis. The inclusion criteria were limited to studies that explicitly investigated carbon-based technologies within water and wastewater treatment applications. Publications focusing exclusively on removal of emerging contaminants in water and wastewater treatment system using carbon-based technologies. This systematic filtering process ensured that only publications fully aligned with the defined scope of emerging trends of carbon-based materials and composite technologies for water and wastewater treatment applications were included in the present analysis.

3. Result and discussion

3.1 General publication trend

The annual publication output and citation frequency related to the removal of emerging contaminants from water and wastewater treatment systems using carbon-based technologies between 2010 and 2025 is illustrated in Fig. 1.1. It can be observed that the number of publications increased gradually from 15 articles in 2010 to about 50 publications by 2014, indicating an initial exploratory phase of research activity. A steady increase is observed from 2015 to 2018 where annual publications rose from around 85 to nearly 160, followed by a pronounced escalation after 2019, with publication counts exceeding 296 in 2020 and reaching more

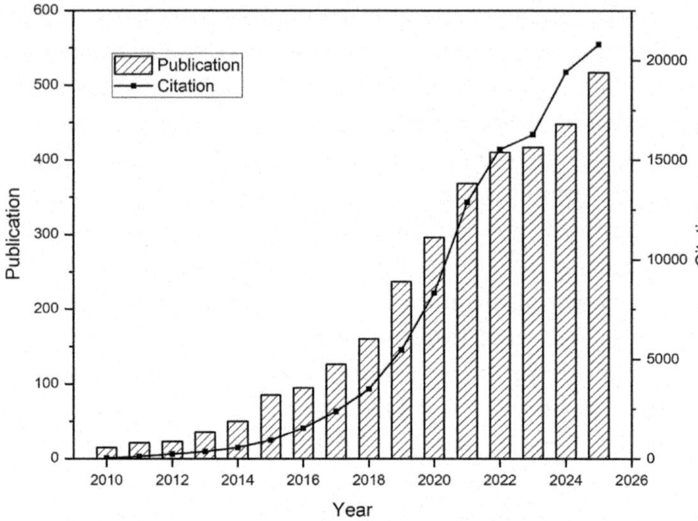

Figure 1.1 Annual number of studies on carbon-based materials and composites in water and wastewater treatment applications from 2010 to 2025.

than 517 by 2025. In parallel, citation counts followed a more accelerated growth pattern, remaining relatively low before 2014 but increasing sharply thereafter, surpassing 5000 citations by 2019 and exceeding 20,000 citations by 2025, reflecting the rising scientific influence of this research area. The observed increase in publication output and citation frequency can be attributed to the growing recognition of emerging contaminants, such as pharmaceuticals, antibiotics, endocrine-disrupting compounds, and industrial additives, as persistent and hazardous pollutants in aquatic environments [10].

Several studies have reported that the increasing detection of these contaminants at trace concentrations, enabled by advances in analytical techniques, has significantly expanded research interest in advanced treatment technologies [11,12]. Furthermore, the sharp rise observed after 2018 coincides with the rapid development of advanced carbon-based materials, including biochar, graphene and its derivatives, carbon nanotubes, and chemically or magnetically functionalized carbon composites, which demonstrate enhanced adsorption capacity, tunable surface chemistry, and improved regeneration performance compared to conventional activated carbon [13–15]. Recent reviews have further highlighted that citation growth is strongly associated with the emergence of engineered and sustainable carbon materials targeting emerging contaminants [16,17].

This consistent trend confirms that the research field has transitioned from fundamental adsorption studies toward a mature, high-impact, and application-oriented phase. Overall, the strong alignment between publication output and citation growth underscores the establishment of a robust knowledge base and highlights carbon-based technologies as a dominant and continuously evolving platform for emerging contaminant removal in water and wastewater treatment systems.

Table 1.1 presents the major funding agencies supporting research on carbon-based materials and composites for water and wastewater treatment applications over the period 2010–25. As shown in the table, the National Natural Science Foundation of China (NSFC) is the most prominent funding body, supporting 584 research projects, which is substantially higher than other agencies. This is followed by the Spanish Government with 211 projects, the National Council for Scientific and Technological Development (CNPq), Brazil with 202 projects, the National Polytechnic Institute of Mexico with 192 projects, and the Brazilian Federal Agency for Support and Evaluation of Graduate Education (CAPES) with 181 projects. The funding periods indicate long-term and continuous investment, particularly in China, Spain, and Brazil, highlighting sustained national commitment to research in carbon-based materials for water and wastewater treatment.

This funding landscape provides important context for interpreting the annual publication and citation trends. The steady and long-term financial support from major national agencies closely corresponds with the progressive increase in research output observed after 2010. In particular, the dominance of NSFC funding aligns with the transition from a gradual

Table 1.1 Top funding agencies for carbon-based materials and composites in water and wastewater treatment applications research.

Funding agencies	Year	No. of funded project
National natural science foundation of China	2010–25	584
Spanish government	2010–25	211
National council for scientific and technological development, Brazil	2013–25	202
National polytechnic institute, Mexico	2013–25	192
Brazilian federal agency for support and evaluation of graduate education	2013–25	181

growth phase in publications between 2010 and 2014 to a more accelerated expansion after 2015. The intensified funding support appears to have facilitated expanded research capacity, advanced material synthesis, and broader international dissemination, contributing to the sharp rise in annual publications after 2019.

The influence of sustained funding is even more evident when examining citation trends. While publication growth reflects increasing research activity, the exponential rise in citations after 2018 suggests that funding-backed studies have achieved substantial scientific impact. This pattern indicates that long-term investment has not only increased the quantity of research output but has also enhanced its quality, visibility, and relevance. The strong citation performance is consistent with the emergence of advanced carbon-based materials, such as biochar, graphene-based composites, carbon nanotubes, and functionalized carbon systems, which have attracted significant attention due to their superior adsorption performance and adaptability for emerging contaminant removal. Furthermore, the contributions from Spain, Brazil, and Mexico correspond with the formation of influential regional research clusters, particularly in biomass-derived carbon materials, adsorption mechanisms, and sustainable composite development. This suggests that national funding priorities have played a decisive role in shaping both the thematic focus and the geographical distribution of high-impact research [18].

3.2 Journal analysis

Table 1.2 presents the top five most productive journals publishing research on carbon-based materials and composites for water and wastewater treatment applications during the period 2010–25. As shown in the table, Chemical Engineering Journal is the leading outlet, contributing 212 publications, followed by the Journal of Environmental Chemical Engineering with 176 publications, Journal of Hazardous Materials with 125 publications, Science of the Total Environment with 123 publications, and Journal of Water Process Engineering with 120 publications. Collectively, these five journals account for a cumulative total of 756 publications, indicating a strong concentration of scholarly output within a limited number of high-impact journals.

The dominance of Chemical Engineering Journal can be attributed to its strong interdisciplinary scope encompassing chemical engineering and environmental engineering, as well as its emphasis on advanced materials,

Table 1.2 Top five productive journals on carbon-based materials and composites in water and wastewater treatment applications research in 2010–25.

Journal	Impact factor	Category	Total percentage (%)	Total cumulative percentage (%)
Chemical engineering journal	13.2	Engineering, chemical (Q1/8/176) Engineering, environmental (Q1/3/83)	212	212
Journal of environmental chemical engineering	7.2	Engineering, chemical (Q1/25/176) Engineering, environmental (Q1/17/83)	176	388
Journal of hazardous materials	11.3	Engineering, environmental (Q1/5/83) Environmental sciences (Q1/19/376)	125	513
Science of the total environment	8.0	Environmental sciences (Q1/39/376)	123	636
Journal of water process engineering	6.7	Engineering, chemical (Q1/28/176) Engineering, environmental (Q1/21/83) Water resources (Q1/9/132)	120	756

adsorption processes, and scalable treatment technologies. Similarly, the high productivity of the Journal of Environmental Chemical Engineering reflects the growing integration of material science with environmental applications, particularly in adsorption, catalysis, and hybrid treatment systems. The consistent presence of Journal of Hazardous Materials and Science of the Total Environment highlights the importance of carbon-based materials in addressing pollutant toxicity, environmental risk, and contaminant fate, while Journal of Water Process Engineering reflects increasing interest in process-oriented and application-driven research.

From a quality perspective, all five journals are classified within Q1 categories across engineering, environmental sciences, and water resources, with impact factors ranging from 6.7 to 13.2. This concentration of publications in top-quartile journals indicates that research on carbon-based materials and composites for water and wastewater treatment has achieved a high level of scientific rigor, relevance, and visibility. The strong cumulative publication percentage further suggests that authors preferentially target journals with broad readership, high citation potential, and strong alignment with emerging contaminant remediation and advanced material development [19]. Overall, the dominance of established Q1 journals underscores the recognition of carbon-based materials and composites as a central research theme in water and wastewater treatment, while also indicating that future advancements in this area are likely to continue being disseminated through these leading publication platforms.

3.3 Contribution and collaboration between countries

Table 1.3 presents the leading countries contributing to research on carbon-based materials and composites for water and wastewater treatment applications from 2010 to 2025. s reported in the table, China is the most prolific contributor, with a total of 901 publications, followed by Spain (427 publications), Brazil (354 publications), India (322 publications), and the United States (233 publications). This distribution reflects a pronounced geographical concentration of research output, with China contributing more than twice the publication volume of the second-ranked country.

The predominance of China can be directly associated with its sustained and large-scale national research funding, particularly through the NSFC, which is identified as the leading funding agency in this field. Continuous

Table 1.3 Top countries focus on carbon-based materials and composites in water and wastewater treatment applications research 2010–25.

Country	Total publication
China	901
Spain	427
Brazil	354
India	322
USA	233

financial support over an extended period has enabled the establishment of large research networks, advanced laboratory infrastructure, and long-term research programs, thereby translating funding intensity into high research productivity and output. Similarly, the substantial contributions from Spain and Brazil align with the strong presence of national funding bodies such as the Spanish Government, CNPq, and CAPES, which have consistently prioritized research on environmental remediation, advanced materials, and water treatment technologies.

Although India does not appear among the top funding agencies listed in Table 1.1, its significant publication output suggests a research landscape driven by application-oriented and cost-effective approaches [20], particularly in the development of biomass-derived carbon materials and adsorption-based treatment systems suited for large-scale and decentralized wastewater treatment. In contrast, the comparatively lower publication output from the United States may reflect a more diversified research portfolio, where carbon-based materials constitute one of several parallel technological pathways for advanced water and wastewater treatment.

When considered alongside the funding distribution and the annual publication and citation trends, the country-level analysis presented in Table 1.3 reveals a strong correspondence between sustained national funding support and research productivity. Countries with long-term, targeted investment strategies—most notably China, Spain, and Brazil—also exhibit higher publication output and increasing scientific impact. Collectively, these findings underscore the pivotal role of national funding frameworks in shaping global research leadership, thematic focus, and scientific influence in the development and application of carbon-based materials and composites for water and wastewater treatment.

Figure 1.2 illustrates the country collaboration network in research on carbon-based materials and composites for water and wastewater treatment

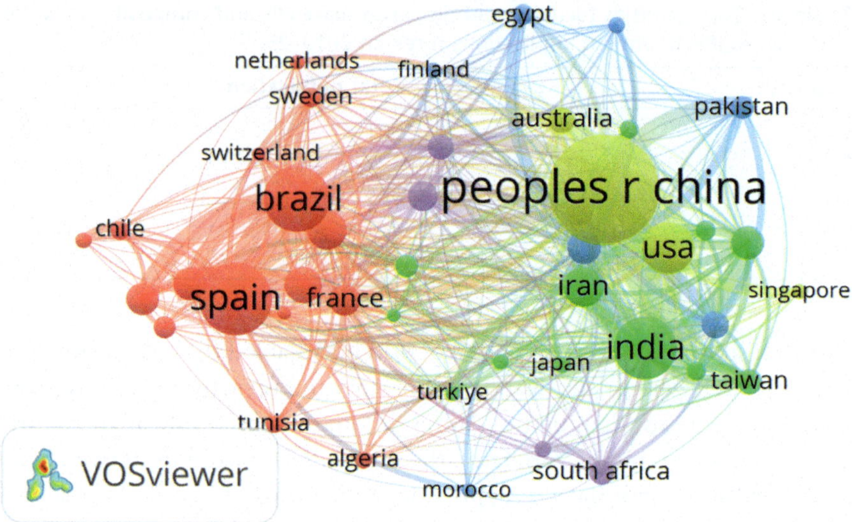

Figure 1.2 Network of keywords pertaining to carbon-based materials and composites in water and wastewater treatment applications research 2010–25 (minimum number of occurrences: 20, 44 out of 112 keywords meet the threshold).

applications from 2010 to 2025, generated using VOSviewer with a minimum occurrence threshold of 20. As shown in the figure, the network is characterized by several well-defined and densely interconnected clusters, indicating strong international collaboration patterns. Among these, China appears as the most dominant and central node, exhibiting extensive collaborative links with multiple countries, including the United States, India, Australia, and Pakistan. The large node size and high link density associated with China indicate both high publication volume and strong international research integration.

A second prominent collaboration cluster is centered around Spain and Brazil, which are closely linked to other European and Latin American countries such as France, Switzerland, the Netherlands, and Chile. This cluster reflects strong regional cooperation and cross-continental collaboration, particularly in areas related to adsorption mechanisms, biomass-derived carbons, and sustainable composite materials. The presence of multiple medium-to-large nodes within this cluster suggests a balanced and collaborative research ecosystem rather than dominance by a single country.

The observed collaboration structure can be explained by sustained national funding strategies and shared research priorities. China's central role aligns closely with its dominant funding support through the NSFC, which has enabled large-scale projects and international partnerships. Similarly, the strong Spain–Brazil collaboration corresponds with consistent funding from national agencies such as the Spanish Government, CNPq, and CAPES, facilitating joint research initiatives and researcher mobility. Countries such as India and the United States, while positioned slightly more peripherally, maintain strong bilateral links with China and European partners, indicating strategic collaboration in advanced material development and application-driven studies. Countries with sustained public investment not only lead in publication volume but also occupy central positions within the global collaboration network, thereby amplifying their scientific visibility and citation impact. This pattern confirms that international collaboration has played a critical role in accelerating knowledge diffusion and advancing high-impact research on carbon-based materials for water and wastewater treatment.

3.4 Keywords

Figure 1.3 presents the keyword co-occurrence network for research on carbon-based materials and composites in water and wastewater treatment applications from 2010 to 2025, generated using a minimum occurrence threshold of 20. The network is dominated by the central keyword "removal", which exhibits strong connections with multiple clusters, highlighting its role as the core objective of carbon-based water and wastewater treatment research. Surrounding this central node, three major thematic clusters can be identified. The blue cluster is primarily associated with adsorption-based processes, characterized by high-frequency keywords such as adsorption, activated carbon, sorption, water, equilibrium, and dyes. This cluster reflects the foundational role of adsorption mechanisms and activated carbon materials in early and ongoing studies, particularly for the removal of organic pollutants and inorganic pollutants from aqueous systems.

The green cluster is centered on degradation and performance-oriented approaches, with prominent keywords including degradation, performance, biochar, composite, catalyst, activation, and advanced oxidation processes. The presence of terms such as visible light, photo-Fenton, hydroxyl radicals, and ozone indicates increasing integration of carbon-based materials with catalytic and oxidative treatment pathways. This cluster signifies a shift toward multifunctional materials that not only adsorb

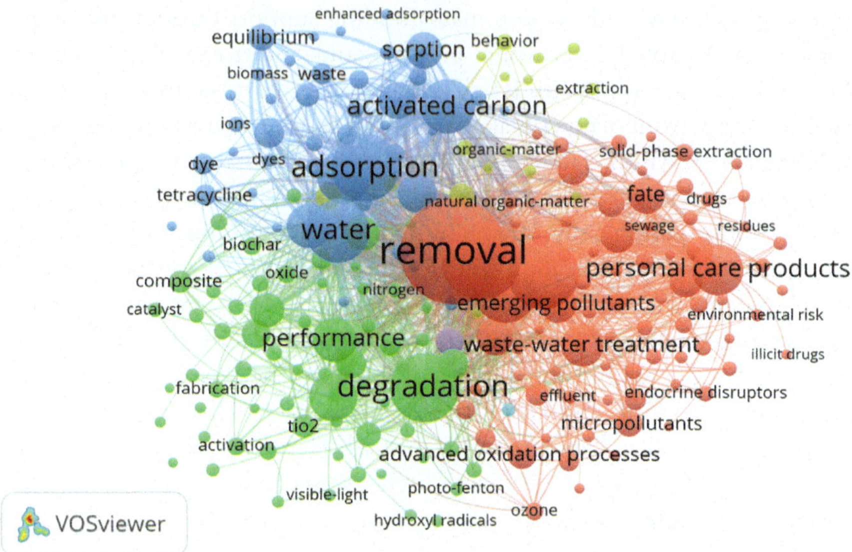

Figure 1.3 Network of keywords pertaining to carbon-based materials and composites in water and wastewater treatment applications research 2010–25 (minimum number of occurrences: 20, 25 out of 4465 keywords meet the threshold).

contaminants but also promote their chemical transformation and mineralization. The red cluster focuses on emerging contaminants and environmental risk, with dominant keywords such as emerging pollutants, personal care products, endocrine disruptors, micropollutants, wastewater treatment, and environmental risk. The strong interconnections within this cluster reflect growing concern over trace-level contaminants and their long-term ecological and health impacts. The linkage between this cluster and both adsorption- and degradation-related clusters suggests that research has increasingly targeted real-world wastewater matrices and complex contaminant mixtures rather than model pollutants alone.

Overall, the keyword network demonstrates the thematic evolution of the field from conventional adsorption studies toward more integrated and application-oriented treatment strategies. The strong connectivity among adsorption, degradation, and emerging contaminant clusters highlights a transition toward hybrid systems that combine carbon-based materials with advanced oxidation and catalytic processes. This evolution reflects the maturation of the research field, where performance enhancement, selectivity, and treatment of complex wastewater streams have become central priorities.

4. Conclusion

In conclusion, this study provides a comprehensive bibliometric assessment of research on carbon-based materials and composites for water and wastewater treatment applications over the period 2010–25, offering a systematic and data-driven overview of the field's evolution, structure, and emerging directions. The analysis reveals a pronounced and sustained growth in both publication output and citation impact, particularly after 2015, reflecting the increasing global concern over emerging contaminants and the growing reliance on carbon-based technologies as effective remediation platforms. The strong alignment between funding intensity, publication productivity, and citation performance highlights the pivotal role of long-term national investment in shaping research leadership and scientific influence. Countries with sustained public funding support—-most notably China, Spain, and Brazil—emerged as dominant contributors, not only in terms of publication volume but also through their central positions within international collaboration networks. These findings underscore the importance of coordinated funding strategies in enabling large-scale research programs, fostering international collaboration, and accelerating knowledge diffusion.

Journal analysis further indicates that research dissemination is highly concentrated within a limited number of top-tier Q1 journals, such as Chemical Engineering Journal, Journal of Environmental Chemical Engineering, and Journal of Hazardous Materials. This concentration reflects both the scientific maturity of the field and the high relevance of carbon-based materials research to interdisciplinary audiences spanning chemical engineering, environmental science, and water resources. Keyword co-occurrence analysis reveals a clear thematic evolution from conventional adsorption studies toward more integrated and application-oriented approaches. While adsorption using activated carbon remains a foundational research theme, recent years have witnessed increasing emphasis on degradation-based mechanisms, advanced oxidation processes, and multifunctional carbon composites targeting complex mixtures of emerging contaminants. The strong interconnectivity between adsorption, degradation, and emerging pollutant clusters highlights a shift toward hybrid treatment strategies capable of addressing real-world wastewater challenges.

Overall, the findings confirm that carbon-based materials and composites have transitioned from traditional adsorbents to versatile and high-impact platforms for emerging contaminant removal in water and wastewater treatment systems. Despite significant progress, future research should prioritize material regeneration, cost-effectiveness, life-cycle assessment, and large-scale implementation to bridge the gap between laboratory studies and practical applications. Expanding international collaboration and broadening research participation from under-represented regions will also be essential to ensure that future advancements address diverse environmental contexts and global water treatment needs.

References

[1] Zachariah, S., Sasi, R., Suchithra, T., 2024. Potential effects and removal strategies for contaminants of emerging concern (CECs) in the circular economy of water. In: Circular economy. CRC Press, pp. 353–372.

[2] Singh, S., Sivaram, N., Dhanjal, D.S., Assefa, H., Singh, J., Ramamurthy, P.C., 2024. Navigating the complexity of emerging contaminants: Sources, impacts, and remediation strategies. J. Indian Inst. Sci. 104, 519–553.

[3] ben Mosbah, M., Mechi, L., Khiari, R., Moussaoui, Y., 2020. Current state of porous carbon for wastewater treatment. Processes 8, 1651.

[4] Ma, Q., Yu, Y., Sindoro, M., Fane, A.G., Wang, R., Zhang, H., 2017. Carbon-based functional materials derived from waste for water remediation and energy storage. Adv. Mater. 29, 1605361.

[5] Benzigar, M.R., Talapaneni, S.N., Joseph, S., Ramadass, K., Singh, G., Scaranto, J., Ravon, U., Al-Bahily, K., Vinu, A., 2018. Recent advances in functionalized micro and mesoporous carbon materials: Synthesis and applications. Chem. Soc. Rev. 47, 2680–2721.

[6] Pérez-Mayoral, E., Matos, I., Bernardo, M., Ventura, M., Fonseca, I.M., 2020. Carbon-based materials for the development of highly dispersed metal catalysts: Towards highly performant catalysts for fine chemical synthesis. Catalysts 10, 1407.

[7] Hall, C.M., 2011. Publish and perish? Bibliometric analysis, journal ranking and the assessment of research quality in tourism. Tour. Manag. 32, 16–27.

[8] Joseph, A.M., Snellings, R., Van den Heede, P., Matthys, S., De Belie, N., 2018. The use of municipal solid waste incineration ash in various building materials: A Belgian point of view. Materials 11, 141.

[9] Aleixandre-Tudo, J.L., Castelló-Cogollos, L., Aleixandre, J.L., Aleixandre-Benavent, R., 2019. Unravelling the scientific research on grape and wine phenolic compounds: A bibliometric study. Scientometrics 119, 119–147.

[10] Zaidi, N., Mir, M.A., Chang, S.K., Abdelli, N., Hasnain, S.M., Ali Khan, M.A., Andrews, K., 2025. Pharmaceuticals and personal care products as emerging contaminants: Environmental fate, detection, and mitigation strategies. Int. J. Environ. Anal. Chem. 1–29.

[11] Kumar, A., Kumar, A., Seema, K., Kumari, K., 2025. Detection and remediation techniques for emerging contaminants. In: Emerging contaminants in water and wastewater: Sources and substances. Springer, pp. 223–253.

[12] Boro, D., Chirania, M., Verma, A.K., Chettri, D., Verma, A.K., 2025. Comprehensive approaches to managing emerging contaminants in wastewater: Identification, sources, monitoring and remediation. Environ. Monit. Assess. 197, 1–29.

[13] Januário, E.F.D., Fachina, Y.J., Wernke, G., Demiti, G.M.M., Beltran, L.B., Bergamasco, R., Vieira, A.M.S., 2022. Application of activated carbon functionalized with graphene oxide for efficient removal of COVID-19 treatment-related pharmaceuticals from water. Chemosphere 289, 133213.
[14] Brito, C.H.V., Gloria, D.C.S., Santos, E.B., Domingues, R.A., Valente, G.T., Vieira, N.C.S., Gonçalves, M., 2023. Porous activated carbon/graphene oxide composite for efficient adsorption of pharmaceutical contaminants. Chem. Eng. Res. Des. 191, 387–400.
[15] Costa, H.P., Duarte, E.D., da Silva, F.V., da Silva, M.G., Vieira, M.G., 2024. Green synthesis of carbon nanotubes functionalized with iron nanoparticles and coffee husk biomass for efficient removal of losartan and diclofenac: Adsorption kinetics and ANN modeling studies. Environ. Res. 251, 118733.
[16] Balakumar, S., Mahesh, N., Kamaraj, M., Saranya, T., Babu, P.S., Aravind, J., Kim, W., Govarthanan, M., 2024. Customized carbon composite nanomaterials for the mitigation of emerging contaminants: A review of recent trends. Carbon Lett. 34, 1091–1114.
[17] Zia-ur-Rehman, M., Usman, M., Aslam, H., Khan, A.A., 2026. Carbon-based nanomaterials for removal of emerging contaminants. In: Advances in treatment methods towards emerging contaminants. Elsevier, pp. 447–460.
[18] Tijssen, R., Winnink, J., 2022. Global and local research excellence in Africa: New perspectives on performance assessment and funding. Sci. Technol. Soc. 27, 368–387.
[19] Tenopir, C., Dalton, E., Fish, A., Christian, L., Jones, M., Smith, M., 2016. What motivates authors of scholarly articles? The importance of journal attributes and potential audience on publication choice. Publications 4, 22.
[20] Karloopia, J.J., Agrawal, R., 2025. Navigating the landscape of design thinking: Evolution, current trends and future research perspectives. Bus. Process Manag. J. 1–26.

CHAPTER 2

Synthesis of carbon-based materials derived from synthetic and natural precursors

Faten Ermala Che Othman
Department of Chemical and Biomolecular Engineering, National University of Singapore, Singapore

1. Introduction

There are various technologies have been developed and applied rapidly in recent years to remove wide range of contaminants or pollutants from the polluted water and wastewater including adsorption, coagulation, advanced oxidation process (AOP), chemical precipitation, electrodialysis (ED), evaporation, flocculation, filtration with coagulation, ion exchange, membrane separation, ozonation, and reverse osmosis (RO). However, most of these methods have been found to be limited as they were often involving high capital and operational costs [1]. For instance, the method such as coagulation and flocculation are highly affected from the large volume of sludge produced. Otherwise, the downside of precipitation process is the generation of harmful sludge that resulting to additional treatment and lack of recyclability is the drawback of ion exchange method. On the other hand, despite of their successful method, ED requires high energy consumption and operating costs. Despite the drawbacks of those techniques, adsorption process by solid sorbents exhibits potential as one of the most efficient methods in the wastewater treatment due to its simple design and inexpensiveness in both initial costs and land requirements [2,3]. Meanwhile, membrane separation has been one of the promising and pervasive method that was in demand due to its numerous strengths such as ease operation, high separation selectivity, low energy demand, good stability, and requires no chemical additives.

Carbon-Based Materials and Composites for Water and Wastewater Treatment
ISBN 978-0-443-33681-2
https://doi.org/10.1016/B978-0-443-33681-2.00002-8

1.1 Adsorbents

Adsorption has become one of the most efficient and sustainable strategy to remediate the contaminants in wastewater in recent years due to its easy handling, simple operation, ease regeneration, low-cost, availability of different adsorbents, especially using of low-cost adsorbent materials from natural materials, agricultural wastes and industrial wastes [1,4,5]. In addition, this method does not cause secondary pollution from the generation of byproducts [6]. In general, adsorption is a surface phenomenon with common mechanism for organic and inorganic pollutants removal. The organic pollutants get sorbed onto the surface of adsorbents due to surface interaction process. In this context, high surface area and excellent pore size distribution are good characteristics for adsorbents to effectively reduce pollutants from water and wastewater. Adsorption can be categorized into two main mechanisms based on type of attraction between adsorbate and adsorbent; physical adsorption/physisorption and chemical adsorption/chemisorption. Physisorption occurs due to van der Waals force of attraction, hydrogen bonding, and dipole interaction, meanwhile chemisorption occurs due to the covalent bonding resulting to electrostatic attraction as shown in Fig. 2.1 [3].

1.2 Membranes

In the last couple of decades, membrane technology has emerged as a favorite choice due to its efficiency and adaptability it offers in water and wastewater treatment, such as high-quality filtration through selective removals, scalable, and small footprint that is space-efficient, low energy usage and reduced chemical dependency. There are various methods in

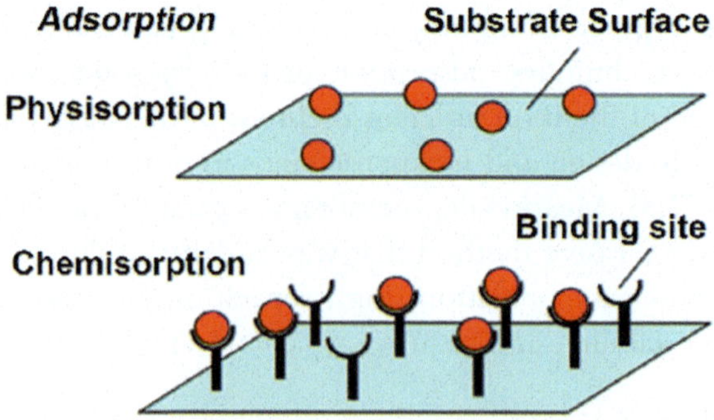

Figure 2.1 Depiction of physisorption and chemisorption interaction between the substrate surface (adsorbent) and the adsorbate molecules [7].

membrane technology and the most commonly used are membrane filtration (microfiltration, nanofiltration, and ultrafiltration), RO, forward osmosis (FO), membrane distillation (MD), and ED. This membrane technology offers many prospects in wastewater treatment including low capital cost and energy requirements, significant reduction in the size of equipment, low chemical utilization, as well as environmental friendliness. Basically, membrane is a barrier that separates two phases from each other by restricting movement of components through it in a selective style. In the context of membrane filtration, for instance, in nanofiltration, membranes only restrain particles with a sizes of 0.1 nm–0.001 μm, in ultrafiltration, particles with sizes of 0.001–0.1 μm, and in microfiltration, particles with size of 0.1–1.0 μm. Meanwhile, in RO, a strong hydrostatic pressure is generated to overcome the intrinsic osmotic pressure of the feed through a semipermeable membrane. In contrast to RO, FO follows the natural osmosis process where water molecules are drawn from one solution to the other through semipermeable membrane, where highly concentrated draw solution used the concentration gradient to draw water molecules from the feed solution until equilibrium is achieved. MD is a technique that involving thermal treatment to create higher pressure in separating the substances based on their volatilities, where only steam molecules can pass through the hydrophobic membrane. ED is a membrane-based separation process that uses an electric field to drive ions through selective ion-exchange membranes, effectively separating salts and other dissolved ions from water. In this process, alternating layers of cation-exchange and anion-exchange membranes are stacked between electrodes. When an electric current is applied, positively charged ions (cations) migrate toward the cathode, passing through cation-exchange membranes, while negatively charged ions (anions) move toward the anode, passing through anion-exchange membranes.

2. Precursors of carbon-based materials

Commonly used carbon-based materials such as coal, activated carbon (AC), carbon nanotubes (CNTs), carbon fibers (CFs), graphene, graphene oxide (GO), and so on are potentially reliable in wide range of applications, including energy storage, diagnosis, drug delivery, electronics, sensors, as well as water purification due to their excellent chemical, electrical, mechanical, and thermal properties. These carbon-based materials are chemically and thermally stable, large specific surface area, low-cost, regenerable, and possess minimum chemical impact on the environment which make them favorable in various applications [8]. These materials can be applied in their pure forms, or they may act as complex hybrid materials. In this section, different precursors of carbon-based materials will be further discussed and those precursors can be anything that is high in carbon backbone in their structure, either naturally or synthetically.

2.1 Synthetic-based precursors

Synthetic precursors of carbon-based materials can be obtained from synthetic polymers such as polymethyl methacrylate (PMMA), polyimide, and polyacrylonitrile (PAN), waste byproducts such as plastic wastes and waste clothes, phenolic resin, and pitch resins such as petroleum and coal tar. Polymers are ideal carbon-rich raw materials in the fabrication of carbon materials due to their abundant availability, diverse structure, and affordable. The adaptable chemical compositions, unique macromolecular and self-assembled structures, and universal synthesis methods are the strengths of these polymer-based precursors, which can be used to fabricate carbon materials with various morphologies and structures using suitable and specific polymers [9]. Phenol formaldehyde resins or phenolic resins are another carbon-based precursor that is produced by the polymerization reaction between phenols and aldehydes or several of their derivatives, which possess excellent mechanical and thermal properties. These phenolic resins have high potential to produce high microporous structures at carbonization temperature of 700–800°C; however, these micropores reduced at higher carbonization temperature, which is the main drawback of this precursor [10]. Phenolic resin is a typical nongraphitized carbon precursor that has 80 wt.% of their original carbon in their chemical structure, whereas the furan resins have 75 wt.% [11]. Meanwhile, pitches as low-cost petrochemical byproducts are a mixture of aromatic hydrocarbons and heterocyclic compounds attained from distillation of petroleum fractions and coal tar. Pitches are comprised of polycyclic aromatic hydrocarbon molecules (aromatic core and side chains), which will form carbon sheets when being slightly heated and form graphite at higher temperatures [12].

Table 2.1 tabulated different types of synthetic precursors including the PAN-based, phenolic- and pitch-resins-based, and graphene-based as adsorbents or membranes or their composites for polluted water and wastewater treatment. The studies listed in Table 2.1 show that PAN-based precursors exhibit high potential to be used as both adsorbents and

Table 2.1 Synthetic precursor of carbon-based materials in water and wastewater treatment.

Type	Precursor	Carbon materials	Pollutants and removal efficiency	Ref.
Polymer-based	PAN + TiO_2	ACF	p-Nitrophenol (98.05%)	[13]
	PAN	ACF	Emulsified oil (97%); p-cresol (91%)	[14]
	PAN + phenolic resin + NaCMC	CF membrane	Emulsified oil (94.2%)	[15]
	PAN	ACF	Phosphate (90%)	[16]
	PMMA + SiO_2	HPC	RB (281.97 mg/g) MB (143.07 mg/g) BY (221.20 mg/g)	[17]
Phenolic resins	Weaved cloth	ACF	Bisphenol A (277.78 mg/g); bisphenol S (426.67 mg/g)	[18]
	Fabrics	NH_3-ACNF and ACF/CNF	Phenol (290 mg/g) Pb^{2+} (40 mg/g)	[19]
	Waste phenolic resin	AC	Methylene blue (485.4 mg/g)	[20]
	Phenolic resin/coal	Carbon membrane	Emulsified oil (99.1%)	[21]
	Phenolic resin + polyurethane	MWCNT-loaded carbon foam	Arsenic (90.5 mg/g)	[22]
	Phenolic resin	CNT-PEG	Phenol (~100%)	[23]
	Phenolic resin/ethanol	Graphite/CNT composite	Humic acid (70%); Cu^{2+} (25%)	[24]

Continued

Table 2.1 Synthetic precursor of carbon-based materials in water and wastewater treatment.—cont'd

Type	Precursor	Carbon materials	Pollutants and removal efficiency	Ref.
Pitch resins	Coal tar pitch	AC	Phenol (250 mg/g)	[25]
	Pitch-based ACF	Urea/nitric acid ACF	Formaldehyde (79.28 mg/g)	[26]
	Coal tar pitch + vinasse	AC	Phenol (57.5 mg/g)	[27]
	Coal tar pitch	ACF	Cu^{2+}(32.41%); Ni^{2+}(43.03%); Cr^{6+} (17.31%)	[28]
Graphene-based	GO powder	rGO	Bisphenol A (193.5 mg/g); 2-nitrophenol (213 mg/g); 2-chlorophenol (341.8 mg/g)	
	GO + EDA + HPEI	GO membrane	Pb^{2+} (95.7%); Ni^{2+} (96.0%); Zn^{2+} (97.4%); Cd^{2+} (90.5%)	[29]
	GO + Ag + P + BWO	GO photocatalyst	TOC (100%)	[30]
	GO + PS	GO membrane	BP-3 (99%); CBZ (38%); DCF (85%); BP-4 (20%); OFLOX (95%); Rh (97%); TRX (100%)	[31]
	GO powder	GO	Emulsified oil (~93%)	[32]

Ag, Silver; *BP-3*, benzophenone-3; *CBZ*, Benzyl chloroformate; *DCF*, diclofenac; *EDA*, Ethylenediamine; *HPEI*; Hyperbranched polyethyleneimine; *NaCMC*, Sodium carboxymethyl cellulose; *OFLOX*, Ofloxacin; *PAN*, Polyacrylonitrile; *PEG*, Polyethylene glycol; *PS*, Polystyrene; *Rh*, rhodamine b; *TiO2*, Titanium dioxide; *TRX*, triton X-100.

membrane materials in the wastewater treatment, but in different forms and structures. Martins et al. [33] have successfully produced PAN-based CFs incorporated with Ag nanoparticles for removal of bacteria in the water such as *Escherichia coli, Candida albicans*, and *Aspergillus niger* with removal efficiencies of 97.8%, 100%, and 99.4%, respectively. Other PAN-based CFs and AC fibers as well as their composites also have been used for removal of other potential water pollutants such as p-nitrophenol, p-cresol, emulsified oil, as well as phosphate. A study by Qi et al. [17] has used PMMA as precursor and SiO_2 as hard-template for fabrication of hierarchical porous carbon (HPC) containing interconnected macro and mesopores for adsorption of bulky dyes in water as shown in Fig. 2.2. The results obtained exhibit that HPC have the highest adsorption capacity toward rhodamine B (RB) of 281.97 mg/g, followed by brilliant yellow (BY) (221.20 mg/g), and methylene blue (MB) (143.07 mg g). This is due to the different sizes of adsorbate molecules of each dye that can pass through certain pore sizes, resulting in different rates of diffusion of dye molecules to the inner surface of the adsorbents [17].

A study by Hu et al. [20] shows high MB adsorption up to 485.4 mg/g by using an adsorbent made of waste phenolic resins-based AC. Despite the dye adsorption, adsorption of bisphenol using phenolic weaved cloth-derived ACFs displays good adsorption capacity; bisphenol A is 277.78 mg/g and bisphenol S is 426.67 mg/g [18]. Meanwhile, for coal tar pitch-based AC studied by Yao et al. [25] exhibits good adsorption capacity toward phenol which is up to 250 mg/g; however, coal tar pitch/vinasses-derived AC shows significant lower adsorption amount of phenol of 57.5 mg/g [27] as compared to Yao et al. Despite their utilization as adsorbents, GO also preferably to be used as membrane-based materials for separation or removal of organic pollutants in the wastewater. For instance, Zambianchi et al. [31] have successfully fabricated GO/PS membrane for

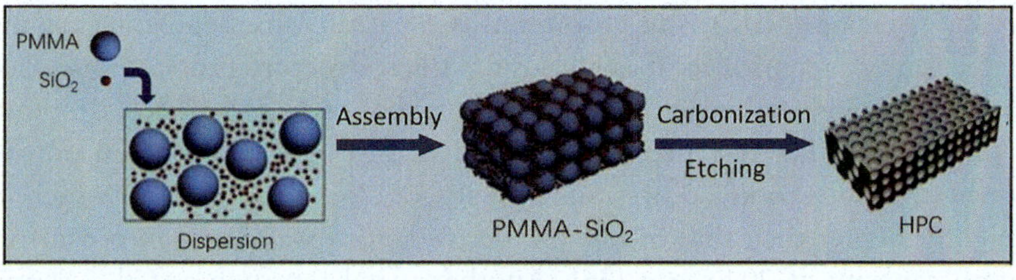

Figure 2.2 Schematic procedure of fabrication of hierarchical porous carbon from dual-template process using PMMA and SiO_2 [17].

removal of different pollutants such as ofloxacin (OFLOX), benzophenone-3 (BP-3), benzyl chloroformate (CBZ), rhodamine b (Rh), diclofenac (DCF), and Triton X-100 (TRX). The removal efficiency of OFLOX is 95%, BP-3 is 99%, CBZ is 38%, Rh is 97%, DCF is 85%, and TRX is 100%.

2.2 Natural-based precursors

Natural precursors of carbon-based materials can be obtained naturally from any materials with carbon structures. Up to now, natural hydrocarbon precursors such as acetylene, benzene, methane, toluene, xylene, and fossil fuels have been commonly used; however, these precursors are nonrenewable and in limited supply, especially fossil fuels, which makes the synthesis of carbon-based materials from these precursors more complicated and expensive. Therefore, in order to develop a more competitive carbon materials, it is necessary to consider developing carbon-based materials from renewable and cheaper precursors such as lignocellulosic materials (e.g., any plant-derived materials and wastes that are high in cellulosic contents) and also biowastes (e.g., industries, agricultural and municipal solid wastes). The study of ultra-structures of plant material and carbon produced by them has revealed that plant-derived precursors possess some basic hydrocarbon compound skeletal structures. In comparison to fossil gaseous and liquid hydrocarbons, the market price of these precursors are very low and cheap [34]. Utilization of biomass can be considered as a waste-to-wealth and sustainable approach due to its cost-effective and less emission of carbon dioxide as compared to deriving carbon-based materials from coal. Fig. 2.3 shows schematic diagram of conversion of AC and CNFs or biofuels and chemicals from biomass materials.

Table 2.2 listed a few natural precursors of carbon-based materials for the treatment of polluted water and wastewater. The removal efficiency or adsorption capacity of heavy metal ions such as lead (Pb^{2+}), cadmium (Cd^{2+}), nickel (Ni^{2+}), and chromium (Cr^{6+}) can vary depending on the selection of precursor itself as well as the microstructure properties (surface area, total pore volume, micropore and mesopores) and their adsorption performance. For instance, the removal of Pb^{2+} by lignin-based porous carbon shows good adsorption capacity of 250.5 mg/g [36], which is slightly higher than that of biochar derived from waste cotton precursors with capacity of 204 mg/g [38]. Another metal ion such as Cd^{2+} shows quite low adsorption capacity on the similar lignin-based porous carbon,

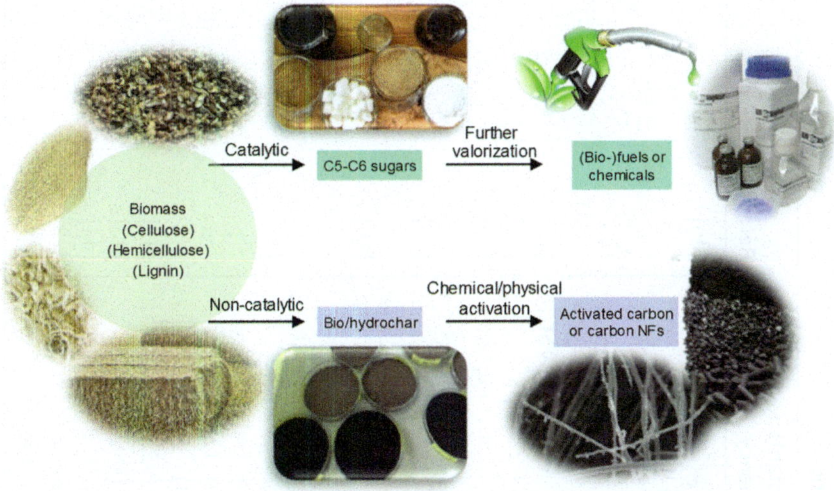

Figure 2.3 Schematic diagram of conversion of biomass into biofuels or activated carbon or carbon nanofibers [35].

with only 126.4 mg/g. However, in comparison to another precursor, such as sunflower stem-based carbon with only adsorption amount of only 0.32 mM/g [40], the finding by Wang et al. [38] shows better results toward Cd^{2+} adsorption and removal. A study by Tao et al. [39] have shown excellent Pb^{2+} removal efficiency up to 135.5% using sewage sludge/sugarcane bagasse AC as compared to hazelnut shell- and walnut shell-based biochar which can remove Pb^{2+} of 83% and 70%, respectively [37].

Other than heavy metals, dyes also another major concerns in the wastewater treatment technologies. The most common dye such as methylene blue (MB) can be effectively removed from the wastewater by using carbon-based materials through different kind of processes. Mahamad et al. [43] have successfully produced pineapple leaves-derived AC to remove MB with adsorption capacity of 288.3 mg/g which is doubled the adsorption performance by other precursor fabricated by Fan et al. [36] using the polyacrylic acid grafted lignosulfonate (SLS)/carboxymethyl cellulose (CMC)/AC composites with value of 113 mg/g. As shown in the list, another type of dye such as Congo red have also been studied in order to study the effectiveness of adsorption capacity of the graphene-like carbon derived from commercial cellulose. As reported by Kim et al. [44]; adsorption capacity of Congo red on the graphene-like carbon shows quite promising results, which is 220.7 mg/g. Instead of Congo red, the resultant graphene-like carbon also been used for adsorption of RhB,

Table 2.2 Natural precursors of carbon-based materials in water and wastewater treatment.

Type	Precursor	Carbon materials	Pollutants & removal efficiency	Ref.
Lignocellulosic and biowastes	Waste cotton + sodium lignin sulfonate (SLS) + carboxymethyl cellulose (CMC)	Biochar	Pb^{2+} (204 mg/g); MB (113 mg/g)	[36]
	Hazelnut shell; walnut shell	Biochar	Pb^{2+} (83% for WSC; 70% for HSC)	[37]
	Lignin-based	Porous carbon	Pb^{2+} (250.5 mg/g); Cd^{2+} (126.4 mg/g)	[38]
	Sewage sludge + sugarcane bagasse	AC	Pb^{2+} (135.5%)	[39]
	Sunflower stem	Carbon	Ni^{2+} (0.45 mM/g); Cd^{2+} (0.32 mM/g); Cr^{6+} (0.28 mM/g)	[40]
	Oil palm leaves	Carbon nanospheres	Malachite green (91%); brilliant green (99%); Congo red (88%)	[41]
	Citrus limetta peels	AC	MB (93%)	[42]
	Pineapple leaves	AC	MB (288.3 mg/g)	[43]
	Commercial cellulose	Graphene-like carbon	RhB (572.8 mg/g); Congo red (220.7 mg/g)	[44]

which results in excellent adsorption performance of 572.8 mg/g. From these findings, it can be concluded that, the adsorption capacity and removal efficiency of pollutants in wastewater can be determined based on the adsorbents' physical and microstructure properties including high specific surface area as well as a wide range of pore size distributions [45], which are the major contributors for high adsorption performance toward specific pollutants.

3. Carbon-based materials as adsorbents and as membranes

3.1 Activated carbon, carbon fibers, and activated carbon fibers

AC, CFs, and activated carbon fibers (ACFs) are all carbon-based materials with unique structures and properties suited to different applications. AC is highly porous with a large internal surface area, generally in the form of granules or powder that are produced by carbonizing organic materials (e.g., coconut shells, coal) and then activating them (typically using heat or chemicals) to create extensive micropores. Meanwhile, long, thin fibers with a high degree of carbon crystallinity, giving them high strength and stiffness are called CFs. CFs are produced by carbonizing precursor fibers (usually PAN) under controlled conditions, without additional activation processes to increase porosity and are known for their mechanical properties, ACFs are similar to CFs but have an activated, highly porous surface, offering a high surface area like AC. ACFs were made by carbonizing and activating precursor fibers (e.g., rayon, PAN), similar to the production of CFs, but with additional activation steps to create micropores. AC is highly porous and mainly used for adsorption, CFs are prized for their mechanical strength and used in structural applications. ACFs combine both high adsorption capacity and fiber form, making them versatile for applications that need both structural integrity and high surface area.

AC has become a good alternative for adsorption due to its porosity and internal pore structure that contribute to its high surface area and make AC applicable for water pollutant treatment. However, the costly commercial AC has initiated a current trend of developing AC from various wastes, agrowastes, and biomaterials such as biomass. Mandal et al. [4] prepared mesoporous AC obtained from kenaf core (KAC) that exhibited ultra-rapid removal of anionic dyes and heavy metal ions. The resultant KAC exhibited a BET surface area of 843.3 m^2/g that is significantly higher than raw kenaf (3.2 m^2/g). This high surface area contributed to their enhanced adsorption capacities with maximum removal efficiency toward Cu^{2+}, Pb^{2+}, and Congo red (CR) dye of 65.1%, 74.6%, and 83.9%, respectively at 47°C. Another similar study conducted by Yunus et al. [46] used honeydew peel-derived AC (surface area up to 1272.4 m^2/g) for removal of metal ions such as Cr^{3+} and Zn^{2+} from mining effluents in tile and electroplating industries with adsorption capacities of 83.5% and 88.9%, respectively. Meanwhile, Rahmani-Sani et al. [47] successfully prepared a novel magnetized AC from chicken feather and eggshell for removing Pb^{2+}, Cd^{2+}, Cu^{2+}, Zn^{2+}, and Ni^{2+} from water. They have reported that the maximum percentage of heavy metal ions removal was 96.1% for Pb^{2+}, 93.8% for Cd^{2+}, 83.3% for Cu^{2+}, 69.6% for Zn^{2+}, and 50.0% for Ni^{2+}.

Despite of AC, CFs, and ACFs are microporous carbon materials that possess larger surface areas, uniform pore size distribution, and faster adsorption rates compared to granular or powdered AC. Additionally, CFs and ACFs can be found in various forms including fabric, felt, thread, and yarn which making them suitable for various applications. For example, a study by Plakas et al. used porous ACFs deposited with titanium dioxide (TiO_2) to enhance their photocatalytic performance for water treatment. The results revealed that ACF/TiO_2 is able to efficiently remove phenol in water that is up to 96% [48]. On the other hand, Silva et al. [49] produced high surface area ACFs (1975 m^2/g) from banana tree pseudo-stem fibers with high adsorption capacity of MB dye (728.9 mg/g). Other wastes such as plastic wastes also alternative precursor for ACFs with potential of removing of thymol blue dye approximately 98.05% as reported by Kumari et al. [50] In term of heavy metal removal, ACFs demonstrate effective adsorption capacity toward Pb^{2+} which is 183.3 mg/g as reported by Hashem et al. [51].

On the other hand, in order to overcome the poor hydrophilicity and surface roughness of CFs and ACFs, the modification of CFs or ACFs as biofilm supports using ferrous oxalate (FeC_2O_4) has been conducted by Xu and Jiang [52]. From their study, it has been reported that the average removal efficiency of chemical oxygen demand (COD) is 97%, ammonia nitrogen (NH_4^+-N) is 96%, and total phosphorus (TP) is 97% which were increased by 7.18%, 10.30%, and 9.40%, respectively as compared to un-modified CFs. Alternatively, a study by Tofighy and Mohammadi [53] successfully synthesized a positively charged membrane produced from charcoal and hyperbranched polyethyleneimine (HPEI) as crosslinkers for removal of divalent heavy metal ions from contaminated water. The results exhibited superior removal performance with a rejection sequence of

R(Zn^{2+}: 95.65%) > R(Cu^{2+}: 92.56%) > R(Ni^{2+}: 92.34%) > R(Pb^{2+}: 90.51%).

Despite dyes and heavy metals, increasing concentrations of oily contaminants in wastewater that pose serious risks to biological ecosystems and their life and need to be reduced. In order to develop promising candidate materials for an efficient and rapid oil/water separation, functional materials with opposite wetting properties for oil and water, such as either superhydrophilic–superoleophobic or superhydrophobic–superoleophilic, need to be considered. For instance, a research by Pakdel et al. [54] developed a functional nonwoven from CF wastes and modified the surface with fluorine-free coatings containing polydimethylsiloxane (PDMS) and zeolitic imidazolate framework-8 (ZIF-8) particles for oil-water separation. They reported an excellent oil–water separation efficiency of >99% and this is due to superior water contact angle (WCA) of 153.5° that contributed by the ZIF-8 particles that boost the superhydrophobicity and absorption capacity. Another study on a superhydrophobic CF membrane by Sun et al. [55] displayed a flexible and high specific surface area, with high WCA of 155.9°, high oil flux of 3590 $L/m^2/h$, and excellent oil–-water separation of more than 98%. Moreover, the resultant CF membrane showed good stability in harsh conditions including strong alkali and acid solutions, low or high temperatures, salt solutions, and organic solvents.

3.2 Biochar

Biochar is a form of charcoal produced by heating organic materials (like agricultural waste, wood chips, or manure) in a low-oxygen environment, a process called pyrolysis. This material has a porous structure, making it highly effective for various environmental and agricultural applications. Its use in carbon capture, soil enhancement, and water filtration are particularly notable. Due to its inherent porosity and chemical properties, biochar can be directly applied or further processed to improve its effectiveness as an adsorbent or membrane material in various applications. A review by Qiu et al. [56] focuses on using biochar as a cost-effective adsorbent for removing heavy metals like lead (Pb), cadmium (Cd), nickel (Ni), and chromium (Cr) from water. It discusses various biochar feedstocks, preparation methods, and activation processes to enhance biochar's adsorption capacity. For instance, biochar produced from rice husk showed a lead (Pb) removal efficiency of up to 90% under optimal conditions, and activated biochar derived from wood waste achieved cadmium (Cd) adsorption capacities of up to 100 mg/g, which are comparable to commercial ACs.

Despite adsorbents, there are various studies that explore the use of biochar as a membrane material, focusing on its application in water purification. For instance, Ghaffar et al. [57] introduced a novel biochar composite membrane, integrating wood-derived biochar with polyvinylidene fluoride (PVdF) using thermal phase inversion. Membranes with varied biochar content demonstrate strong mechanical properties, high water flux, and hydrophilicity, particularly at higher pyrolysis temperatures. The membranes effectively adsorbed Rhodamine B dye (47–187 mg/g) and removed *E. coli* (74%–93%) through filtration. Importantly, cleaning restores their adsorption and filtration abilities, making them practical for pollutant management. This scalable membrane design provides a multifunctional alternative to particle-form biochar for environmental applications.

3.3 Carbon nanotubes

CNTs and carbon allotropes with a cylindrical nanostructure, can be applied for eliminating the pharmaceutical-derived contaminants or their metabolites; they are superior nanosorbents for the removal of organic/inorganic pollutants in comparison to common sorbents, for example, zeolite, clay, AC, and diatomite. CNTs, as good adsorbents, have well-defined cylindrical hollow structures, stronger physicochemical interactions, high aspect ratios, large surface area, high sorbent capacity, hydrophobic side and easily modifiable exteriors.

CNTs and CNT-based composites exhibit the capacity for dye adsorption from aqueous systems, which include multi-walled CNTs (MWCNTs) and single-walled CNTs (SWCNTs); later demonstrates superior adsorption capacity than MWCNTs. Though, CNTs have interesting edge over AC, their large-scale applications for sizable wastewater treatment is not favorable due to the excessive manufacturing costs. In addition to high surface area, CNTs have modifiable surfaces and are endowed with significantly measurable adsorption sites located on their outer and/or inner layer surfaces; theoretically a favorable third generation of carbon-based adsorbents. CNT-based composites exhibit remarkable wastewater remediation capabilities for multiple organic/inorganic and biological water contaminants. CNTs function as scaffolds in water purification due to their hydrophobic surface; however, they should be stabilized in aqueous suspensions to avoid aggregation, which can reduce their active surface area.

Song et al. [58] reported the application of magnetic MWCNTs (MMCNTs) for effective removal of five typical cationic dyes from aqueous solutions. In their study, they have focused on the varieties of adsorbate to investigate the difference in adsorption performance due to different structure. For instance, the hydrophilicity and steric hindrance between the adsorbent and the positive charge center (C+ and N+) of cationic dyes could lead to the adsorption affinity or force difference. Moreover, with different molecular structures or sizes, the adsorption sites have an unequal affinity to dyes which leads to the difference in the adsorption amounts. Thus, MWCNTs exhibit maximum adsorption capacities at 23°C toward malachite green oxalate, auramine O, neutral red, crystal violet, and rhodamine B were 442.2, 295.2, 183.4, 165.3, and 143.0 mg/g, respectively. These results proven that varieties in structure, size, and properties of dyes led to the difference adsorption performance. Meanwhile, in another study by Wang et al. [59] efficiently removed chromium (VI) more than 95% with maximum adsorption capacity of 445.1 mg/g from wastewater by using spherical MMCNTs coupled with deep eutectic solvents (DES).

Despite the dyes and metal ions, the oily wastewater also another pollutant that commonly found in wastewater, which can be treated by using the MWCNTs membranes. For instance, Zarghami et al. [60] successfully synthesized amino-functionalized MWCNTs anchored onto the polydopamine (PDA)-coated polyether sulfone (PES) (PES/PDA/N-MWCNTs) membrane with high oil rejection of more than 99% and total fouling ratio of 22.35%, which is significantly lower than that of the unmodified PES membrane (98.38%). This resultant modified membranes shows that the clean water concentration (5 ppm) after oily wastewater treatment met the discharge limit. In 2023, Khalil and co-researchers have fabricated high performance mixed matrix membrane for removal of organic foulant bovine serum albumin (BSA) and humic acid (HA) in raw wastewater from a wastewater treatment plant. They modified polylactic acid (PLA) membrane by incorporating self-assembled functionalized GO carboxylic MWCNTs (f-GO/MWCNTs-COOH). Incorporation of f-GO/MWCNTs-COOH up to 3 wt.% into the mixed matrix membrane exhibited the highest percentage of BSA and HA removal of 96% and 98%, respectively. Moreover, the resultant membrane displayed excellent fouling resistance of ~92% when treating the wastewater, with a flux recovery ratio (FRR) [61].

3.4 Graphene and its derivatives

One of important branches of carbon-based material is graphene derivative, namely GO. This particle consists of an oxygen-containing functional group attached to the graphene layer. Such a structure, especially a high density of oxygen functional groups (e.g., carboxyl, hydroxyl, carbonyl, and epoxy) in the carbon lattice, enables GO to form stable suspensions in aqueous solution. This property enables practical applications that are specific to processes occurring in the water environment. Graphene nanomaterials have been applied for treating wastewater as they show significant relevant physicochemical characteristics that include high surface area, simple functionalization, and hydrophobicity. Graphene nanocomposites can be utilized as nanosorbents for eliminating hazardous and toxic mix of multiple pollutants, and for their subsequent catalytic degradation using catalytic, photocatalytic, electrocatalytic, and photoelectrocatalytic oxidation, and reduction processes via their commercial production, and use are still a formidable challenge.

For instance, uniform nanoporous graphene sheets can be applied for filtrating and desalinating water by varying their pore size and hence the pressure, the main drawback being their mechanical stability after increasing the pore number. In 2019, a review article by Yin et al. [62] have summarized the potential of CNT/graphene-based nanomaterials in the adsorption of organic wastewater treatment. This is due to their mesoporous structure, high specific surface area, tunable surface properties, and high chemical stability which is helpful in enduring harsh wastewater conditions such as acidic, basic, and salty conditions at high temperatures or high concentrations.

In textile industries, dyes are among the main pollutants that possessed complex molecular structure which are believed to be carcinogenic, poisonous, and environmentally perilous. These pollutants in the wastewater can be removed by using graphene-based materials or their derivatives via absorption. In a study by Fallah et al. [63], they have used GO/acrylamide/sodium alginate (GO/AM/SA) hydrogel composites for removal of MB dye from aqueous media with efficiency up to 95.76%, which is better than dye removal against magnetic hydrogels (91%). This result is due to the high surface area of the GO-containing hydrogels and their ability to absorb a large amount of water and water-soluble species. Another study by Guo et al. [64] have reported highly efficient removal of both MB and rhodamine B (RhB) nearly 100% by using

GO/polyethyleneimine (GO/PEI) hydrogels. Tabish et al. [65] have produced highly porous graphene from GO nanosheet via thermal treatment without the use of any template or addition of any catalyst with efficient adsorption capacities toward MB and RhB of about 99%. Polyethersulfone (PES) membrane was modified by Marjani et al. [66] with the incorporation of GO also shown good dye removal performance. The dye rejection results displayed the PES/GO membrane is ~90% for methyl orange (MO) and ~70% of MB.

As one of the hazardous water pollutants, heavy metals accumulation in the environment and in living organisms can cause serious threat to the environment and health issues. Graphene is in high surface area (which can be up to 3000 m^2/g) and amenable to surface modification to tailor metal binding sites to adsorb and/or sense metal ions as compared to other carbon-based materials, which AC (ranging from 500 to 2000 m^2/g) and biochar (ranging between 10 and 400 m^2/g). Malhotra et al. [67] have successfully produced an environmentally friendly biographene made of graphite powder and BSA protein for adsorption of metal ions for applications in environmental sensing and wastewater treatment. The metal ion binding of the resultant biographene with metal ions such as Co^{2+}, Cr^{3+}, Cu^{2+}, Ni^{2+}, and Fe^{2+} were computationally evaluated by density functional theory (DFT). The maximum mass of metal chloride bound per unit mass of biographene is known as B_{max}, where for Co^{2+} is 47.0 ± 3.3, Cr^{3+} is 25.8 ± 1.9, Cu^{2+} is 22.3 ± 1.0, Ni^{2+} is 37.4 ± 2.4, and Fe^{2+} is 95.5 ± 24.6. Heavy metal removal such as Zn, Cu, and Cd ions by using PES/GO membrane produced by Marjani et al. [66] shows value of ~87%, ~70%, and ~68%, respectively.

4. Conclusion

Carbon-based materials are capable of being synthesized from a wide range of natural and synthetic precursors, with the selection of precursor being crucial to the final material's properties and applications. Common carbon-based materials, like AC, CFs, CNTs, and graphene, necessitate customized features to fulfill certain application demands. These materials display varied structures, geometries, and functions, which substantially affect their performance across multiple industrial and scientific domains. Recent years have seen an increasing focus on the development of sustainable and economically viable carbon-based materials, motivated by the necessity for environmental sustainability and financial practicality. Consequently, natural precursors have garnered significant interest owing to their availability, renewability, and cost-effectiveness. This trend underscores the transition to utilizing more sustainable resources in the production of advanced carbon-based materials, in accordance with worldwide initiatives to diminish reliance on nonrenewable resources.

References

[1] Rashed, M.N., 2013. Adsorption technique for the removal of organic pollutants from water and wastewater. In: Organic Pollutants-Monitoring, Risk and Treatment. IntechOpen.
[2] Tee, G.T., Gok, X.Y., Yong, W.F., 2022. Adsorption of pollutants in wastewater via biosorbents, nanoparticles and magnetic biosorbents: A review. Environ. Res. 212, 113248.
[3] El-Baz, A.A.A., Hendy, I.A., Dohdoh, A.M., Srour, M.I., 2020. Adsorption technique for pollutants removal; current new trends and future challenges–A review. Egypt. Int. J. Eng. Sci. Technol. 32, 1–24.
[4] Mandal, S., Calderon, J., Marpu, S.B., Omary, M.A., Shi, S.Q., 2021. Mesoporous activated carbon as a green adsorbent for the removal of heavy metals and Congo red: Characterization, adsorption kinetics, and isotherm studies. J. Contam. Hydrol. 243, 103869.
[5] Tichem, T.M., Wang, Y., Gameli, R.B., Mbage, B., Li, B., 2023. Multicomponent adsorption of pollutants from wastewater using low-cost eco-friendly iron-modified rice husk biochar in the era of green chemistry. Sustainability 15 (23), 16348.
[6] Younas, F., Mustafa, A., Farooqi, Z.U.R., Wang, X., Younas, S., Mohy-Ud-Din, W., Ashir Hameed, M., Mohsin Abrar, M., Maitlo, A.A., Noreen, S., 2021. Current and emerging adsorbent technologies for wastewater treatment: Trends, limitations, and environmental implications. Water 13 (2), 215.
[7] Kennedy, K.K., Maseka, K.J., Mbulo, M., 2018. Selected adsorbents for removal of contaminants from wastewater: Towards engineering clay minerals. Open J. Appl. Sci. 8 (8), 355–369.
[8] Ani, J., Akpomie, K., Okoro, U., Aneke, L., Onukwuli, O., Ujam, O., 2020. Potentials of activated carbon produced from biomass materials for sequestration of dyes, heavy metals, and crude oil components from aqueous environment. Appl. Water Sci. 10 (2), 69.
[9] Zhao, Z., Sun, L., Li, Y., Feng, W., 2023. Polymer-derived carbon materials for energy storage devices: A mini review. Carbon 210, 118066.
[10] Lenghaus, K., Qiao, G.G., Solomon, D.H., Gomez, C., Rodriguez-Reinoso, F., Sepulveda-Escribano, A., 2002. Controlling carbon microporosity: The structure of carbons obtained from different phenolic resin precursors. Carbon 40 (5), 743–749.
[11] Renda, C.G., Bertholdo, R., 2018. Study of phenolic resin and their tendency for carbon graphitization. J. Polym. Res. 25 (11), 241.
[12] Daulbayev, C., Kaidar, B., Sultanov, F., Bakbolat, B., Smagulova, G., Mansurov, Z., 2021. The recent progress in pitch derived carbon fibers applications. A review. S. Afr. J. Chem. Eng. 38, 9–20.
[13] Yue, Y., Wang, Y., Qu, C., Xu, X., 2021. Modification of polyacrylonitrile-based activated carbon fibers and their p-nitrophenol adsorption and degradation properties. J. Environ. Chem. Eng. 9 (4), 105390.

[14] Alkathiri, D., Sabri, M., Ibrahim, T., ElSayed, Y., Jumean, F., 2020. Development of activated carbon fibers for removal of organic contaminants. Int. J. Environ. Sci. Technol. 17 (12), 4841–4852.
[15] Li, H., Zhang, B., Hong, X., Wu, Y., Wang, T., 2022. Optimizing the microstructure and properties of microfiltration carbon membranes enabled with PAN fibers for emulsified oil removal from wastewater. Chem. Eng. Res. Des. 184, 566–576.
[16] Sakamoto, T., Amano, Y., Machida, M., 2020. Phosphate ion adsorption properties of PAN-based activated carbon fibers prepared with K2CO3 activation. SN Appl. Sci. 2 (4), 702.
[17] Qi, C., Xu, L., Zhang, M., Zhang, M., 2019. Fabrication and application of hierarchical porous carbon for the adsorption of bulky dyes. Microporous Mesoporous Mater. 290, 109651.
[18] Srivastava, A., Singh, M., Karsauliya, K., Mondal, D.P., Khare, P., Singh, S., Singh, S. P., 2020. Effective elimination of endocrine disrupting bisphenol A and S from drinking water using phenolic resin-based activated carbon fiber: Adsorption, thermodynamic and kinetic studies. Environ. Nanotechnol. Monit. Manag. 14, 100316.
[19] Chakraborty, A., Deva, D., Sharma, A., Verma, N., 2011. Adsorbents based on carbon microfibers and carbon nanofibers for the removal of phenol and lead from water. J. Colloid Interface Sci. 359 (1), 228–239.
[20] Hu, W., Cheng, S., Xia, H., Zhang, L., Jiang, X., Zhang, Q., Chen, Q., 2019. Waste phenolic resin derived activated carbon by microwave-assisted KOH activation and application to dye wastewater treatment. Green Process. Synth. 8 (1), 408–415.
[21] Li, H., Zhang, B., Wu, Y., 2024. Highly efficient removal of emulsified oil from oily wastewater by microfiltration carbon membranes made from phenolic resin/coal. Environ. Technol. 45 (18), 3692–3705.
[22] Agrawal, P.R., Singh, N., Kumari, S., Dhakate, S.R., 2018. Multiwall carbon nanotube embedded phenolic resin-based carbon foam for the removal of As (V) from contaminated water. Mater. Res. Express 5 (3), 035601.
[23] Bin-Dahman, O.A., Saleh, T.A., 2020. Synthesis of carbon nanotubes grafted with PEG and its efficiency for the removal of phenol from industrial wastewater. Environ. Nanotechnol. Monit. Manag. 13, 100286.
[24] Zghal, S., Jedidi, I., Cretin, M., Cerneaux, S., Abdelmouleh, M., 2020. One-step synthesis of highly porous carbon graphite/carbon nanotubes composite by in-situ growth of carbon nanotubes for the removal of humic acid and copper (II) from wastewater. Diam. Relat. Mater. 101, 107557.
[25] Yao, P., Cen, J., Fang, M., Wang, T., Wang, Q., 2018. A study on the preparation of pitch-based high-strength columnar activated carbon and mechanism of phenol adsorption from aqueous solution. RSC Adv. 8 (31), 17558–17568.
[26] Ryu, D.-Y., Shimohara, T., Nakabayashi, K., Miyawaki, J., Park, J.-I., Yoon, S.-H., 2019. Urea/nitric acid co-impregnated pitch-based activated carbon fiber for the effective removal of formaldehyde. J. Ind. Eng. Chem. 80, 98–105.
[27] Gao, M., Wang, X., Xia, C., Song, N., Ma, Y., Wang, Q., Yang, T., Ge, S., Wu, C., Lam, S.S., 2021. Phenol removal via activated carbon from co-pyrolysis of waste coal tar pitch and vinasse. Kor. J. Chem. Eng. 38 (1), 64–71.
[28] Choi, B.K., Ko, J.K., Park, S.-J., Seo, M.-K., 2017. Adsorption characteristics of heavy metals ions by physical activation on coal tar pitch-based activated carbon fibers. Carbon Lett. 22, 96–100.
[29] Zhang, Y., Zhang, S., Chung, T.-S., 2015. Nanometric graphene oxide framework membranes with enhanced heavy metal removal via nanofiltration. Environ. Sci. Technol. 49 (16), 10235–10242.

[30] Ma, Q., Ming, J., Sun, X., Liu, N., Chen, G., Yang, Y., 2022. Visible light active graphene oxide modified Ag/Ag2O/BiPO4/Bi2WO6 for photocatalytic removal of organic pollutants and bacteria in wastewater. Chemosphere 306, 135512.
[31] Zambianchi, M., Durso, M., Liscio, A., Treossi, E., Bettini, C., Capobianco, M.L., Aluigi, A., Kovtun, A., Ruani, G., Corticelli, F., 2017. Graphene oxide doped polysulfone membrane adsorbers for the removal of organic contaminants from water. Chem. Eng. J. 326, 130–140.
[32] Diraki, A., Mackey, H.R., McKay, G., Abdala, A., 2019. Removal of emulsified and dissolved diesel oil from high salinity wastewater by adsorption onto graphene oxide. J. Environ. Chem. Eng. 7 (3), 103106.
[33] Martins, I.M., Sampaio, A.G., Lima, G.M., Oliveira e Campos, M.A., Rodgher, S., Rodrigues-Siqueli, A.C., Baldan, M.R., Marcuzzo, J.S., Koga-Ito, C.Y., 2023. Application of textile (PAN-based) activated carbon fibers decorated with silver nanoparticles in water treatment. Front. Environ. Sci. 10, 1100583.
[34] Kumar, R., Singh, R.K., Singh, D.P., 2016. Natural and waste hydrocarbon precursors for the synthesis of carbon based nanomaterials: Graphene and CNTs. Renew. Sustain. Energy Rev. 58, 976–1006.
[35] Zhang, X., Wang, Y., Cai, J., Wilson, K., Lee, A.F., 2020. Bio/hydrochar sorbents for environmental remediation. Energy Environ. Mater. 3 (4), 453–468.
[36] Fan, X., Peng, L., Wang, X., Han, S., Yang, L., Wang, H., Hao, C., 2022. Efficient capture of lead ion and methylene blue by functionalized biomass carbon-based adsorbent for wastewater treatment. Ind. Crop. Prod. 183, 114966.
[37] Kaya, N., Arslan, F., Yildiz Uzun, Z., 2020. Production and characterization of carbon-based adsorbents from waste lignocellulosic biomass: Their effectiveness in heavy metal removal. Fullerenes, Nanotub. Carbon Nanostruct. 28 (10), 769–780.
[38] Wang, A., Zheng, Z., Li, R., Hu, D., Lu, Y., Luo, H., Yan, K., 2019. Biomass-derived porous carbon highly efficient for removal of Pb (II) and Cd (II). Green Energy Environ. 4 (4), 414–423.
[39] Tao, H.-C., Zhang, H.-R., Li, J.-B., Ding, W.-Y., 2015. Biomass based activated carbon obtained from sludge and sugarcane bagasse for removing lead ion from wastewater. Bioresour. Technol. 192, 611–617.
[40] Jain, M., Garg, V., Kadirvelu, K., Sillanpää, M., 2016. Adsorption of heavy metals from multi-metal aqueous solution by sunflower plant biomass-based carbons. Int. J. Environ. Sci. Technol. 13 (2), 493–500.
[41] Krishnappa, B., Bhat, V.S., Ancy, V., Joshi, J.C., Jyothi, M.S., Naik, M., Hegde, G., 2022. Biowaste-derived, highly efficient, reusable carbon nanospheres for speedy removal of organic dyes from aqueous solutions. Molecules 27 (20), 7017.
[42] Singh, S., Sidhu, G.K., Singh, H., 2019. Removal of methylene blue dye using activated carbon prepared from biowaste precursor. Indian Chem. Eng. 61 (1), 28–39.
[43] Mahamad, M.N., Zaini, M.A.A., Zakaria, Z.A., 2015. Preparation and characterization of activated carbon from pineapple waste biomass for dye removal. Int. Biodeterior. Biodegrad. 102, 274–280.
[44] Kim, S.-H., Kim, D.-S., Moradi, H., Chang, Y.-Y., Yang, J.-K., 2023. Highly porous biobased graphene-like carbon adsorbent for dye removal: Preparation, adsorption mechanisms and optimization. J. Environ. Chem. Eng. 11 (2), 109278.
[45] Wang, B., Lan, J., Bo, C., Gong, B., Ou, J., 2023. Adsorption of heavy metal onto biomass-derived activated carbon. RSC Adv. 13 (7), 4275–4302.
[46] Yunus, Z.M., Al-Gheethi, A., Othman, N., Hamdan, R., Ruslan, N.N., 2020. Removal of heavy metals from mining effluents in tile and electroplating industries using honeydew peel activated carbon: A microstructure and techno-economic analysis. J. Clean. Prod. 251, 119738.

[47] Rahmani-Sani, A., Singh, P., Raizada, P., Lima, E.C., Anastopoulos, I., Giannakoudakis, D.A., Sivamani, S., Dontsova, T.A., Hosseini-Bandegharaei, A., 2020. Use of chicken feather and eggshell to synthesize a novel magnetized activated carbon for sorption of heavy metal ions. Bioresour. Technol. 297, 122452.
[48] Plakas, K.V., Taxintari, A., Karabelas, A.J., 2019. Enhanced photo-catalytic performance of activated carbon fibers for water treatment. Water 11 (9), 1794.
[49] Silva, M.C., Spessato, L., Silva, T.L., Lopes, G.K., Zanella, H.G., Yokoyama, J.T., Cazetta, A.L., Almeida, V.C., 2021. H3PO4–activated carbon fibers of high surface area from banana tree pseudo-stem fibers: Adsorption studies of methylene blue dye in batch and fixed bed systems. J. Mol. Liq. 324, 114771.
[50] Kumari, M., Chaudhary, G.R., Chaudhary, S., Umar, A., 2022. Transformation of solid plastic waste to activated carbon fibres for wastewater treatment. Chemosphere 294, 133692.
[51] Hashem, A., Aniagor, C.O., Nasr, M.F., Abou-Okeil, A., 2021. Efficacy of treated sodium alginate and activated carbon fibre for Pb (II) adsorption. Int. J. Biol. Macromol. 176, 201–216.
[52] Xu, S., Jiang, Q., 2018. Surface modification of carbon fiber support by ferrous oxalate for biofilm wastewater treatment system. J. Clean. Prod. 194, 416–424.
[53] Tofighy, M.A., Mohammadi, T., 2020. Divalent heavy metal ions removal from contaminated water using positively charged membrane prepared from a new carbon nanomaterial and HPEI. Chem. Eng. J. 388, 124192.
[54] Pakdel, E., Wang, J., Varley, R., Wang, X., 2021. Recycled carbon fiber nonwoven functionalized with fluorine-free superhydrophobic PDMS/ZIF-8 coating for efficient oil-water separation. J. Environ. Chem. Eng. 9 (6), 106329.
[55] Sun, X., Bai, L., Li, J., Huang, L., Sun, H., Gao, X., 2021. Robust preparation of flexibly super-hydrophobic carbon fiber membrane by electrospinning for efficient oil-water separation in harsh environments. Carbon 182, 11–22.
[56] Qiu, B., Tao, X., Wang, H., Li, W., Ding, X., Chu, H., 2021. Biochar as a low-cost adsorbent for aqueous heavy metal removal: A review. J. Anal. Appl. Pyrolysis 155, 105081.
[57] Ghaffar, A., Zhu, X., Chen, B., 2018. Biochar composite membrane for high performance pollutant management: Fabrication, structural characteristics and synergistic mechanisms. Environ. Pollut. 233, 1013–1023.
[58] Song, G., Shi, Y., Wang, H., Li, A., Li, W., Sun, Y., Ding, G., 2022. Effective sorptive removal of five cationic dyes from aqueous solutions by using magnetic multi-walled carbon nanotubes. Water Sci. Technol. 85 (7), 1999–2014.
[59] Wang, L., Zhu, Y., Ma, L., Hai, X., Li, X., Yang, Z., Gao, Y., Yuan, M., Xiong, H., Chen, M., 2024. Efficient removal of chromium (VI) from wastewater based on magnetic multiwalled carbon nanotubes coupled with deep eutectic solvents. Chemosphere 362, 142732.
[60] Zarghami, S., Mohammadi, T., Sadrzadeh, M., Van der Bruggen, B., 2020. Bio-inspired anchoring of amino-functionalized multi-wall carbon nanotubes (N-MWCNTs) onto PES membrane using polydopamine for oily wastewater treatment. Sci. Total Environ. 711, 134951.
[61] Khalil, H., Wadi, V.S., Hegab, H.M., Nassar, L., Naddeo, V., Yousef, A.F., Banat, F., Hasan, S.W., 2023. High-performance f-GO/MWCNTs-COOH nanohybrid-based polylactic acid mixed matrix membrane for wastewater treatment. J. Water Process Eng. 53, 103784.
[62] Yin, Z., Cui, C., Chen, H., Duoni, Yu, X., Qian, W., 2020. The application of carbon nanotube/graphene-based nanomaterials in wastewater treatment. Small 16 (15), 1902301.

[63] Fallah, S., Mamaghani, H.R., Yegani, R., Hajinajaf, N., Pourabbas, B., 2020. Use of graphene substrates for wastewater treatment of textile industries. Adv. Compos. Hybrid Mater. 3 (2), 187–193.
[64] Guo, H., Jiao, T., Zhang, Q., Guo, W., Peng, Q., Yan, X., 2015. Preparation of graphene oxide-based hydrogels as efficient dye adsorbents for wastewater treatment. Nanoscale Res. Lett. 10 (1), 272.
[65] Tabish, T.A., Memon, F.A., Gomez, D.E., Horsell, D.W., Zhang, S., 2018. A facile synthesis of porous graphene for efficient water and wastewater treatment. Sci. Rep. 8 (1), 1817.
[66] Marjani, A., Nakhjiri, A.T., Adimi, M., Jirandehi, H.F., Shirazian, S., 2020. Effect of graphene oxide on modifying polyethersulfone membrane performance and its application in wastewater treatment. Sci. Rep. 10 (1), 2049.
[67] Malhotra, M., Puglia, M., Kalluri, A., Chowdhury, D., Kumar, C.V., 2022. Adsorption of metal ions on graphene sheet for applications in environmental sensing and wastewater treatment. Sens. Acutators Rep. 4, 100077.

CHAPTER 03

Carbon-based adsorbent materials and composites in water and wastewater treatment

Yuvrajsinh B. Rajput[1], Vishwajit Chavda[1,6], Mansi Gandhi[2], Kenit Acharya[1], Rawesh Kumar[3], Sapna Raghav[4] and Pallavi Jain[5]

[1]Department of Chemistry, Institute of Science, Humanities and Liberal Studies, Indus University, Ahmedabad, Gujarat, India; [2]Department of Chemistry, Gujarat Technological University, Ahmedabad, Gujarat, India; [3]Chemistry Department, Patna University, Patna, Bihar, India; [4]Department of Chemistry, Shri Jagdishprasad Jhabarmal Tibbrewala (JJT) University, Jhunjhunu, Rajasthan, India; [5]Department of Chemistry, SRM Institute of Science and Technology, Delhi NCR Campus, Modinagar, Uttar Pradesh, India; [6]Department of Environmental Engineering, Kyungpook National University, Daegu, South Korea

1. Introduction

Due to rapidly growing populations, the rise of urbanization, multiyear shortages, and increasing demand for water from nature worldwide, there is currently an inadequate supply of water that is safe to consume; it is highly recommended that water undergo treatment immediately. Various workable strategies to produce more usable water are proportionate to the problem of diminishing water resources. Moreover, the largest practical water resources are categorized as effluent from industry and agriculture [1–3]. There are inherent drawbacks regarding every water purification technique, especially those concerning expenses, efficacy, and application. Implementing adsorption to treat wastewater from industry or alternative sources to boost its quality for consumption in agricultural and industrial processes afterward offers a tempting prospect. During the last few years, numerous investigations have been undertaken to investigate the characteristics of materials as adsorbents for facilitating the elimination of undesirable and toxic substances [2–8]. Water quality can be improved by adsorbing contaminants from liquid forms, like coloring agents, metals, phenol compounds, and medicinal products, thereby clarifying products and recovering essential chemical compounds. preventing long-term

Carbon-Based Materials and Composites for Water and Wastewater Treatment
ISBN 978-0-443-33681-2
https://doi.org/10.1016/B978-0-443-33681-2.00003-X

medical impacts on humans is vital [9]. The current research intends to minimize standard organic and inorganic compounds that are potentially dangerous whenever they become apparent in either liquid or gas media. Carbon-based materials will define the features associated with the adsorbent-adsorbate interaction. In the past several decades, we have made considerable progress in understanding carbon-based materials' adsorption, processes, and isotherms. The primary focus of this chapter is the synthesis and modification of carbon-based materials using varied materials, along with the associated overview of the adsorption capability and processes used to remove pollutants from aqueous solutions.

1.1 Adsorbents

Adsorption processes are affected by many factors, including the structure of the adsorbent, the properties of the fluid, the structure of the contaminants, the operating conditions, and the way the system is set up. Adsorbents have been utilized to adsorb pollutants from environmental water, including zeolites, biomasses, clays, silica gel, composites, alumina-activated carbon (AC), and biological and polymeric materials [10]. Within the past 10 years, there has been an important spike in scientific investigation of nanomaterials made from carbon as adsorption agents. Carbon-based structures have unique properties and are remarkably different from one another. This field also grows because it opens new research areas in many chemistry, physics, and engineering fields. There are diverse ways that carbon-based materials can soak up diverse types of water pollutants, such as metalloids, pesticides, medicines, hazardous heavy metal ions, and organic-inorganic chemicals [11,12]. The typical pollutants are concentrated and transferred to other phases by absorption in a water treatment process. The interactions of adsorbent and adsorbate molecules define and depend on the adsorption phenomena. A carbon-based adsorbent's ability to absorb chemical compounds is determined by the qualities of the adsorbate (pK_a, polarity, functionality, size, and molecular weight), the adsorbent (functional groups, pore size, and structure), and the solution parameters (temperature, pH, and ionic strength) [13,14]. Chemical material binding and accumulating on various adsorbents in the water medium is regulated by the dipole-dipole, induced-dipole, van der Waals, and H-binding interactions. Among the several interactions, the hydrophobic effect, electrostatic and covalent interactions, p–p bonds, hydrogen, and others are crucial for adsorption. Carbon aerogels (CAs), graphene, biochar (BC), CNFs, CNTs, and AC are the components of carbon-based materials [15–17]. The numerous varieties of carbon-based materials utilized in the adsorption procedure are shown in Fig. 3.1.

1.2 Equilibrium isotherm of adsorption

An adsorbent's thermodynamic characteristics and adsorption capacity can be determined by the equilibrium isotherms, which additionally provide an opportunity to determine how the adsorbate-adsorbent interacts. Since equilibrium isotherms connect the adsorption capacity (q_e) to the adsorbate's equilibrium concentration (C_e), they serve as useful resources for describing the adsorption process. In addition, isotherms like liquid-solid adsorption may be utilized to determine the adsorption process and get an increased awareness of the features of the adsorbent surface and the type of adsorbate [18]. Based on their initial slope, their S, H, C, and L curves make up the four different types of equilibrium values [19]. Under the

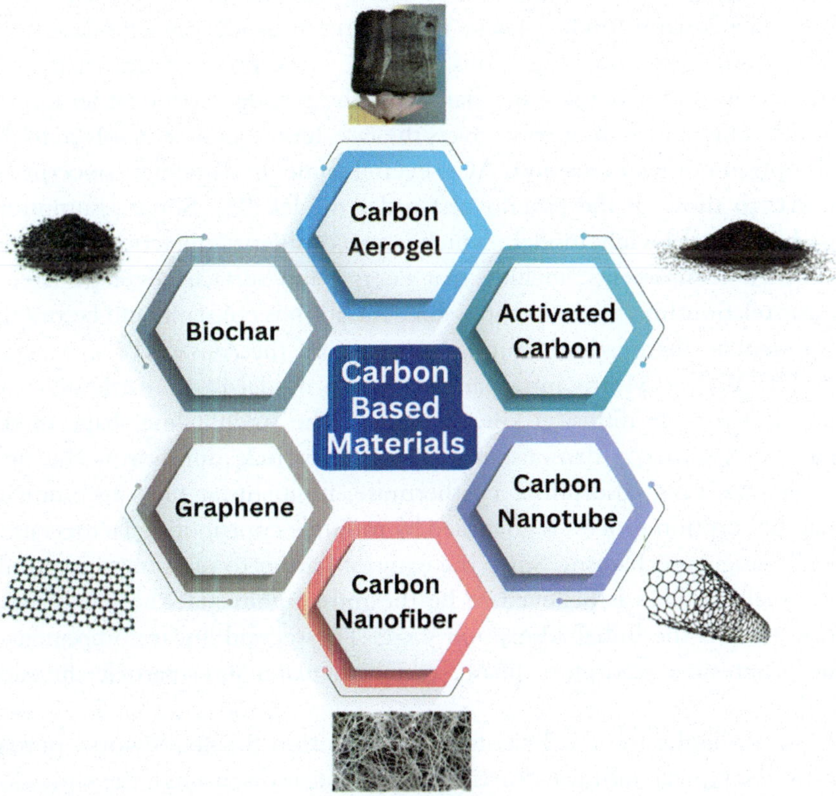

Figure 3.1 Adsorption processes in different applicable carbon-based materials.

isotherm methods, the fact that could potentially be implemented for modeling the results of adsorption is statistical particle physics models, the BET isotherm (multilayer), Dubinin–Radushkevich (D–R), Henry's rule, Temkin, Redlich–Peterson (R–P), Freundlich, and Langmuir (monolayer) [20].

1.3 Kinetics of adsorption model

Reviewing the adsorption kinetics, which may be produced by physical or chemical means, can help define the kind of adsorption by examining the mechanism and rate of adsorption processes and their potential uses.

Adsorption reaches its maximum when a single complete layer of adsorbate molecules forms on the surface of the adsorbent. The concentration of adsorbates remains constant, and adsorption occurs only at specific localized sites without any interactions between the adsorbed molecules. Additionally, the adsorption energy remains unaffected by the extent of surface coverage, and the adsorption process follows a pseudo-first-order kinetic model. These five assumptions are the foundation for the pseudo-first-order equations [21,22]. Adsorption is anticipated to follow a second-order rate mechanism in the pseudo-second-order kinetic model. This approach mainly covers the rate-limiting phase involved in the adsorption of metal ions onto ACs, even if some fundamental concerns are similar to those of the pseudo-first-order model [23]. Some assumptions used in the Elovich model, which also considers interactions between adsorbed components, include that energy and surface coverage have a linear relationship, the adsorbate's concentration is constant, and before the exponential, there is limited ion absorption on the ACs [24].

The Weber–Morris intra-particle kinetic model is the name given to the intra-particle diffusion kinetic model. The straight-line shape of the graph q_t versus $t_{1/2}$ demonstrates that intra-particle diffusion is the only factor regulating adsorption. Furthermore, multi-linear plots recommend that the sorption process is governed by multiple contributing factors rather than a single mechanism. Since the exterior barrier to mass transfer around the small particles is believed to be the most significant, the first sharper stage reflects the initial adsorption stage. The second linear component is the progressive adsorption phase, which regulates intra-particle diffusion [25].

Accordingly, the coefficients t_1 and t_2 within this framework approximate the typical diffusion durations for propagation through the saturation shell and the exterior liquid film [26]. The mass equilibrium of a fixed bed and the breakthroughs curve in a constant process of adsorption could possibly be analyzed by a range of laboratory-based mathematical models, including the Adams, Thomas, Wolborska, Bohart–, and Yoon–Nelson models [20].

1.4 Mechanisms of adsorption

The overview explains the mechanics underlying carbon-based nanomaterials' removal of aqueous contaminants. The bonding between contaminants and the functionalized groups of carbon-based materials is generally complex. The chemical and physical characteristics of the carbon-based substance's surface, chemical composition of adsorbate, and the characteristics of the water-based solution significantly affects the effectiveness of adsorption works.

Electrostatic interaction (physisorption), physical adsorption, ion exchange, surface complexation, and precipitation (chemisorption) are two adsorption methods onto carbon adsorbents [27]. Chemisorption is generally more effective than physisorption at removing contaminants from aqueous solutions. In certain conditions, the surface-modified groups function successfully in ion exchange, surface complexation, and electrostatic interaction, amongst additional procedures [28,29].

2. Adsorption on carbonaceous materials

The existence of carbon in many distinct kinds of types is mainly because of its distinctive electronic structure, enabling the emergence of strong chemical bonds in a range of forms. The uniqueness of carbon arises from its ability to establish single, double, and triple covalent bonds. In addition, carbon atoms readily undergo catenation, allowing them to link with each other to generate linear or branched chains as well as cyclic structures, while also forming stable bonds with various non-metal elements [30,31]. Carbon's L shell contains four valence electrons, two in the 2s orbital and two in the 2p orbital, while the inner K shell holds two electrons. Through sp^3, sp^2, or sp hybridization, the 2s and 2p orbitals enable the formation of single, double, and triple bonds, respectively.

This unique bonding capability creates carbon frameworks in 1D, 2D, and 3D structures. The previously remarkable materials generated by carbon atoms enduring sp^2 hybridization include porous carbon, graphene nanoribbons, fullerene, and CNT. Natural carbon allotropes, such as diamond and graphite, feature crystal structures that exhibit various defects, such as each additional crystalline phase. These reported lattice imperfections may be classified into four distinct categories based on their dimensions: Zero dimension (point), one dimension (line), two dimensions (planar), and three dimensions (volume). Structural geological embarking during which crystals were formed and modified can frequently be seen by the lattice flaws that regulate their morphology.

Crystalline graphite, diamond, and amorphous carbon are the three allotropes occurring naturally. In their pure (100% carbon) state, several amorphous carbon materials, including coal, lampblack, charcoal, coke, and carbon black, have not been observed. They should all be considered allotropes because they have diverse features. Yet, they are preferred to be grouped as amorphous allotropes because their structures are not manipulable like those of crystalline allotropes; amorphous forms of carbon are often not appealing alternatives for researchers [32]. Since carbon-based adsorbents have been used for so long, it is hard to pinpoint their actual origin. A partly de-volatilized carbonaceous substance, wood char, or coal char, was used as an adsorbent before the advent of what is now known as carbon-based adsorbents. Using carbonaceous adsorbents to remove pollutants from synthetic materials, water, soil, and air has been extensively studied in previous research. These adsorbents are commonly classified using CNTs, BC, AC, CAs, GO, and CNFs symbols. Adsorption mechanisms involve the binding of molecules or atoms (adsorbates) onto carbon surfaces (adsorbents) through chemical interactions and/or surface forces (electrostatic or dispersive) in liquid and gas media. The large specific surface area of carbon-based material adsorbents serves as one of their most significant qualities. For an outcome, researchers tried to look at the progression of these materials' applications, modifications in structure, and capabilities over various kinds of applications in the industry.

2.1 Activated carbon: Structure, properties, and adsorptive applications

Due to its remarkable surface area (500–3000 $m^2\,g^{-1}$), affordable price, superior charge holding ability, and availability of many functional groups, activated carbon (AC) is a highly porous carbonaceous nanomaterial which is frequently utilized as an adsorbent. These elements could effectively eliminate various impurities from gaseous or aqueous media [33]. In water purification processes, multiple states of chemical forms of AC, such as granular in nature, powdered form, and fiber forms, are frequently used [34–36].

2.1.1 Synthesis and activation

The method of raw material synthesis, carbonization, pelletizing, and activation were all the basic steps comprising AC preparation [37,38]. To generate a porous structure from nanomaterials possessing low surface area, two principal activation approaches are commonly applied: thermal activation and chemical activation [39]. Physical (thermal) activation requires carbonization at 400–500°C to remove volatile substances and partial gasification by an oxidizing gas, including CO, flue gas, or steam (800–1000°C) to generate functional groups, porosity, and surface area following the initial treatment and pelletizing [40,41]. Activation of chemicals is another method, which entails adding ingredients to the synthesis process before carbonization [42]. The cellulose material undergoes degradation and dehydration by the current additions during carbonization at 250–650°C. Activators such as sulfuric acid, phosphoric acid, or others are mixed with lignin in a raw material after carbonizing at temperatures as high as 900°C [43]. Different structures, functional groupings, and activities can be provided based on the activation process and material sources [44].

A fluidized bed and a rotary kiln are crucial in manufacturing AC materials. A revolving kiln is a device employed in pyro processing to enhance the raw materials' ambient temperature. The kiln is a cylindrical sphere that shifts and rotates slowly on its horizontal axis. This system's fundamental component is a cylindrical device featuring a refractory lining that rotates at a speed of 1–5 rpm and is positioned at a slight horizontal inclination of about 5–10°.

Another method for achieving the appropriate temperature for kiln systems is auxiliary heating. The material progressively descends to the bottom end of the kiln as it spins, undergoing some stirring and mixing. Hot gases move around the kiln to improve heat transmission. Post-activation, activation, and pre-activation are the three stages of the AC manufacturing procedure (with a rotary kiln) [45,46]. The quality of the material is evaluated before activation, and the required size characteristics (also referred to as kiln-feed) are acquired.

- The process of activation converts the kiln input into AC. Within certain designs, the rotary kiln employs steam and elevated temperatures based on the kiln-feed qualities.
- The post-activation process involves quality verifying the kiln output material's specifications.

A fluidized bed reactor can produce AC by carefully optimizing various process parameters, including fluidizing velocity, reaction duration, fluidizing medium, activation temperature, particle size, and static bed height.

2.1.2 Origins of activated carbon

Since lignocelluloses are frequently discovered in anthracite, wood, bituminous coal, and biomass, these substances constitute vital sources of AC; however, more precursor materials are now widely accessible and may be used to create inexpensive carbons [47,48]. AC can come from various general waste goods, including plastics and agricultural waste. Table 3.1 lists several significant and extremely prevalent carbon-based adsorbents for pollutant removal from wastewater.

2.1.3 Modification techniques for activated carbon

Research on AC modification is used to evaluate the activity of ACs in an adsorption approach for a variety of contaminants, which has increased in prominence as a consequence of the discovery of further sites on ACs that improve their affinity for particular contaminants and accelerate their efficient elimination from a variety of fields, particularly wastewater effluent. Identifying the key factors influencing the adsorption capacity of AC is essential before its use or modification through physical and chemical processes designed to enhance adsorption efficiency for organic and inorganic compounds and metals in aqueous environments [88].

Pore volume (PV), pore-size distribution (PS), surface area (SSA), and the existence of functionalized groups at the surface represent the fundamental and unique features of AC, which are vital to adsorption ability [89]. These qualities can be enhanced further after being modified with more nanostructured materials or shrinking to the nanoscale. While PS is closely related to the nature of the unique raw material, the degree of activation throughout the manufacturing phase, the regeneration cycle, and the physical characteristics of the AC, adsorption capacity increases with a particular surface area because of the accessibility of adsorption sites. Much attention is focused on creating surface-modified carbons using different techniques to boost AC's capacity to remove certain contaminants. Following the activation step, the surface of AC is often altered. Three types of modifications might be distinguished: chemical, biological, and physical. There are additional reports in the literature on oxidant [90] and non-oxidative [91] approaches to treating AC.

Table 3.1 Details of adsorption capacities by carbon-based nanomaterials.

Nanomaterials	Contaminants	Capacity of adsorbent (mg g^{-1})	Refs.
Bismuth sulfide–silver sulfide nanocomposite	Auramine-O (Fluorescent basic dye)	202.43	[49]
	Ponceau-4R (Red azo food dye)	191.57	
Gold–magnetite nanocomposite	Rhodamine123 (Cationic fluorescent dye)	71.46	[50]
	Disulfine blue (Anionic dye)	76.38	
Zinc oxide	Safranin O (Basic red dye)	32.06	[51]
Chromium ferrite	Methyl violet (MV) (Cationic triarylmethane dye)	65.67	[52]
Multi-walled carbon nanotubes	Acid red 18 (Anionic sulfonated azo dye)	166.67	[53]
Platinum	Congo red (CR) (Anionic diazo dye)	43.478	[54]
Carbon nanotubes/Activated carbon fabric	Basic violet 10 (Basic dye)	220	[55]
Copper:Doped zinc sulfide	Methyl orange (MO) (Anionic azo dye)	44.65	[56]
	Sunset yellow (SY) (Orange azo food dye)	50.54	

Continued

Table 3.1 Details of adsorption capacities by carbon-based nanomaterials.—cont'd

Nanomaterials	Contaminants	Capacity of adsorbent (mg g^{-1})	Refs.
Functionalized-multi-walled carbon nanotubes	Congo red (CR)	148	[57]
	Reactive green HE4BD (Reactive green dye)	152	
	Golden yellow MR (Azo dye)	141	
Multi-walled carbon nanotubes	Reactive red M-2BE	335.7	[58]
Carbon nanotubes	C.I. Direct yellow 86 (Direct yellow textile dye)	56.2	[59]
	C.I. Direct red 224 (Direct red textile dye)	61.3	
Multi-walled carbon nanotubes/Glycine/β-cyclodextrin	Methylene blue (MB)	90.90	[60]
	Acid blue 113	172.41	
	Methyl orange (MO) (Anionic azo dye)	96.15	
	Disperse red 1	500	
Carbon Nanotubes/Zinc:Zinc oxide@Nickel phosphide	Azure B (Cationic thiazine dye)	38.60	[61]
Multi-walled carbon nanotubes/Iron carbide	Direct red 23	172	[62]
Sulfur-functionalized carbon nanotubes (CNT-Ss)	Mercury(II) ions	151.51	[63]
Chitosan-coated carbon nanotubes treated with acid	Lead(II) ions	158.7	[64]

Multi-walled carbon nanotubes	Cadmium(II) ions	181.81	[65]
Multi-walled carbon nanotubes	Nickel(II) ions	3.7	[66]
Carbon nanotubes	Chromium(VI) ions	1.0	[67]
Iron oxide/Carbon nanotube composite	Copper(II) ions	23.4	[68]
Multi-walled carbon nanotubes–Iron oxide	Mercury(II) ions	238.87	[69]
Graphene oxide/Iron oxide	Methylene blue (MB) (Cationic thiazine dye)	39	[70]
Magnetite single-walled carbon nanotubes with cobalt sulfide	Mercury(II) ions	1666	[71]
Graphene oxide	Cobalt(II) ions	21.28	[72]
Deep eutectic solvent functionalized graphene oxide @ titanium dioxide	Methylene blue (MB) (Cationic thiazine dye)	45	[73]
Graphene oxide	Acid orange 8	29.0	[74]
	Direct red 23	15.3	
Deep eutectic solvent functionalized graphene oxide @ zirconium dioxide	Methylene blue (MB) (Cationic thiazine dye)	35	[75]
Graphene oxide	Basic yellow 28	68.5	[76]
	Basic red 46	76.9	

Continued

Table 3.1 Details of adsorption capacities by carbon-based nanomaterials.—cont'd

Nanomaterials	Contaminants	Capacity of adsorbent (mg g^{-1})	Refs.
Carboxylic modified tubular carbon nanofibers	Uranium(VI) ions	1928.59	[77]
Biochar	Congo red (CR) (Anionic diazo dye)	404.4	[78]
	Malachite green (MG) (Triphenylmethane cationic dye)	2468	
Biochar/Nanoscale zero-valent iron	Malachite green (MG) (Triphenylmethane cationic dye)	515.77	[79]
Carbon nanofiber web	Chromium(VI) ions	56.9	[80]
Biochar	Malachite green (MG) (Triphenylmethane cationic dye)	1738.6	[81]
	Crystal violet (CV) (Triphenylmethane-based cationic dye)	175.6	
	Congo red (CR) (Anionic diazo dye)	86.9	
Carbonaceous nanofibers	Uranium(VI) ions	125	[82]
	Europium(III) ions	91	
Magnetic tubular CNFs	Copper(II) ions	375.93	[83]
Cellulose acetate	Cadmium(II) ions	15.53	[84]
Cellulose acetate	Copper(II) ions	55.25	[85]
Cellulose acetate	Strontium(II) ions	24.63	[86]
Cellulose acetate	Lead(II) ions	34.72	[87]

2.1.4 Application

The primary industrial applications of AC include air treatment, decolourization, solvent recovery, water treatment, and military defense against toxic chemical attacks like mustard gas and nuclear reactors. Household items include air purifiers, cooker hoods, refrigerator deodorizers, cigarette filters, and deep-fat fryer cartridges. Gold and silver can be recovered from a cyanide-based solution as precious metals. ACs are also useful as supports for catalysis. The enormous hydrophobic interior surface of ACs allows them to absorb chemicals, making them accessible to reactants due to their inert porous nature. A variety of forms of AC are made in various shapes and sizes due to their intended application, like powdered AC (PAC), granular AC (GAC), extruded AC (EAC), AC-molecular sieves, and AC fibers. For gas separation, carbon molecular sieves have grown increasingly common in the past few years for nitrogen and oxygen [92–96]. These carbons exhibit high adsorption efficiency due to the presence of micropores whose dimensions are comparable to those of the adsorbate molecules. The separation capability of carbon molecular sieves, particularly for gas mixtures, is governed by the molecular size of the species and their ability to diffuse into the pore network. The separation effectiveness has been enhanced using AC fibers for molecular sieves in gas filtration since they frequently possess higher capacities and a quicker adsorption rate than conventional granular ACs. The alterations of nanoparticle (NP) surface by AC have additionally been the focus of several studies to establish surfaces that might improve the eradication of chemical species. Several different nanostructures have been identified on ZnO [51], $CrFeO_3$ [52], AC: Bi_2S_3–Ag_2S [49], Au–Fe_3O_4 [50], Cu_2O [97], Pt [54], ZnS: Cu [56], Cu: ZnS [98], CdSe, CdTe [99], TiO_2 [100], $Cd(OH)_2$ [101], Fe_3O_4 [102], Cu $(OH)_2$ [103]. Table 3.1 shows how these nanostructures can adsorb contaminants on AC.

Applications of ACs can be broadly divided into two categories from a unique perspective: gas-phase and liquid-phase. Granular or shaped ACs are used in gas-phase applications for catalytic supports, separation, and storage. The nanocomposites Ni–Co-NPs–gAl_2O_3 and Cu–Zn–Ni-NPs–AC were developed by Daneshyar et al. [104] and used to recover H_2S molecules from natural gas, as seen in Fig. 3.2. Cu–Zn–Ni/AC was the most effective in extracting compounds.

Liquid-phase carbons exhibit a significantly larger pore area within the macropore range, allowing liquids to penetrate the pores faster. This constitutes the main difference among ACs used for liquid-phase and gas-phase

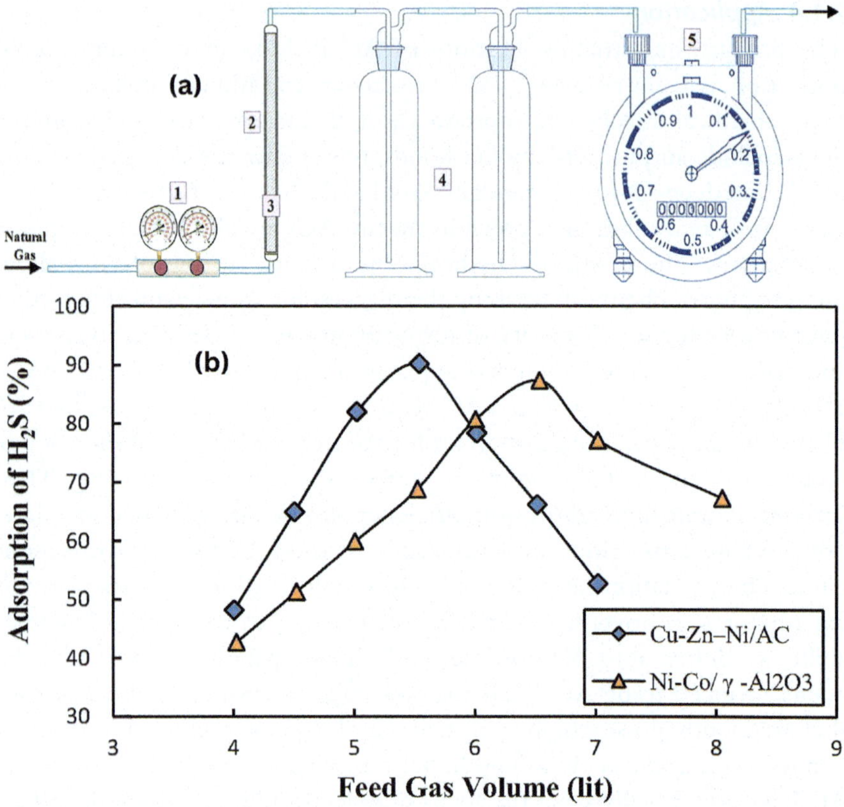

Figure 3.2 (a) The practical setup for the adsorption-based removal of H_2S; (b) the effect of feed gas volume on H_2S adsorption. Reproduced with the license granted by Elsevier from Ref. [104], Copyright 2017.

carbons. Higher adsorption of large molecules, which could indicate pollutants or products in an array of liquid-phase applications, is an additional advantage of larger pores. Using powdered AC (PAC), Zhang et al. [105] studied those 28 antimicrobial agents containing tetracyclines (TCs), quinolones (QNs), sulfonamides (SAs), penicillins (PNs), macrolides (MCs), and chloramphenicol (CPs), which were separated. Under ideal conditions, the elimination efficiency for surface water was 99.6%, and deionized water was 99.9%, with a 20 mg L^{-1} PAC dosage during a 120 min interaction period. The findings indicated that any other antimicrobial showed more potential for adsorption. The potential of metallic ions like Co [106], Cu [107], Mn [108], Hg [109], Pd [110], Cd [111], V [112] and Pb [113,114] to bind to

ACs via a wide range of patterns of interaction is currently presented. Pb^{2+} adsorption onto the surface of AC derived from cow bones undergoing chemical alteration with HNO_3 was studied [115]. The pH of the solution influences Pb^{2+} adsorption on AC; metal hydroxide precipitation dominates at pH 4–6.0. At 50%, the metal Pb^{2+} recovery was consistent across all cycles.

The adsorption of Cr (VI), both in the presence and absence of humic acid (HA), was best represented by the pseudo-second-order kinetic model, while the equilibrium data showed good agreement with the Freundlich isotherm model [105]. The adsorption of Cr (III)–HA complexes occurs due to the interactions of binding in HA and Cr (VI), the HA-induced reduction of Cr (VI), and the complexation of Cr (III) by HA attached to the surface, which may be the cause of the enhanced removal of Cr (VI). Since they have special qualities, phenolic organic compounds are employed in many industrial-based products. However, they are classified as priority pollutants because they are hazardous to species, even at low levels.

2.2 Carbon nanotubes: Structure, types, and key physicochemical properties

CNTs, also referred to as buckytubes, are cylindrical carbon allotropes having distinctive characteristics that render them suitable for an array of applications in industry.

First developed in 1991 by Ijima via an arc-discharge procedure, CNTs were prominent components of the fullerene family and exhibited unique electrical, mechanical, and thermal characteristics and exceptional durability [116–118]. A high melting point, extremely broadband gap, thermal conductivity, and high tensile strength are just a few of its exceptional characteristics. CNTs can be classified based on the number of concentric tubes in their structure [119,120]. CNTs are generally categorized into three types: single-walled CNTs (SWCNTs), double-walled CNTs (DWCNTs), and multi-walled CNTs (MWCNTs).

One graphene sheet (1–2 nm diameter) can be rolled up on itself to develop SWCNTs, although the preparation process can alter the length of the sheets [121]. SWCNTs may typically form bundles or ropes, which are hexagonally arranged to resemble crystals. The outer tube of DWCNTs encloses the inner tube, which is composed of two concentric CNTs [122]. They combine the distinctive characteristics of SWCNT, which render it preferable to SWCNT in terms of unique inner and outer tube doping or higher stability and stiffness.

Multiple graphene layers are wrapped up on each other to form MWCNT [123]. The thickness of these layers ranges from 2 to 50 nm, depending on the count of graphene tubes, with an inter-layer spacing of approximately 0.34 nm [124].

2.2.1 Synthesis and modification

Five typical ways may be utilized to produce CNTs in enormous quantities: chemical vapor deposition (CVD), electrolysis [125], sonochemical or hydrothermal processes [126,127], laser ablation [128], and carbon arc-discharge technology [128]. These methods can be used to create various kinds of CNT. The first methods for producing SWCNTs on a large scale (gram) were laser ablation and carbon arc discharge. Since the basis of both SWCNTs and MWCNTs is the condensation of hot gaseous carbon atoms formed when solid carbon evaporates, these can be developed [129,130]. SWCNTs are grown utilizing a metal catalyst in the arc-discharge technique. Moreover, the laser ablation method may generate 1–10 g of superior SWCNTs [129]. These methods are less beneficial for the manufacture of nanotubes due to their limitations, such as the significant use of energy and the expensive equipment required. Arc discharge has several drawbacks, including the need for purification, elevated temperatures, and twisted nanotubes. In contrast, laser-based methods only apply in lab settings and require further raw material purification. MWCNTs, nano-onions, nano-rods, nanowires, and nano-belts are among the additional carbonaceous nano-architectures that may be constructed using the sonochemical/hydrothermal technique [127]. Their stability at room temperature, their straightforwardness of commencing material preparation, and their ability for operation at low temperatures (150–180°C) with a requirement of hydrocarbons or carrier gas represent a few of the benefits of this method over alternative approaches [131].

2.2.2 Application

Healthcare, nanotechnologies, production, electronic devices, exceptionally strong composite materials, hydrogen-storing media, and electronic devices [132], adsorbents, catalysts [119], actuators, energy integration devices [133], nano-probes, and sensors [121] are merely a few of the various uses that are suitable for nanotubes. The adsorption abilities of graphene and carbon nanotube nanomaterials are more effective compared to those exhibited by conventional adsorbents such as zeolite, silica, TiO_2, chitosan, zero-valent iron, iron oxide, and polymers because of their substantial surface area, adaptable production, non-corrosive characteristics, tunable surface chemistry and surface containing oxygen functional groups. Consequently, CNTs, in their unmodified or modified form, are currently employed as adsorbents to extract contaminants from gaseous or aqueous environments. In a fluidized bed reactor, MWCNTs are produced by the CVD technique with a Co–Mo/MgO nanocatalyst present while H_2 and CH_4 gases are swirling. The allylamide-grafted multiwall CNTs produced from MWCNTs can be used for effective H_2S elimination [134]. According to a report, this material's effective nitrogen functions give it a high capacity for H_2S elimination. To extract MB from a water-based solution, oxidized multiwalled carbon nanotube (OMWCNT)–iron oxide and OMWCNT–k-carrageenan–iron oxide nanocomposites have been generated by Duman et al. [135] using chemical vapor deposition. The adsorption procedure removed antibiotics such as quinolones, macrolides, sulfonamides, and tetracyclines. Based on experimental data, MWCNT–Pd-NPs perform better than MWCNTs as adsorbents in terms of adsorbent mass and decreased contact period in removing a specific amount of SY and AZ. With ultrasonic [136], it has been shown that dyes like SY and azur(II) (AZ) are capable of being adsorbed over MWCNT and MWCNT–Pd-NPs. Table 3.1 gives details of the pollutant adsorption capabilities of CNTs.

Balarak et al. studied eliminating contaminants from wastewater and industrial effluents using acid blue 92 dye adsorption onto SWCNT [137]. The existence of -ve charged atoms in the dye and functionalized groups with H-atoms on the surface of the adsorbent was additionally demonstrated by the FTIR analysis, in addition to H-bonding, induced London scattering, and dipole by dipole interactions within the adsorption methods. The result was caused by the fact that this dye failed to exist in its ionized state below the pH range of 3, corresponding to the ideal pH level. The physical adsorption procedure might be clarified further by the low level of the usual Gibbs free energy fluctuation detected in the thermodynamic studies.

Fard et al. recently suggested using ethylenediamine to functionalize SWCNT to adsorb lead and cadmium ions from synthetic wastewater [138]. The surface of CNTs following modification proved to have greater porosity and roughness and, therefore, may have improved metal ion adsorption. This contributed to analyzing the adsorbent material's morphology using the FESEM. Comparisons between modified and non-functionalized SWCNT are required to show whether functionalizing SWCNT increases its efficacy and absorption capacity.

2.3 Graphene

A single layer of atoms of carbon termed graphene forms a closely packed, 2D crystalline honeycomb structure [139–142]. A vital aspect of graphene's sp^2 hybridization of the two 2p and 2s orbitals of the carbon atom was its hexagonal form. The discovery that graphene is a mono-layered 2D hexagonal crystal, attributable to the electron confinement in 2D materials, might explain the conventional quantum Hall effect [143,144].

2.3.1 Synthesis and modification techniques

In science, graphene's fascinating properties, such as its strong Young's modulus, excellent thermal conductivity, rapid charge carrier mobility, and enormous surface area, could be extensively and thoroughly investigated [145]. The primary techniques employed for making graphene involve exfoliation and cleavage, plasma-enhanced CVD, thermal CVD, unzipping CNT, thermal degradation on substrates, and chemical methods. Extensive investigations have been conducted on various graphene derivatives, including one-dimensional graphene nanoribbons (GNRs), two-dimensional graphene nanosheets (GNSs), and graphene quantum dots (GQDs), which exhibit diverse thickness ranges.

There are four main types of graphene fabrication using MWCNT unzipping techniques: oxidative unzipping, electrochemical unzipping, reductive-intercalation-assisted approach, and an unidentified fourth category of methodologies. Alkali metals' widely recognized capability to intercalate graphite with Z-axis expansion is essential to the first approach. Due to the extreme strain this lattice growth produces on the concentric walls, the tubes burst (open longitudinally). The resulting GNRs are multi-layered and foliated despite their high conductivity. Their affinity to one another prevents them from exfoliating into single-layer ribbons. By treating MWCNTs in acidic oxidative environments, the oxidative approach produces graphene oxide (GO) from graphite using a formulation like that employed in the Hummers process. GO nanoribbons (GONRs) are the end outcome. Unlike GNRs generated by the reductive-intercalation process, GONRs may grow as single-layered structures and readily exfoliate in aqueous solution.

2.3.2 Application

Transistors, fuel cells, ultracapacitors, clear electrodes, batteries, biosensors, gas sensors, extraction, drug delivery, hydrogen storage, adsorbents, and photocatalysts are possible uses for graphene. Among the most significant fields of discussion in chemistry is graphene's potential application in the adsorption of organic moieties, which are prevalent toxic elements. The capacity of various potentially dangerous chemicals to be effectively and adaptively eliminated from water-based solutions has also been extensively investigated. Chemical pesticides [146–148], volatile organic chemicals [149,150], oil [151–153], phenolic compounds [154–156], antimicrobial agents [157–159], gases [160], metallic materials [161–163], and colors [164–166] include examples. Table 3.1 summarizes graphene's and compound graphene's pollutant adsorption capabilities.

The removal of Zn(II), Pb(II), Cd(II), Co(II), Ni(II), and Cu(II) metal ions can be accomplished by GO/cellulose membranes [167]. The metal ions under study sorb against membranes in a monolayer coating based on adsorption kinetics and isotherms. Yu and colleagues developed a GO nanosheet to effectively eliminate four common antibiotic-resistant genes (ARGs) with two distinct chemical structures: double-stranded (ds)-ARGs and cyclic (c)-ARGs. In addition, cobalt (II), nickel (II), zinc (II), cadmium (II), and lead (II) ions exhibited maximal adsorption effectiveness of 15.5, 14.3, 26.6, 16.7, 26.8, and 107.9 mg g^{-1} during acidic conditions at pH 4.5.

The maximum adsorption equilibrium reached in just 15 min demonstrated how effective GO is at removing ARGs [168]. At a rate of 2.88 logs toward ds-ARGs and 3.11 logs toward c-ARGs, four ARGs have been eliminated from 300 mg mL^{-1} GO liquid because of increased adsorption sites, including oxygen-containing groups and conjugate *P* region sites. The adsorption schematics and adsorption kinetics of both c- and ds-ARGs on graphene oxide (GO) nanosheets are shown in Fig. 3.3. Guo et al. used a carbon–GO nanofiber composite to study treatment of volatile organic compounds (VOCs), like butanone and benzene [169]. At 20°C, the maximum adsorption capacities of butanone and benzene onto the GO/carbon composite nanofibers were 130.5 cm^3 g^{-1} and 83.2 cm^3 g^{-1}, respectively. Consequently, GO concurrently enhances the nanofibers' adsorption capabilities for polar VOCs. Bisphenol A (BPA) was removed from the water-based solution by Xu et al. with graphene [170]. A Langmuir isotherm showed that graphene's maximum BPA adsorption capacity was 182 mg g^{-1} at 302.15 K. BPA adsorption onto graphene

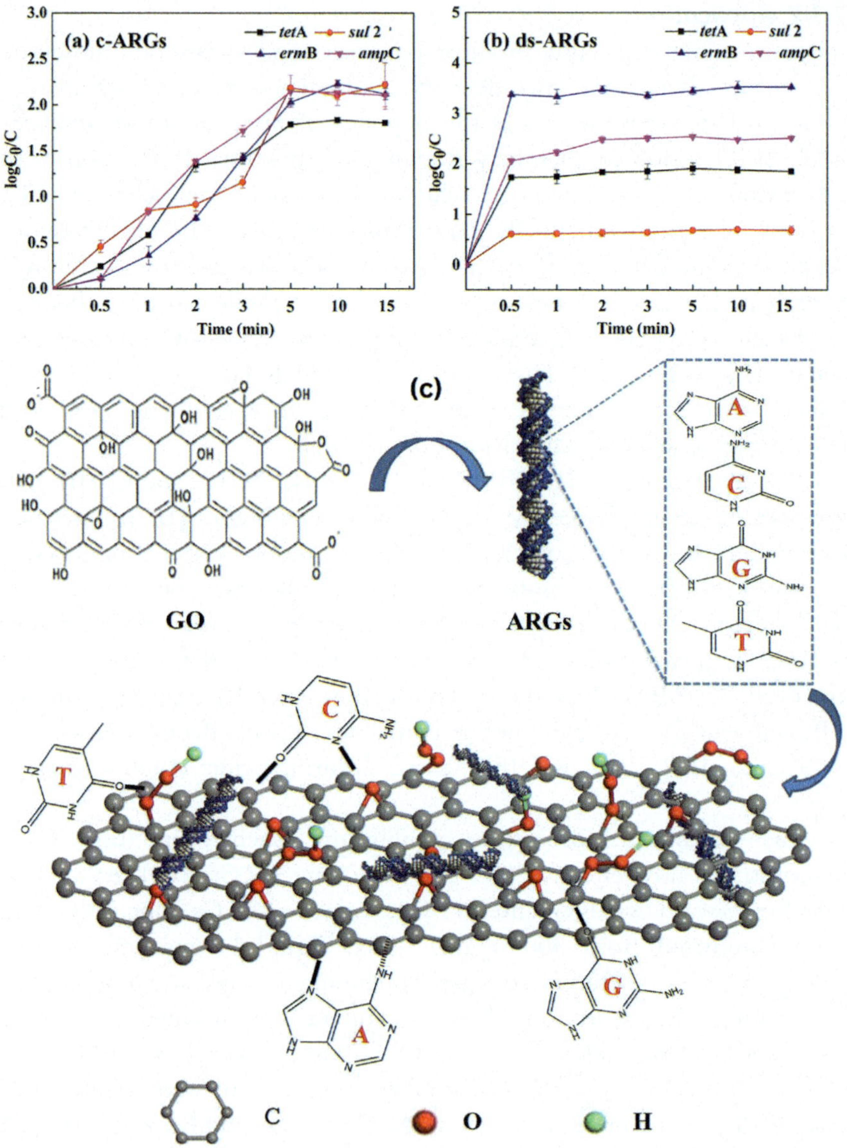

Figure 3.3 Adsorptive interaction of (a) c-ARGs and (b) ds-ARGs with graphene oxide nanosheets, together with (c) a schematic representation illustrating the binding of circular antibiotic resistance genes and double-stranded antibiotic resistance genes on graphene oxide surfaces, where A, C, G, and T correspond to adenine, cytosine, guanine, and thymine, respectively. Reproduced with the license granted by Elsevier from Ref. [168], Copyright 2017.

involves p–p interactions and hydrogen bonding. Gong et al. synthesized PNIPAM-grafted GO (GO–PNIPAM) by grafting poly-(N-isopropylacrylamide) (PNIPAM) onto graphene oxide using Brodie's technique to adsorb phenol contaminants. The results demonstrate that the effectiveness of adsorption is strongly influenced by temperature, suggesting that phenol is adsorbed selectively because of various interactions, including hydrogen bonds. According to Sharif and the team, GO/iron oxide (rGO–IO) adsorbents were created utilizing a solvothermal technique and utilized to eliminate MB. Increasing the IO loading to 75 wt% reduced adsorption performance to 26 mg g^{-1}, while 60 wt% IO enabled an improved adsorption capacity of 39 mg g^{-1}. For the effective treatment of MB, a magnetic β-cyclodextrin/chitosan with GO composite (MCCG) was additionally applied [171].

Chavda et al. used reline to modify GO, enabling ZrO_2 attachment. FTIR confirmed functionalization, with a 2900 cm^{-1} peak indicating choline-derived alkyl chains. Elemental analysis detected chlorine, verifying DES incorporation. TGA showed improved thermal stability, reducing surface oxygen decomposition. The nanocomposite achieved ~100% MB adsorption (20 mg/L) in 5 min [75]. Later, Chavda et al. functionalized GO@TiO_2 with reline, enhancing adsorption. FE-SEM showed aggregated graphene with uniformly distributed TiO_2. TGA revealed nitrogen and chlorine decomposition (300–550°C). Due to DES coverage, the BET analysis showed a surface area drop (94.12–107.30 m^2 g^{-1}). This nanocomposite removed 100% MB in 3 min (Fig. 3.4) [73].

2.4 Carbon nanofibers

2.4.1 Fabrication and modification

Using electrospinning technology, CNFs originated and are now used as intriguing materials in various industries. The two approaches that could form CNFs of average diameters of 50–200 nm include vapor deposition fabrication, which has been used since [172,173], and electrospinning, which has been applied increasingly for CNF synthesis [174–176]. This technique is frequently used to produce the progenitors of CNFs using polymer nanofibers. The involved polymers include polyacrylonitrile (PAN) [177], lignin [178], phenolic resins [179–181], pitches, poly (vinylidene fluoride) (PVDF) [182–184], polybenzimidazole (PBI) [185,186], polyimides (PIs) [187–189], and poly (vinyl alcohol) (PVA) [190]. CNF composites' primary characteristics include acceptable

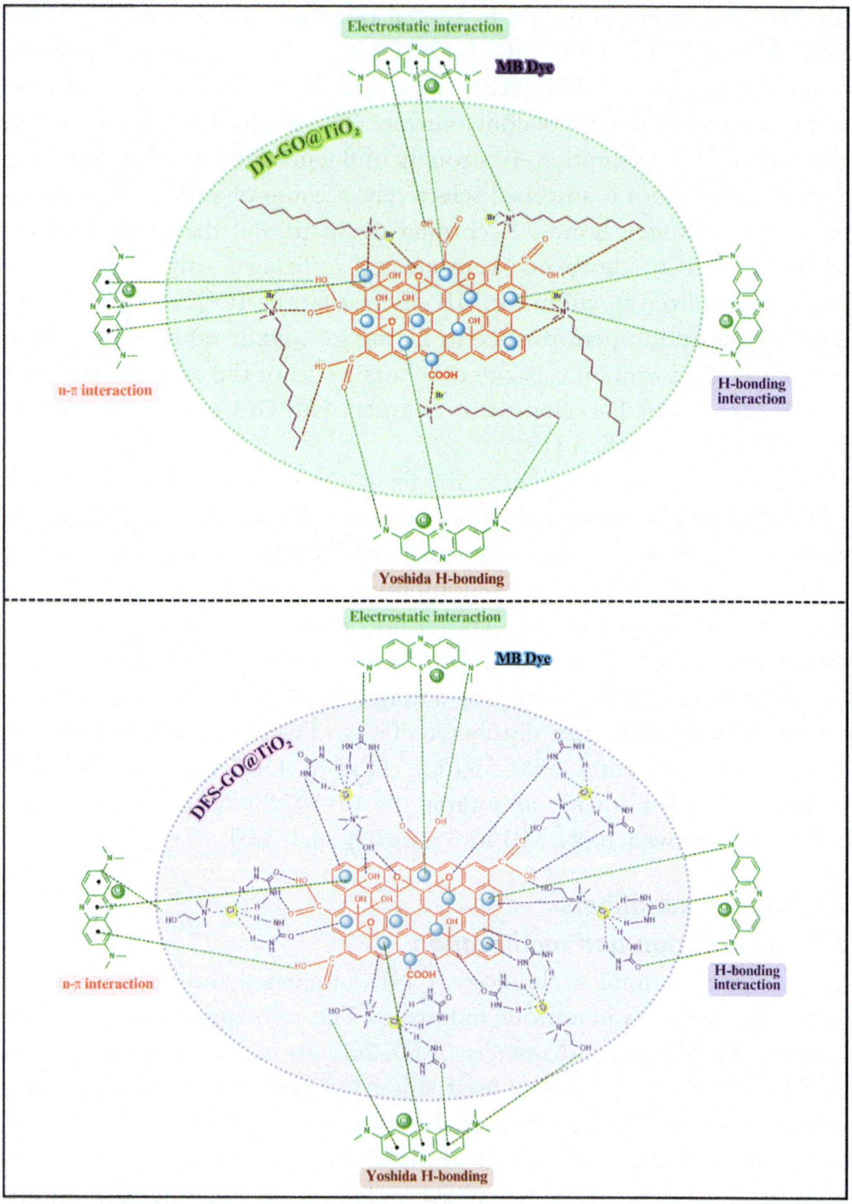

Figure 3.4 Illustrates methylene blue adsorption on DT-GO@TiO_2 and DES-GO@TiO_2 nanocomposites through various interaction mechanisms. Reproduced with the license granted by the American chemical society from Ref. [73], Copyright 2024.

mechanical, electrical, and thermal conductivity qualities. Their thermal and mechanical transport phenomena have also been extensively researched [191–193]. The second method used to validate the discovery of CNT is effectively exposing nanofibers to thermal processes to carbonize them at different temperatures and ambient conditions. These nanofibers' diameter decreases because of the volume and weight changes typically occurring throughout the carbonization process [194,195].

2.4.2 Application

CNFs and their derived compounds have been effectively used to remove organic solvents, hazardous industrial chemicals, gaseous persistent organic pollutants [196], organic contaminants such as pesticides and phenolic compounds, VOCs [197,198], dyes [199,200], and metals from the environment. Because of their electron transport, functional groups, and huge surface area, AC nanofibers have been used to effectively eliminate a wide variety of contaminants. These features make them excellent catalytic materials and remarkable adsorbents [201,202]. Table 3.1 lists the CNFs' capacity for adsorption to remove different contaminants from wastewater.

The nitrogen concentration of activated carbon nanofibers derived from electrospinning polyacrylonitrile nanofibers has a significant effect on formaldehyde adsorption [203]. These nanofibers exhibit carbonization when exposed to steam containing various nitrogen-containing functional groups. In humid environments, the ACNFs exhibit more formaldehyde adsorption capacity than the typical thick ACFs. CNFs were developed using electrospun PAN polymer and then thermally stabilized, carbonized, and activated. They were then used to adsorb organic compounds such as 2-chlorophenol (2-CP), bisphenol (BPA), and ciprofloxacin (CIP) [204]. The CIP > BPA > 2-CP order can be used to organize their maximal adsorption capacity and micro/mesoporous shape. These CNTs possess an extensive surface area of 2326 $m^2\ g^{-1}$. Consequently, the highest adsorption capacities of these CNFs for pollutants follow this order: CIP > BPA > 2-CP. Compared to commercial AC, these CNFs exhibit improvements of 1.1 for 2-CP, 1.6 for BPA, and 2.6 for CIP. A two-step procedure consisting of solvothermal and electrospinning techniques is used to create Fe_3O_4/PAN composite NFs [205]. The analysis showed the emergence of uniform NP coatings on PAN nanofibers. The findings were adapted by a pseudo-second-order kinetics model when tetracycline (TC) was eliminated from Fe_3O_4/PAN composite NFs. According to the Langmuir isotherm model, the maximum adsorption capacity reached 257.07 $mg\ g^{-1}$

at pH 6. These nanofibers exhibit high renderability, low cost, and an acceptable, ecologically friendly connection in adsorption/desorption cycles. The steps for creating Fe_3O_4/PAN nanocomposite NFs and their adsorption kinetics and capacities are shown in Fig. 3.5.

With AC nanofibers (ACNFs), the adsorption equilibrium isotherms linked to VOCs, such as ethanol and benzene, have been thoroughly investigated [206]. After electrospinning, steam activated the ACNFs (PAN/DMF 10 wt%). Their micropores and smaller diameters should allow for additional adsorption sites. The findings indicate that ACNFs exhibit greater VOC adsorption capabilities at exceptionally low relative

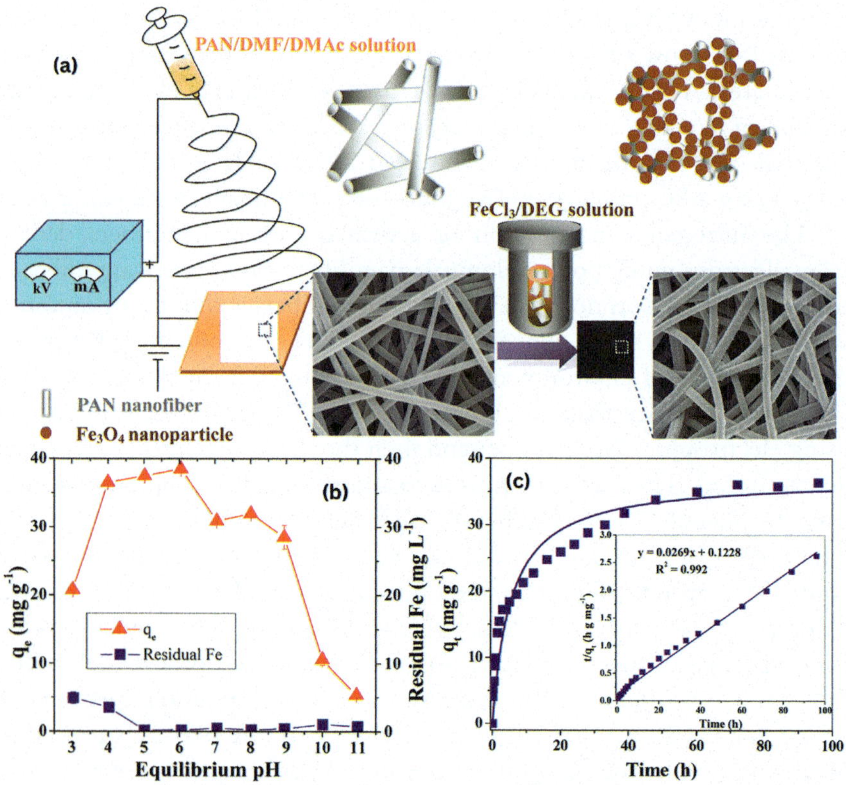

Figure 3.5 (a) Synthesis of Fe_3O_4/PAN nanocomposite nanofibers, (b) adsorption capacity of the pollutant TC on Fe_3O_4/PAN nanocomposite nanofibers and the remaining Fe content in solution at different pH levels, and (c) adsorption rate evaluation of tetracycline on iron (II, III) oxide/polyacrylonitrile nanocomposite nanofibers. Reproduced with the license granted by the American chemical society from Ref. [205], Copyright 2015.

pressures than AC fiber. Combining oxidizing acids (HNO_3 and H_2SO_4) to generate AECNFs, the comparable study additionally carried out surface oxidation [207]. The level of acid oxidation increases with the quantity of oxygen functionalized groups on the nanofibers, increasing the oxidized nanofibers' adsorption potential. Both pure and oxidized CNFs' adsorption ability has been assessed by applying the isotherms model with organic chemicals such as ethanol (polar), butanone, and benzene (non-polar). Furthermore, compared to benzene, the AECNFs had superior water adsorption capabilities and an increased capacity for ethanol and butanone to adsorb on nanofibers. Recently, researchers used a novel adsorbent composed of CNFs and manganese oxide to study the adsorption of Pb^{2+} ions for contaminated water treatment [208]. Polyacrylonitrile functioned as a precursor, enabling the electrospinning method employed in this research to generate the CNFs. CNFs were produced by calcining this molecule, carbonizing it in a furnace, and then peroxidizing it. The SEM morphology investigations show that the diameter value (277 nm) is acceptable.

2.5 Biochar

Various agricultural and industrial biomass materials can undergo pyrolysis at different temperatures under low-oxygen conditions, forming BC, a carbon-rich solid residue. Charcoal, AC, and BC are the three types of carbon that may be made from organic compounds [209,210]. They differ in their practical applications even though their compositions and production methods are remarkably similar. AC always connects with filtering, while BC is linked with water remediation and soil conditioning. Wood is usually extracted to make charcoal, which is used as fuel. Several investigations have claimed that BC requires an activation mechanism and often arises at temperatures that are lower compared to those of AC. The possible advantages of BC in improving soil quality and eliminating environmental pollutants [211] have drawn much attention [212,213]. Because it is less expensive, BC can be used instead of AC, a renowned adsorbent, because of its ability to remove a broad range of contaminants [214]. Recently, BC research was started to build upon this approach to recycle and reduce organic waste resources.

2.5.1 *Synthesis of biochar*

The organic portion of animal manure, forestry waste, municipal solid waste, by-products of industrial biomass, and agricultural wastes are among the raw materials used to synthesize BC. Biomass-derived products, such as BC, are made from various biomass feedstock types through pyrolysis at temperatures between 300 and 700°C, either with or without oxygen [215]. BC's own physicochemical characteristics, including particular surface area, pore characteristics, surface charges, and unique functional groups, influence its adsorption capacity. The material type and the pyrolysis conditions, like the temperature and time of the pyrolysis, the heating rate, and the pyrolysis furnace, determine the features of BC [211]. The use of surface oxidation, nanoscale metals, surface reduction, and impregnation with mineral components constitutes typical techniques for BC modification [214].

2.5.2 Practical implications

The performance of the prepared BC was determined through its adsorption-based removal efficiency toward different pollutants, including antimicrobial agents [216], heavy metals [217,218], agrochemicals [219], and industrial colorants [220]. Table 3.1 shows an overview of BC's adsorbent capabilities for pollutants. Biochar generated from lemons with zerovalent iron support was created for the rapid adsorption of MB dye in a study by Abd El-Monaem et al. [221]. Magnetization was crucial in the process of biosorption, as in earlier studies. However, the kinetic investigation showed that, unlike the work's equilibrium period of 15 min, equilibrium was established in just 5 min due to the Langmuir isotherm. In this study, an intriguing test was conducted to determine whether the adsorbent could be reused. Its recyclability was verified, allowing for seven cycles of adsorption and desorption.

However, biochar has the drawback of low reactivity and limited selectivity [222]. One potential solution to address this drawback is to alter the pyrolysis process's environment, which could result in particles smaller than 10–600 μm [223]. Therefore, the benefits of nanotechnology and biochar are combined to create nanobiochar [224]. Improved surface-volume ratio increases superficial energy, and more adsorption sites come from this size reduction, improving dye adsorption efficiency [225]. NCB-TA-PC nanocomposite was recently created by functionalizing corncob with triethylenetetramine and treating it with sulfuric acid. This nanobiochar effectively removes harmful anionic dyes like Sunset Yellow and tartrazine [226]. The kernels and husks of the corn cob were heated at a 50°C min^{-1} rate to 400°C for 1 h to create the nanobiochar. The final material was rinsed with deionized water and stirred for 40 min. After filtering, it was oven-dried for 48 h at 50°C. SEM and HR-TEM were used for morphological analysis to describe it. The nanometric scale of the NCB-TAPC nanocomposite was verified by HR-TEM and SEM, which revealed diameters of 11.0 and 18.6 nm, respectively.

2.6 Nanocomposites

Nanocomposites are solids with heterogeneous structures that incorporate clusters of nanocrystals or NPs dispersed throughout a matrix. These materials typically consist of two phases: a continuous phase known as the matrix and a dispersed phase of NPs [227]. Carbon-based nanocomposites have been utilized to filter contaminants out of textiles. Almoisheer and the researcher formed a nanocomposite of CNT, GO, and silica to eliminate Congo red dye from artificial wastewater [228]. GO, silica NPs, and SWCNT were shaken, mixed, and sonicated before being placed in a hydrothermal reactor to form the nanocomposite. Adsorption experiments were conducted at pH 2 and 10, utilizing 10 mg of adsorbent and 20 mL of dye solution. The dye was reported to have the highest adsorption capacity of 456 mg g^{-1} at 3.0 pH. Three varying temperatures, 20, 30, and 40°C, were employed for thermodynamic investigations and to generate adsorption isotherms. The range of Congo red concentrations was 50–500 mg L^{-1}. With a coefficient of determination of 0.955, the Langmuir approach was the isotherm that most effectively represented the dye adsorption process for the objectives of the adsorption isotherm study.

MO dye was extracted from an aqueous background utilizing a different innovative nanocomposite made of chitosan and nanodiamonds [229]. The nanocomposite production used different nanodiamond concentrations (0.1%, 0.5%, 1%, and 1.5% by weight). Adsorption studies were performed to assess chitosan nanodiamonds and pure chitosan adsorption capacities. Even at exceptionally low concentrations of nanodiamonds, the nanocomposite's adsorption capacity was twice that of pure chitosan. The required amount of the adsorption material has been studied with a 100 mg L^{-1} initial dye level. The dye removal percentage increased from 25% to 88% at 0.05 and 1 g L^{-1} of the adsorbent quantity. Since wastewater contains increased concentrations of dissolved surfactants and salts, which can influence the dye adsorption procedure, the impact of ionic strength was assessed using KCl and NaCl at an acidic pH of 4. It showed that if the level of both salts increases, the rate of dye adsorption is reduced. Using actual samples, like industrial effluents, could result in a minor change in process efficiency.

Utilizing an adsorbent of magnetic chitosan-GO, Sohni and team investigated removing azo-type dyes from textile wastes [230]. The material was made using the following methods: chitosan crosslinking, graphene oxide synthesis, and magnetite NP development. According to adsorption experiments, 93% of the bromophenol blue dye and 90% of the acid red-17 dye were swiftly eliminated in the first phase of the process. The longer the contact period, the quicker the dye removal approached equilibrium. With a coefficient of regression of 0.99 for both dyes, the pseudo-second-order kinetic model perfectly describes the experimental kinetic data results. This nanocomposite eliminated around 79% of the bromophenol blue from industrial effluent and acid red dye-17 (86%) from river waters. The nanocomposite's ability to adsorb the colors under investigation in extremely chemically complicated matrices was shown through this application.

Hexavalent chromium has also been extracted from aqueous solutions using GO to make a ternary nanocomposite with diethylenetriamine and magnetite NPs. The hydrothermal method was used to synthesize this nanocomposite, and XRD, TEM, FESEM, and FTIR were used to describe the material. The adsorption experiments were conducted using a potassium dichromate solution and $0.2\ g\ L^{-1}$ of adsorbent. Inductively coupled plasma mass spectrometry determined the total amount of chromium in the upper phase after 12 h of stirring to reach equilibrium. The highest agreement between the adsorption isotherms and the Langmuir model was obtained at 298, 308, and 318 K. This shows that monolayer hexavalent chromium was adsorbed with adsorption energies that were comparable to those of the adsorbent. To eliminate hexavalent chromium, Zhang et al. also employed graphene chitosan; however, they changed the nanocomposite by adding disodium ethylenediaminetetraacetate [231]. Since chromium in solution requires various chemical compounds depending on its pH and attains its highest adsorption capacity at pH 2, the effect of pH on the adsorption process was studied. Thermodynamic studies also showed the adsorption process to be spontaneous, endothermic, probable, and unreliable.

2.7 Additional carbon-based materials

There are additionally carbon-containing materials compared to AC, CNTs, graphene, CNFs, CAs, and BC, which can facilitate the adsorption of naturally harmful substances. Porous organic frameworks consisting of carbon are known as covalent organic frameworks (COFs) and metal-organic frameworks (MOFs) [7,8]. MOFs and COFs are two different forms of adsorbents employed to eliminate various pollutants efficiently. Both materials' remarkable porosity and ease of pore size shift from the scale of microporous to mesoporous are distinctive features that have received much attention nowadays [232,233].

3. Conclusion and future perspectives

In conclusion, this chapter explores the important uses of carbon-based adsorbents, including AC, graphene, CNFs, CNTs, carbon aerogels, and BC, for removing hazardous contaminants from aquatic backgrounds. Additionally, it compiles the influence of different manufacturing processes on the physicochemical properties of carbon-based adsorbents and their corresponding impact on contaminant adsorption. Carbon-based materials are the most appropriate options for managing contaminants due to their remarkable features and straightforwardness of production and based on the field's entire phenomenal expansion. Carbon-based materials with specific morphologies, topologies, and functionalized groups are promising for broad environmental uses. These materials may absorb significant amounts of chemical contaminants, such as phenolic, pesticides, dyes, oils, and VOCs, antimicrobial agents, and gas impurities. Because of their superior adsorption capacity and excellent selectivity, carbon nanomaterials are the most excellent carriers for adsorption-based pollution separation of both organic and inorganic contaminants. Improving and streamlining the preparation process was necessary for studies on carbon-based adsorbents for the elimination of contaminants. Many harmful substances can be eliminated from water-based solutions by interacting with functionally related groups within carbon molecules. The physicochemical behavior of adsorption in nanomaterials composed of carbon is significantly influenced by synthesis, functionalization, modification, and doping and coating techniques. For the purpose of enhancing the formation, mechanisms, and potential uses of carbon-based adsorbents for the removal of frequently discovered toxic substances, the present study is anticipated to assist in driving future studies. Hence, the discovery of carbon-based adsorbents is required to discard various pollutants cost-effectively. While studies have tried to produce various carbon-based materials as adsorbents, it is acknowledged that synthesizing carbon adsorbents necessitates a range of comprehensive and extensive procedures. Materials preparation employing various techniques should also be discovered in detail with the goal of minimizing and eliminating additional pollutants induced by the physicochemical fabrication of carbon-based materials. Because freshwater includes a range of contaminants with complicated combinations, further knowledge is necessary in real-life large-scale situations. A significant amount of research exclusively evaluates carbon-based materials in laboratory environments.

References

[1] Tortajada, C., van Rensburg, P., 2020. Drink more recycled wastewater. Nature 577, 26–28. https://doi.org/10.1038/d41586-019-03913-6.

[2] Anjum, M., Miandad, R., Waqas, M., Gehany, F., Barakat, M.A., 2019. Remediation of wastewater using various nano-materials. Arab. J. Chem. 12, 4897–4919. https://doi.org/10.1016/j.arabjc.2016.10.004.

[3] Yang, J., Hou, B., Wang, J., Tian, B., Bi, J., Wang, N., et al., 2019. Nanomaterials for the removal of heavy metals from wastewater. Nanomaterials 9, 424. https://doi.org/10.3390/nano9030424.

[4] Shah, P.A., Chavda, V., Hirpara, D., Sharma, V.S., Shrivastav, P.S., Kumar, S., 2023. Exploring the potential of deep eutectic solvents in pharmaceuticals: Challenges and opportunities. J. Mol. Liq. 390, 123171. https://doi.org/10.1016/j.molliq.2023.123171.

[5] Shah, P.A., Chavda, V., Hirpara, D., Pokar, R., Kumar, S., Shrivastav, P.S., 2024. Advances in deep eutectic solvent-based synthesis of nanomaterials for environmental remediation. J. Mol. Liq. 416, 126482. https://doi.org/10.1016/j.molliq.2024.126482.

[6] Mishra, S.R., Gadore, V., Chavda, V., Panda, S., Roy, S., Sahoo, P., et al., 2025. Black phosphorene: A versatile allotrope revolutionizing environmental, energy, and biomedical applications. Coord. Chem. Rev. 524, 216345. https://doi.org/10.1016/j.ccr.2024.216345.

[7] Mishra, S.R., Chavda, V., Roy, S., Sarkar, N., Gadore, V., Ahmaruzzaman, Md, et al., 2025. Covalent organic framework nanocomposites for superior lithium-ion storage: Bridging molecular design and battery technology. J. Energy Storage 108, 115026. https://doi.org/10.1016/j.est.2024.115026.

[8] Mishra, S.R., Gadore, V., Roy, S., Chavda, V., Ahmaruzzaman, M., 2024. Advancements in Ni-MOFs: Unravelling the photocatalytic tapestry for organic contaminant degradation. Int. J. Environ. Anal. Chem. 1–19. https://doi.org/10.1080/03067319.2024.2343106.

[9] Abbo, H.S., Gupta, K.C., Khaligh, N.G., Titinchi, S.J.J., 2021. Carbon nanomaterials for wastewater treatment. Chem. Bio. Eng. Rev. 8, 463–489. https://doi.org/10.1002/cben.202100003.

[10] Rafatullah, M., Sulaiman, O., Hashim, R., Ahmad, A., 2010. Adsorption of methylene blue on low-cost adsorbents: A review. J. Hazard. Mater. 177, 70–80. https://doi.org/10.1016/j.jhazmat.2009.12.047.

[11] Liu, P., Yan, T., Zhang, J., Shi, L., Zhang, D., 2017. Separation and recovery of heavy metal ions and salt ions from wastewater by 3D graphene-based asymmetric electrodes via capacitive deionization. J. Mater. Chem. A Mater. 5, 14748–14757. https://doi.org/10.1039/C7TA03515B.

[12] Björklund, K., Li, L.Y., 2017. Adsorption of organic stormwater pollutants onto activated carbon from sewage sludge. J. Environ. Manage. 197, 490–497. https://doi.org/10.1016/j.jenvman.2017.04.011.

[13] Aldegs, Y., Elbarghouthi, M., Elsheikh, A., Walker, G., 2008. Effect of solution pH, ionic strength, and temperature on adsorption behavior of reactive dyes on activated carbon. Dyes Pigments 77, 16–23. https://doi.org/10.1016/j.dyepig.2007.03.001.

[14] Haghseresht, F., Nouri, S., Finnerty, J.J., Lu, G.Q., 2002. Effects of surface chemistry on aromatic compound adsorption from dilute aqueous solutions by activated carbon. J. Phys. Chem. B 106, 10935–10943. https://doi.org/10.1021/jp025522a.

[15] Yang, K., Xing, B., 2010. Adsorption of organic compounds by carbon nanomaterials in aqueous phase: Polanyi theory and its application. Chem. Rev. 110, 5989–6008. https://doi.org/10.1021/cr100059s.

[16] Yang, K., Wu, W., Jing, Q., Zhu, L., 2008. Aqueous adsorption of aniline, phenol, and their substitutes by multi-walled carbon nanotubes. Environ. Sci. Technol. 42, 7931–7936. https://doi.org/10.1021/es801463v.

[17] Dehgani, Z., Sedghi asl, M., Ghaedi, M., Sabzehmeidani, M.M., Adhami, E., 2020. Removal of paraquat from aqueous solutions by a bentonite modified zero-valent iron adsorbent. New J. Chem. 44, 13368–13376. https://doi.org/10.1039/D0NJ02259D.

[18] Gas Adsorption Equilibria, 2005. Kluwer Academic Publishers, Boston. https://doi.org/10.1007/b102056.

[19] Giles, C.H., MacEwan, T.H., Nakhwa, S.N., Smith, D., 1960. 786. Studies in adsorption. Part XI. A system of classification of solution adsorption isotherms, and its use in diagnosis of adsorption mechanisms and in measurement of specific surface areas of solids. J. Chem. Soc. 3973. https://doi.org/10.1039/jr9600003973.

[20] Bonilla-Petriciolet, A., Mendoza-Castillo, D.I., Reynel-Ávila, H.E. (Eds.), 2017. Adsorption processes for water treatment and purification. Springer International Publishing, Cham. https://doi.org/10.1007/978-3-319-58136-1.

[21] Kast, W., 1985. Principles of adsorption and adsorption processes. Chem. Eng. Process. Process Intensif. 19, 118. https://doi.org/10.1016/0255-2701(85)80013-1.

[22] Wang, H., Shen, H., Shen, C., Li, Y., Ying, Z., Duan, Y., 2019. Kinetics and mechanism study of mercury adsorption by activated carbon in wet oxy-fuel conditions. Energy Fuels 33, 1344–1353. https://doi.org/10.1021/acs.energyfuels.8b03610.

[23] Harrache, Z., Abbas, M., Aksil, T., Trari, M., 2019. Thermodynamic and kinetics studies on adsorption of indigo carmine from aqueous solution by activated carbon. Microchem. J. 144, 180–189. https://doi.org/10.1016/j.microc.2018.09.004.

[24] Largitte, L., Pasquier, R., 2016. A review of the kinetics adsorption models and their application to the adsorption of lead by an activated carbon. Chem. Eng. Res. Des. 109, 495–504. https://doi.org/10.1016/j.cherd.2016.02.006.

[25] El-Sharkawy, R., El-Ghamry, H.A., 2019. Multi-walled carbon nanotubes decorated with Cu(II) triazole schiff base complex for adsorptive removal of synthetic dyes. J. Mol. Liq. 282, 515–526. https://doi.org/10.1016/j.molliq.2019.02.137.

[26] Pakdehi, S.G., Babaee, S., Azizi, H.R., 2017. Kinetic study and optimization of dehydration of dimethyl amino ethyl azide (DMAZ) using response surface methodology. Bull. Chem. Soc. Jpn. 90, 1325–1332. https://doi.org/10.1246/bcsj.20170253.

[27] Yang, X., Wan, Y., Zheng, Y., He, F., Yu, Z., Huang, J., et al., 2019. Surface functional groups of carbon-based adsorbents and their roles in the removal of heavy metals from aqueous solutions: A critical review. Chem. Eng. J. 366, 608–621. https://doi.org/10.1016/j.cej.2019.02.119.

[28] Zhang, L., Pan, F., Liu, X., Yang, L., Jiang, X., Yang, J., et al., 2013. Multi-walled carbon nanotubes as sorbent for recovery of endocrine disrupting compound-bisphenol F from wastewater. Chem. Eng. J. 218, 238–246. https://doi.org/10.1016/j.cej.2012.12.046.

[29] Gupta, V.K., Moradi, O., Tyagi, I., Agarwal, S., Sadegh, H., Shahryari-Ghoshekandi, R., et al., 2016. Study on the removal of heavy metal ions from industry waste by carbon nanotubes: Effect of the surface modification: a review. Crit. Rev. Environ. Sci. Technol. 46, 93–118. https://doi.org/10.1080/10643389.2015.1061874.

[30] Loos, M., 2015. Allotropes of Carbon and carbon nanotubes. Carbon nanotube reinforced composites. Elsevier, pp. 73–101. https://doi.org/10.1016/B978-1-4557-3195-4.00003-5.

[31] Nasir, S., Hussein, M., Zainal, Z., Yusof, N., 2018. Carbon-based nanomaterials/allotropes: A glimpse of their synthesis, properties and some applications. Materials 11, 295. https://doi.org/10.3390/ma11020295.

[32] Okwundu, O.S., Aniekwe, E.U., Nwanno, C.E., 2018. Unlimited potentials of carbon: Different structures and uses (A review). Metall. Mater. Eng. 24, 145–171. https://doi.org/10.30544/388.

[33] Ioannidou, O., Zabaniotou, A., 2007. Agricultural residues as precursors for activated carbon production—A review. Renew. Sustain. Energy Rev. 11, 1966–2005. https://doi.org/10.1016/j.rser.2006.03.013.

[34] Dąbrowski, A., Podkościelny, P., Hubicki, Z., Barczak, M., 2005. Adsorption of phenolic compounds by activated Carbon—A critical review. Chemosphere 58, 1049–1070. https://doi.org/10.1016/j.chemosphere.2004.09.067.

[35] Pasa, S., Yılmaz, N., Bulduk, İ., Alagöz, O., 2025. Effective adsorption performance of hemp root-derived activated carbon for tamoxifen-contaminated wastewater. J. Mol. Liq. 425, 127219. https://doi.org/10.1016/j.molliq.2025.127219.

[36] Pan, X., Zhao, N., Shi, H., Wang, H., Ruan, F., Wang, H., et al., 2025. Biomass activated carbon derived from golden needle mushroom root for the methylene blue and methyl orange adsorption from wastewater. Ind. Crops. Prod. 223, 120051. https://doi.org/10.1016/j.indcrop.2024.120051.

[37] Deng, H., Li, G., Yang, H., Tang, J., Tang, J., 2010. Preparation of activated carbons from cotton stalk by microwave assisted KOH and K2CO3 activation. Chem. Eng. J. 163, 373–381. https://doi.org/10.1016/j.cej.2010.08.019.

[38] Alslaibi, T.M., Abustan, I., Ahmad, M.A., Foul, A.A., 2013. A review: Production of activated carbon from agricultural byproducts via conventional and microwave heating. J. Chem. Technol. Biotechnol. 88, 1183–1190. https://doi.org/10.1002/jctb.4028.

[39] Jafari, M., Rahimi, M.R., Ghaedi, M., Dashtian, K., 2017. ZnO nanoparticles loaded different mesh size of porous activated carbon prepared from Pinus eldarica and its effects on simultaneous removal of dyes: Multivariate optimization. Chem. Eng. Res. Des. 125, 408–421. https://doi.org/10.1016/j.cherd.2017.07.011.

[40] Yamashita, A., Shikada, J., Oshima, T., Baba, Y., 2014. Preparation of activated carbon with high surface area using biomass wastes and its application to methane storage technology. Kagaku Kogaku Ronbunshu 40, 98–103. https://doi.org/10.1252/kakoronbunshu.40.98.

[41] Figueiredo, J.L., Pereira, M.F.R., Freitas, M.M.A., Órfão, J.J.M., 1999. Modification of the surface chemistry of activated carbons. Carbon N Y 37, 1379–1389. https://doi.org/10.1016/S0008-6223(98)00333-9.

[42] Dias, J.M., Alvim-Ferraz, M.C.M., Almeida, M.F., Rivera-Utrilla, J., Sánchez-Polo, M., 2007. Waste materials for activated carbon preparation and its use in aqueous-phase treatment: A review. J. Environ. Manage. 85, 833–846. https://doi.org/10.1016/j.jenvman.2007.07.031.

[43] Prahas, D., Kartika, Y., Indraswati, N., Ismadji, S., 2008. Activated carbon from jackfruit peel waste by H3PO4 chemical activation: Pore structure and surface chemistry characterization. Chem. Eng. J. 140, 32–42. https://doi.org/10.1016/j.cej.2007.08.032.

[44] Allen, S.J., Whitten, L., Mckay, G., 1998. The production and characterisation of activated carbons: A review. Dev. Chem. Eng. Miner. Process. 6, 231–261. https://doi.org/10.1002/apj.5500060501.
[45] Andas, J., Rahman, M.L.A., Yahya, M.S.M., 2017. Preparation and characterization of activated carbon from palm kernel shell. IOP Conf. Ser. Mater. Sci. Eng. 226, 012156. https://doi.org/10.1088/1757-899X/226/1/012156.
[46] Promdee, K., Chanvidhwatanakit, J., Satitkune, S., Boonmee, C., Kawichai, T., Jarernprasert, S., et al., 2017. Characterization of carbon materials and differences from activated carbon particle (ACP) and coal briquettes product (CBP) derived from coconut shell via rotary kiln. Renew. Sustain. Energy Rev. 75, 1175–1186. https://doi.org/10.1016/j.rser.2016.11.099.
[47] Shoaib, M., Al-Swaidan, H.M., 2015. Optimization and characterization of sliced activated carbon prepared from date palm tree fronds by physical activation. Biomass Bioenergy 73, 124–134. https://doi.org/10.1016/j.biombioe.2014.12.016.
[48] Nabais, J.M.V., Laginhas, C.E.C., Carrott, P.J.M., Ribeiro Carrott, M.M.L., 2011. Production of activated carbons from almond shell. Fuel Process. Technol. 92, 234–240. https://doi.org/10.1016/j.fuproc.2010.03.024.
[49] Dashtian, K., Porhemat, S., Rezvani, A.R., Ghaedi, M., Sabzehmeidani, M.M., 2018. Adsorption of semisoft pollutants onto Bi 2 S 3/Ag 2 S-AC under the influence of ultrasonic waves as external filed. J. Ind. Eng. Chem. 60, 390–400. https://doi.org/10.1016/j.jiec.2017.11.026.
[50] Bagheri, S., Aghaei, H., Ghaedi, M., Asfaram, A., Monajemi, M., Bazrafshan, A.A., 2018. Synthesis of nanocomposites of iron oxide/gold (Fe3O4/Au) loaded on activated carbon and their application in water treatment by using sonochemistry: Optimization study. Ultrason. Sonochem. 41, 279–287. https://doi.org/10.1016/j.ultsonch.2017.09.031.
[51] Sharifpour, E., Ghaedi, M., Nasiri Azad, F., Dashtian, K., Hadadi, H., Purkait, M.K., 2018. Zinc oxide nanorod-loaded activated carbon for ultrasound-assisted adsorption of safranin O: Central composite design and genetic algorithm optimization. Appl. Organomet. Chem. 32. https://doi.org/10.1002/aoc.4099.
[52] Bazrafshan, A.A., Hajati, S., Ghaedi, M., Asfaram, A., 2018. Synthesis and characterization of antibacterial chromium iron oxide nanoparticle-loaded activated carbon for ultrasound-assisted wastewater treatment. Appl. Organomet. Chem. 32. https://doi.org/10.1002/aoc.3981.
[53] Shirmardi, M., Mesdaghinia, A., Mahvi, A.H., Nasseri, S., Nabizadeh, R., 2012. Kinetics and equilibrium studies on adsorption of acid red 18 (Azo-Dye) using multiwall carbon nanotubes (MWCNTs) from aqueous solution. J. Chem. 9, 2371–2383. https://doi.org/10.1155/2012/541909.
[54] Ghaedi, M., Hossaini, S.J., Ramezani, S., 2012. Platinum nanoparticles loaded on activated carbon as novel adsorbent for the removal of Congo red. J. Chem. 9, 63–74. https://doi.org/10.1155/2012/257584.
[55] Wang, J.-P., Yang, H.-C., Hsieh, C.-T., 2010. Adsorption of phenol and basic dye on carbon nanotubes/carbon fabric composites from aqueous solution. Sep. Sci. Technol. 46, 340–348. https://doi.org/10.1080/01496395.2010.508066.
[56] Askari, H., Ghaedi, M., Dashtian, K., 2017. Comparative studies on removal of methyl orange and sunset yellow dyes using ZnS:Cu–NPs–AC: Optimization, equilibrium, kinetic and thermodynamic studies. Desalination Water Treat. 77, 355–365. https://doi.org/10.5004/dwt.2017.20360.
[57] Mishra, A.K., Arockiadoss, T., Ramaprabhu, S., 2010. Study of removal of azo dye by functionalized multi walled carbon nanotubes. Chem. Eng. J. 162, 1026–1034. https://doi.org/10.1016/j.cej.2010.07.014.

[58] Machado, F.M., Bergmann, C.P., Fernandes, T.H.M., Lima, E.C., Royer, B., Calvete, T., et al., 2011. Adsorption of reactive red M-2BE dye from water solutions by multi-walled carbon nanotubes and activated carbon. J. Hazard. Mater. 192, 1122–1131. https://doi.org/10.1016/j.jhazmat.2011.06.020.
[59] He, B.Y., Wang, X.C., Xue, H.Y., Hao, X.L., 2011. Adsorption of direct dyes from aqueous solutions by gelatin microspheres: Kinetics and equilibrium. Adv. Mat. Res. 332–334, 1577–1581. https://doi.org/10.4028/www.scientific.net/AMR.332-334.1577.
[60] Mohammadi, A., Veisi, P., 2018. High adsorption performance of β-cyclodextrin-functionalized multi-walled carbon nanotubes for the removal of organic dyes from water and industrial wastewater. J. Environ. Chem. Eng. 6, 4634–4643. https://doi.org/10.1016/j.jece.2018.07.002.
[61] Alipanahpour, D.E., Ghaedi, M., Asfaram, A., Mehrabi, F., Sadeghfar, F., 2019. Efficient adsorption of Azure B onto CNTs/Zn:ZnO@Ni2P-NCs from aqueous solution in the presence of ultrasound wave based on multivariate optimization. J. Ind. Eng. Chem. 74, 55–62. https://doi.org/10.1016/j.jiec.2018.12.050.
[62] Konicki, W., Pełech, I., Mijowska, E., Jasińska, I., 2012. Adsorption of anionic dye direct red 23 onto magnetic multi-walled carbon nanotubes-Fe3C nanocomposite: Kinetics, equilibrium and thermodynamics. Chem. Eng. J. 210, 87–95. https://doi.org/10.1016/j.cej.2012.08.025.
[63] Gupta, A., Vidyarthi, S.R., Sankararamakrishnan, N., 2014. Enhanced sorption of mercury from compact fluorescent bulbs and contaminated water streams using functionalized multiwalled carbon nanotubes. J. Hazard. Mater. 274, 132–144. https://doi.org/10.1016/j.jhazmat.2014.03.020.
[64] Gong, J., Feng, J., Liu, J., Jiang, Z., Chen, X., Mijowska, E., et al., 2014. Catalytic carbonization of polypropylene into cup-stacked carbon nanotubes with high performances in adsorption of heavy metallic ions and organic dyes. Chem. Eng. J. 248, 27–40. https://doi.org/10.1016/j.cej.2014.01.107.
[65] Bhanjana, G., Dilbaghi, N., Kim, K.-H., Kumar, S., 2017. Carbon nanotubes as sorbent material for removal of cadmium. J. Mol. Liq. 242, 966–970. https://doi.org/10.1016/j.molliq.2017.07.072.
[66] Yang, S., Li, J., Shao, D., Hu, J., Wang, X., 2009. Adsorption of Ni(II) on oxidized multi-walled carbon nanotubes: Effect of contact time, pH, foreign ions and PAA. J Hazard. Mater. 166, 109–116. https://doi.org/10.1016/j.jhazmat.2008.11.003.
[67] Ihsanullah, A.-K.F.A., Abu-Sharkh, B., Abulkibash, A.M., Qureshi, M.I., Laoui, T., et al., 2016. Effect of acid modification on adsorption of hexavalent chromium (Cr (VI)) from aqueous solution by activated carbon and carbon nanotubes. Desalination Water Treat. 57, 7232–7244. https://doi.org/10.1080/19443994.2015.1021847.
[68] Yang, Z.-F., Li, L.-Y., Hsieh, C.-T., Juang, R.-S., 2018. Co-precipitation of magnetic Fe3O4 nanoparticles onto carbon nanotubes for removal of copper ions from aqueous solution. J. Taiwan Inst. Chem. Eng. 82, 56–63. https://doi.org/10.1016/j.jtice.2017.11.009.
[69] Sadegh, H., Ali, G.A.M., Makhlouf, A.S.H., Chong, K.F., Alharbi, N.S., Agarwal, S., et al., 2018. MWCNTs-Fe3O4 nanocomposite for Hg(II) high adsorption efficiency. J. Mol. Liq. 258, 345–353. https://doi.org/10.1016/j.molliq.2018.03.012.
[70] Sharif, F., Gagnon, L.R., Mulmi, S., Roberts, E.P.L., 2017. Electrochemical regeneration of a reduced graphene oxide/magnetite composite adsorbent loaded with methylene blue. Water Res. 114, 237–245. https://doi.org/10.1016/j.watres.2017.02.042.

[71] Alijani, H., Shariatinia, Z., 2018. Synthesis of high growth rate SWCNTs and their magnetite cobalt sulfide nanohybrid as super-adsorbent for mercury removal. Chem. Eng. Res. Des. 129, 132–149. https://doi.org/10.1016/j.cherd.2017.11.014.
[72] Lingamdinne, L.P., Koduru, J.R., Roh, H., Choi, Y.-L., Chang, Y.-Y., Yang, J.-K., 2016. Adsorption removal of Co(II) from waste-water using graphene oxide. Hydrometallurgy 165, 90–96. https://doi.org/10.1016/j.hydromet.2015.10.021.
[73] Chavda, V., Hirpara, D., Rao, V., Kumar, S., 2024. Deep eutectic Solvent/surfactant-engineered GO@TiO_2 advanced material for sustained and ultrafast adsorption of hazardous dye from contaminated aqueous environment. ACS Appl. Eng. Mater. 2, 604–617. https://doi.org/10.1021/acsaenm.3c00696.
[74] Konicki, W., Aleksandrzak, M., Moszyński, D., Mijowska, E., 2017. Adsorption of anionic azo-dyes from aqueous solutions onto graphene oxide: Equilibrium, kinetic and thermodynamic studies. J. Colloid Interface Sci. 496, 188–200. https://doi.org/10.1016/j.jcis.2017.02.031.
[75] Chavda, V., Patel, B., Singh, S., Hirpara, D., Rajeswari, V.D., Kumar, S., 2023. A sustainable approach for the adsorption of methylene blue from an aqueous background: An adsorbent based on DES/CGS modified GO@ZrO_2. RSC Sustain. 1, 2038–2057. https://doi.org/10.1039/D3SU00236E.
[76] Konicki, W., Aleksandrzak, M., Mijowska, E., 2017. Equilibrium, kinetic and thermodynamic studies on adsorption of cationic dyes from aqueous solutions using graphene oxide. Chem. Eng. Res. Des. 123, 35–49. https://doi.org/10.1016/j.cherd.2017.03.036.
[77] Ahmad, M., Yang, K., Li, L., Fan, Y., Shah, T., Zhang, Q., et al., 2020. Modified tubular carbon nanofibers for adsorption of Uranium(VI) from water. ACS Appl. Nano Mater. 3, 6394–6405. https://doi.org/10.1021/acsanm.0c00837.
[78] Tran, H.N., 2020. Comments on "High-efficiency removal of dyes from wastewater by fully recycling litchi peel biochar." Chemosphere 257, 126444. https://doi.org/10.1016/j.chemosphere.2020.126444.
[79] Eltaweil, A.S., Ali, M.H., Abd El-Monaem, E.M., El-Subruiti, G.M., 2020. Mesoporous magnetic biochar composite for enhanced adsorption of malachite green dye: Characterization, adsorption kinetics, thermodynamics and isotherms. Adv. Powder Technol. 31, 1253–1263. https://doi.org/10.1016/j.apt.2020.01.005.
[80] Pan, X., Nan, H., Yang, P., Yang, L., Wang, G., Chen, H., et al., 2018. Low-temperature and large-scale synthesis of carbon nanofiber web via electrospinning and their efficient removal of Cr(VI) ions. ChemistrySelect 3, 10543–10548. https://doi.org/10.1002/slct.201802844.
[81] Yang, S.-S., Chen, Y., Kang, J.-H., Xie, T.-R., He, L., Xing, D.-F., et al., 2019. Generation of high-efficient biochar for dye adsorption using frass of yellow mealworms (larvae of Tenebrio molitor linnaeus) fed with wheat straw for insect biomass production. J. Clean. Prod. 227, 33–47. https://doi.org/10.1016/j.jclepro.2019.04.005.
[82] Sun, Y., Wu, Z.-Y., Wang, X., Ding, C., Cheng, W., Yu, S.-H., et al., 2016. Macroscopic and microscopic investigation of U(VI) and Eu(III) adsorption on carbonaceous nanofibers. Environ. Sci. Technol. 50, 4459–4467. https://doi.org/10.1021/acs.est.6b00058.
[83] Ahmad, M., Wang, J., Xu, J., Zhang, Q., Zhang, B., 2020. Magnetic tubular carbon nanofibers as efficient Cu(II) ion adsorbent from wastewater. J. Clean. Prod. 252, 119825. https://doi.org/10.1016/j.jclepro.2019.119825.
[84] Goel, J., Kadirvelu, K., Rajagopal, C., Garg, V.K., 2006. Cadmium(II) uptake from aqueous solution by adsorption onto carbon aerogel using a response surface methodological approach. Ind. Eng. Chem. Res. 45, 6531–6537. https://doi.org/10.1021/ie060010u.

[85] Li, J., Zheng, L., Liu, H., 2017. A novel carbon aerogel prepared for adsorption of copper(II) ion in water. J. Porous Mater. 24, 1575–1580. https://doi.org/10.1007/s10934-017-0397-y.
[86] Zhu, W., Li, Y., Yu, Y., Duan, T., Zhou, D., Wang, L., et al., 2018. Environment-friendly bio-materials based on cotton-carbon aerogel for strontium removal from aqueous solution. J. Radioanal. Nucl. Chem. 316, 553–560. https://doi.org/10.1007/s10967-018-5782-8.
[87] Goel, J., Kadirvelu, K., Rajagopal, C., Garg, V.K., 2005. Removal of mercury(II) from aqueous solution by adsorption on carbon aerogel: Response surface methodological approach. Carbon N Y 43, 197–200. https://doi.org/10.1016/j.carbon.2004.08.002.
[88] Sarı, A., Tuzen, M., 2009. Kinetic and equilibrium studies of biosorption of Pb(II) and Cd(II) from aqueous solution by macrofungus (Amanita rubescens) biomass. J. Hazard. Mater. 164, 1004–1011. https://doi.org/10.1016/j.jhazmat.2008.09.002.
[89] Ribas, M.C., Adebayo, M.A., Prola, L.D.T., Lima, E.C., Cataluña, R., Feris, L.A., et al., 2014. Comparison of a homemade cocoa shell activated carbon with commercial activated carbon for the removal of reactive violet 5 dye from aqueous solutions. Chem. Eng. J. 248, 315–326. https://doi.org/10.1016/j.cej.2014.03.054.
[90] Santiago, M., Stüber, F., Fortuny, A., Fabregat, A., Font, J., 2005. Modified activated carbons for catalytic wet air oxidation of phenol. Carbon N Y 43, 2134–2145. https://doi.org/10.1016/j.carbon.2005.03.026.
[91] Sato, S., Yoshihara, K., Moriyama, K., Machida, M., Tatsumoto, H., 2007. Influence of activated carbon surface acidity on adsorption of heavy metal ions and aromatics from aqueous solution. Appl. Surf. Sci. 253, 8554–8559. https://doi.org/10.1016/j.apsusc.2007.04.025.
[92] Mohamed, A.R., Mohammadi, M., Darzi, G.N., 2010. Preparation of carbon molecular sieve from lignocellulosic biomass: A review. Renew. Sustain. Energy Rev. 14, 1591–1599. https://doi.org/10.1016/j.rser.2010.01.024.
[93] Buonomenna, M.G., Yave, W., Golemme, G., 2012. Some approaches for high performance polymer based membranes for gas separation: Block copolymers, carbon molecular sieves and mixed matrix membranes. RSC Adv. 2, 10745. https://doi.org/10.1039/c2ra20748f.
[94] Wahby, A., Ramos-Fernández, J.M., Martínez-Escandell, M., Sepúlveda-Escribano, A., Silvestre-Albero, J., Rodríguez-Reinoso, F., 2010. High-surface-area carbon molecular sieves for selective CO_2 adsorption. ChemSusChem 3, 974–981. https://doi.org/10.1002/cssc.201000083.
[95] Shafeeyan, M.S., Daud, W.M.A.W., Houshmand, A., Arami-Niya, A., 2011. Ammonia modification of activated carbon to enhance carbon dioxide adsorption: Effect of pre-oxidation. Appl. Surf. Sci. 257, 3936–3942. https://doi.org/10.1016/j.apsusc.2010.11.127.
[96] Zhang, Z., Xu, M., Wang, H., Li, Z., 2010. Enhancement of CO2 adsorption on high surface area activated carbon modified by N2, H2 and ammonia. Chem. Eng. J. 160, 571–577. https://doi.org/10.1016/j.cej.2010.03.070.
[97] Agarwal, S., Tyagi, I., Gupta, V.K., Bagheri, A.R., Ghaedi, M., Asfaram, A., et al., 2016. Rapid adsorption of ternary dye pollutants onto copper (I) oxide nanoparticle loaded on activated carbon: Experimental optimization via response surface methodology. J. Environ. Chem. Eng. 4, 1769–1779. https://doi.org/10.1016/j.jece.2016.03.002.
[98] Agarwal, S., Tyagi, I., Gupta, V.K., Dastkhoon, M., Ghaedi, M., Yousefi, F., et al., 2016. Ultrasound-assisted adsorption of sunset yellow CFC dye onto Cu doped ZnS nanoparticles loaded on activated carbon using response surface methodology based

on central composite design. J. Mol. Liq. 219, 332–340. https://doi.org/10.1016/j.molliq.2016.02.100.

[99] Ghaedi, M., Hekmati Jah, A., Khodadoust, S., Sahraei, R., Daneshfar, A., Mihandoost, A., et al., 2012. Cadmium telluride nanoparticles loaded on activated carbon as adsorbent for removal of sunset yellow. Spectrochim. Acta Mol. Biomol. Spectrosc. 90, 22–27. https://doi.org/10.1016/j.saa.2011.12.064.

[100] Bagheri, A.R., Ghaedi, M., Dashtian, K., Hajati, S., Bazrafshan, A.A., 2018. Simultaneous removal of Cu^{2+} and Cr^{3+} ions from aqueous solution based on complexation with eriochrome cyanine-r and derivative spectrophotometric method. Appl. Organomet. Chem. 32. https://doi.org/10.1002/aoc.3918.

[101] Ghaedi, M., Haghdoust, S., Kokhdan, S.N., Mihandoost, A., Sahraie, R., Daneshfar, A., 2012. Comparison of activated carbon, multiwalled carbon nanotubes, and cadmium hydroxide nanowire loaded on activated carbon as adsorbents for kinetic and equilibrium study of removal of safranine O. Spectrosc. Lett. 45, 500–510. https://doi.org/10.1080/00387010.2011.641058.

[102] Bagheri, A.R., Ghaedi, M., Asfaram, A., Bazrafshan, A.A., Jannesar, R., 2017. Comparative study on ultrasonic assisted adsorption of dyes from single system onto Fe3O4 magnetite nanoparticles loaded on activated carbon: Experimental design methodology. Ultrason. Sonochem. 34, 294–304. https://doi.org/10.1016/j.ultsonch.2016.05.047.

[103] Dashamiri, S., Ghaedi, M., Asfaram, A., Zare, F., Wang, S., 2017. Multi-response optimization of ultrasound assisted competitive adsorption of dyes onto Cu (OH) 2-Nanoparticle loaded activated carbon: Central composite design. Ultrason. Sonochem. 34, 343–353. https://doi.org/10.1016/j.ultsonch.2016.06.007.

[104] Daneshyar, A., Ghaedi, M., Sabzehmeidani, M.M., Daneshyar, A., 2017. H2S adsorption onto Cu-Zn–Ni nanoparticles loaded activated carbon and Ni-Co nanoparticles loaded γ-Al2O3: Optimization and adsorption isotherms. J. Colloid Interface Sci. 490, 553–561. https://doi.org/10.1016/j.jcis.2016.11.068.

[105] Zhang, X., Guo, W., Ngo, H.H., Wen, H., Li, N., Wu, W., 2016. Performance evaluation of powdered activated carbon for removing 28 types of antibiotics from water. J. Environ. Manage. 172, 193–200. https://doi.org/10.1016/j.jenvman.2016.02.038.

[106] Abbas, M., Kaddour, S., Trari, M., 2014. Kinetic and equilibrium studies of cobalt adsorption on apricot stone activated carbon. J. Ind. Eng. Chem. 20, 745–751. https://doi.org/10.1016/j.jiec.2013.06.030.

[107] Alslaibi, T.M., Abustan, I., Ahmad, M.A., Foul, A.A., 2014. Kinetics and equilibrium adsorption of iron (II), lead (II), and copper (II) onto activated carbon prepared from olive stone waste. Desalination Water Treat. 52, 7887–7897. https://doi.org/10.1080/19443994.2013.833875.

[108] Esfandiar, N., Nasernejad, B., Ebadi, T., 2014. Removal of Mn(II) from groundwater by sugarcane bagasse and activated carbon (A comparative study): Application of response surface methodology (RSM). J. Ind. Eng. Chem. 20, 3726–3736. https://doi.org/10.1016/j.jiec.2013.12.072.

[109] González, P.G., Pliego-Cuervo, Y.B., 2014. Adsorption of Cd(II), Hg(II) and Zn(II) from aqueous solution using mesoporous activated carbon produced from Bambusa vulgaris striata. Chem. Eng. Res. Des. 92, 2715–2724. https://doi.org/10.1016/j.cherd.2014.02.013.

[110] Sharififard, H., Soleimani, M., Ashtiani, F.Z., 2012. Evaluation of activated carbon and bio-polymer modified activated carbon performance for palladium and platinum removal. J. Taiwan Inst. Chem. Eng. 43, 696–703. https://doi.org/10.1016/j.jtice.2012.04.007.

[111] Hydari, S., Sharififard, H., Nabavinia, M., Parvizi, M reza, 2012. A comparative investigation on removal performances of commercial activated carbon, chitosan biosorbent and chitosan/activated carbon composite for cadmium. Chem. Eng. J. 193–194, 276–282. https://doi.org/10.1016/j.cej.2012.04.057.
[112] Sharififard, H., Soleimani, M., 2015. Performance comparison of activated carbon and ferric oxide-hydroxide–activated carbon nanocomposite as vanadium(<scp>v</scp>) ion adsorbents. RSC Adv. 5, 80650–80660. https://doi.org/10.1039/C5RA14493K.
[113] Dil, E.A., Ghaedi, M., Asfaram, A., Hajati, S., Mehrabi, F., Goudarzi, A., 2017. Preparation of nanomaterials for the ultrasound-enhanced removal of Pb2+ ions and malachite green dye: Chemometric optimization and modeling. Ultrason. Sonochem. 34, 677–691. https://doi.org/10.1016/j.ultsonch.2016.07.001.
[114] Sharifpour, E., Khafri, H.Z., Ghaedi, M., Asfaram, A., Jannesar, R., 2018. Isotherms and kinetic study of ultrasound-assisted adsorption of malachite green and Pb2+ ions from aqueous samples by copper sulfide nanorods loaded on activated carbon: Experimental design optimization. Ultrason. Sonochem. 40, 373–382. https://doi.org/10.1016/j.ultsonch.2017.07.030.
[115] Cechinel, M.A.P., Ulson de Souza, S.M.A.G., Ulson de Souza, A.A., 2014. Study of lead (II) adsorption onto activated carbon originating from cow bone. J. Clean. Prod. 65, 342–349. https://doi.org/10.1016/j.jclepro.2013.08.020.
[116] Iijima, S., 1991. Helical microtubules of graphitic carbon. Nature 354, 56–58. https://doi.org/10.1038/354056a0.
[117] Yu, C., Xing, J., Xu, K., Chen, S., 2024. Enhanced adsorption of micropollutants on oxygen plasma-modified carbon nanotubes (O-CNTs) under electrochemical assistance. J. Water Process Eng. 58, 104854. https://doi.org/10.1016/j.jwpe.2024.104854.
[118] Xie, Y., Zhang, Y., Fan, H., Zhang, J., Hao, Q., Dou, B., et al., 2024. Adsorption mechanism of low-carbon alcohols and acids on carbon nanotubes in fischer-tropsch synthesis wastewater. J. Water Process Eng. 65, 105827. https://doi.org/10.1016/j.jwpe.2024.105827.
[119] Popov, V., 2004. Carbon nanotubes: Properties and application. Mater. Sci. Eng. R Rep. 43, 61–102. https://doi.org/10.1016/j.mser.2003.10.001.
[120] Ying, L.S., bin, M.S.MA, b, Mohamed, Y.H., Abdul Rashid, S.B., Abd, b., Razak, J., 2011. Continuous production of carbon nanotubes – A review. J. Ind. Eng. Chem. 17, 367–376. https://doi.org/10.1016/j.jiec.2011.05.007.
[121] Eatemadi, A., Daraee, H., Karimkhanloo, H., Kouhi, M., Zarghami, N., Akbarzadeh, A., et al., 2014. Carbon nanotubes: Properties, synthesis, purification, and medical applications. Nanoscale Res. Lett. 9, 393. https://doi.org/10.1186/1556-276X-9-393.
[122] Vander Wal, R.L., Berger, G.M., Ticich, T.M., 2003. Carbon nanotube synthesis in a flame using laser ablation for in situ catalyst generation. Appl. Phys. A Mater. Sci. Process. 77, 885–889. https://doi.org/10.1007/s00339-003-2196-3.
[123] Ibrahim, K.S., 2013. Carbon nanotubes-properties and applications: A review. Carbon Lett. 14, 131–144. https://doi.org/10.5714/CL.2013.14.3.131.
[124] Pichler, T., 2007. Carbon ahead. Nat. Mater. 6, 332–333. https://doi.org/10.1038/nmat1898.
[125] Novoselova, I.A., Oliinyk, N.F., Volkov, S.V., Konchits, A.A., Yanchuk, I.B., Yefanov, V.S., et al., 2008. Electrolytic synthesis of carbon nanotubes from carbon dioxide in molten salts and their characterization. Physica E Low Dimens. Syst. Nanostruct. 40, 2231–2237. https://doi.org/10.1016/j.physe.2007.10.069.
[126] Manafi, S., Nadali, H., Irani, H.R., 2008. Low temperature synthesis of multi-walled carbon nanotubes via a sonochemical/hydrothermal method. Mater. Lett. 62, 4175–4176. https://doi.org/10.1016/j.matlet.2008.05.072.

[127] Jeong, S.-H., Ko, J.-H., Park, J.-B., Park, W., 2004. A sonochemical route to single-walled carbon nanotubes under ambient conditions. J. Am. Chem. Soc. 126, 15982–15983. https://doi.org/10.1021/ja0451867.

[128] Prasek, J., Drbohlavova, J., Chomoucka, J., Hubalek, J., Jasek, O., Adam, V., et al., 2011. Methods for carbon nanotubes synthesis—Review. J. Mater. Chem. 21, 15872. https://doi.org/10.1039/c1jm12254a.

[129] Saifuddin, N., Raziah, A.Z., Junizah, A.R., 2013. Carbon nanotubes: A review on structure and their interaction with proteins. J. Chem. 2013. https://doi.org/10.1155/2013/676815.

[130] Dai, H., 2002. Carbon nanotubes: Opportunities and challenges. Surf. Sci. 500, 218–241. https://doi.org/10.1016/S0039-6028(01)01558-8.

[131] Wang, W., Huang, J.Y., Wang, D.Z., Ren, Z.F., 2005. Low-temperature hydrothermal synthesis of multiwall carbon nanotubes. Carbon N Y 43, 1328–1331. https://doi.org/10.1016/j.carbon.2004.12.020.

[132] Hasnain, M.S., Nayak, A.K., 2019. Appl. Carbon Nanotub. 33–36. https://doi.org/10.1007/978-981-15-0910-0_6.

[133] de Heer, W.A., 2004. Nanotubes and the pursuit of applications. MRS Bull. 29, 281–285. https://doi.org/10.1557/mrs2004.81.

[134] Rashidi, A., Abbasabadi, M.K., Khodabakhshi, S., 2016. Allylamide-grafted multi-wall carbon nanotubes as a new type of nanoadsorbent for the H_2S removal from gas stream. J. Nat. Gas Sci. Eng. 36, 13–19. https://doi.org/10.1016/j.jngse.2016.10.001.

[135] Duman, O., Tunç, S., Polat, T.G., Bozoğlan, B.K., 2016. Synthesis of magnetic oxidized multiwalled carbon nanotube-κ-carrageenan-Fe_3O_4 nanocomposite adsorbent and its application in cationic methylene blue dye adsorption. Carbohydr. Polym. 147, 79–88. https://doi.org/10.1016/j.carbpol.2016.03.099.

[136] Karimi, R., Yousefi, F., Ghaedi, M., Dashtian, K., 2016. Back propagation artificial neural network and central composite design modeling of operational parameter impact for sunset yellow and azur (II) adsorption onto MWCNT and MWCNT-Pd-NPs: Isotherm and kinetic study. Chemometr. Intell. Lab. Syst. 159, 127–137. https://doi.org/10.1016/j.chemolab.2016.10.012.

[137] Balarak, D., Zafariyan, M., Igwegbe, C.A., Onyechi, K.K., Ighalo, J.O., 2021. Adsorption of acid blue 92 dye from aqueous solutions by single-walled carbon nanotubes: Isothermal, kinetic, and thermodynamic studies. Environ. Process. 8, 869–888. https://doi.org/10.1007/s40710-021-00505-3.

[138] Fard, E.M., Parvareh, A., Moravaji, M.K., 2022. Optimization of removal of lead and cadmium from industrial wastewater by ethylenediamine-modified single-walled carbon nanotubes. Int. J. Environ. Sci. Technol. 19, 2747–2760. https://doi.org/10.1007/s13762-021-03390-3.

[139] Chavda, V., Patel, H.D., Patel, A.K., Hirpara, D., Acharya, N.K., Kumar, S., 2024. Deep eutectic solvent engineered <scp>GO</scp>/<scp> ZrO_2 </scp> nanofiller for mixed matrix membranes: An efficient hydrogen gas separation. J. Appl. Polym. Sci. 141. https://doi.org/10.1002/app.55938.

[140] Chavda, V., Hirpara, D., Kumar, S., 2022. GO/ionic surfactant inspired photophysical modulation of rhodamine B in reline with or without additives. J. Mol. Liq. 368, 120614. https://doi.org/10.1016/j.molliq.2022.120614.

[141] Li, G., Du, R., Cao, Z., Li, C., Xue, J., Ma, X., et al., 2024. Research progress in graphene-based adsorbents for wastewater treatment: Preparation, adsorption properties and mechanisms for inorganic and organic pollutants. C (Basel) 10, 78. https://doi.org/10.3390/c10030078.

[142] Wang, H., Yi, L., Huang, F., Huang, Q., Zhou, T., 2024. Facile synthesis of graphene nanosheets on wastewater sediments for high efficient adsorption of

methylene blue. Sep. Purif. Technol. 337, 126366. https://doi.org/10.1016/j.seppur.2024.126366.

[143] Wang, H., Maiyalagan, T., Wang, X., 2012. Review on recent progress in nitrogen-doped graphene: Synthesis, characterization, and its potential applications. ACS Catal. 2, 781–794. https://doi.org/10.1021/cs200652y.

[144] Jia, X., Campos-Delgado, J., Terrones, M., Meunier, V., Dresselhaus, M.S., 2011. Graphene edges: A review of their fabrication and characterization. Nanoscale 3, 86–95. https://doi.org/10.1039/C0NR00600A.

[145] Manesh, K.M., Kim, H.T., Santhosh, P., Gopalan, A.I., Lee, K.-P., 2008. A novel glucose biosensor based on immobilization of glucose oxidase into multiwall carbon nanotubes–polyelectrolyte-loaded electrospun nanofibrous membrane. Biosens. Bioelectron. 23, 771–779. https://doi.org/10.1016/j.bios.2007.08.016.

[146] Shrivas, K., Ghosale, A., Nirmalkar, N., Srivastava, A., Singh, S.K., Shinde, S.S., 2017. Removal of endrin and dieldrin isomeric pesticides through stereoselective adsorption behavior on the graphene oxide-magnetic nanoparticles. Environ. Sci. Pollut. Control Ser. 24, 24980–24988. https://doi.org/10.1007/s11356-017-0159-z.

[147] Koushik, D., Sen Gupta, S., Maliyekkal, S.M., Pradeep, T., 2016. Rapid dehalogenation of pesticides and organics at the interface of reduced graphene oxide–silver nanocomposite. J. Hazard. Mater. 308, 192–198. https://doi.org/10.1016/j.jhazmat.2016.01.004.

[148] Agarwal, S., Sadeghi, N., Tyagi, I., Gupta, V.K., Fakhri, A., 2016. Adsorption of toxic carbamate pesticide oxamyl from liquid phase by newly synthesized and characterized graphene quantum dots nanomaterials. J. Colloid Interface Sci. 478, 430–438. https://doi.org/10.1016/j.jcis.2016.06.029.

[149] Yu, L., Wang, L., Xu, W., Chen, L., Fu, M., Wu, J., et al., 2018. Adsorption of VOCs on reduced graphene oxide. J. Environ. Sci. 67, 171–178. https://doi.org/10.1016/j.jes.2017.08.022.

[150] Matsuyama, S., Sugiyama, T., Ikoma, T., Cross, J.S., 2016. Fabrication of 3D graphene and 3D graphene oxide devices for sensing VOCs. MRS Adv. 1, 1359–1364. https://doi.org/10.1557/adv.2016.151.

[151] Zhu, H., Chen, D., Li, N., Xu, Q., Li, H., He, J., et al., 2015. Graphene foam with switchable oil wettability for oil and organic solvents recovery. Adv. Funct. Mater. 25, 597–605. https://doi.org/10.1002/adfm.201403864.

[152] Wu, R., Yu, B., Liu, X., Li, H., Wang, W., Chen, L., et al., 2016. One-pot hydrothermal preparation of graphene sponge for the removal of oils and organic solvents. Appl. Surf. Sci. 362, 56–62. https://doi.org/10.1016/j.apsusc.2015.11.215.

[153] Tran, D.N.H., Kabiri, S., Sim, T.R., Losic, D., 2015. Selective adsorption of oil–water mixtures using polydimethylsiloxane (PDMS)–Graphene sponges. Environ. Sci. 1, 298–305. https://doi.org/10.1039/C5EW00035A.

[154] Gaber, D., Abu Haija, M., Eskhan, A., Banat, F., 2017. Graphene as an efficient and reusable adsorbent compared to activated carbons for the removal of phenol from aqueous solutions. Water Air Soil Pollut. 228, 320. https://doi.org/10.1007/s11270-017-3499-x.

[155] Fu, C.-C., Juang, R.-S., Huq, M.M., Hsieh, C.-T., 2016. Enhanced adsorption and photodegradation of phenol in aqueous suspensions of titania/graphene oxide composite catalysts. J. Taiwan Inst. Chem. Eng. 67, 338–345. https://doi.org/10.1016/j.jtice.2016.07.043.

[156] Bele, S., Samanidou, V., Deliyanni, E., 2016. Effect of the reduction degree of graphene oxide on the adsorption of bisphenol A. Chem. Eng. Res. Des. 109, 573–585. https://doi.org/10.1016/j.cherd.2016.03.002.

[157] Xia, P., Zhu, B., Yu, J., Cao, S., Jaroniec, M., 2017. Ultra-thin nanosheet assemblies of graphitic carbon nitride for enhanced photocatalytic CO_2 reduction. J. Mater. Chem. A Mater. 5, 3230–3238. https://doi.org/10.1039/C6TA08310B.
[158] Rostamian, R., Behnejad, H., 2018. A comprehensive adsorption study and modeling of antibiotics as a pharmaceutical waste by graphene oxide nanosheets. Ecotoxicol. Environ. Saf. 147, 117–123. https://doi.org/10.1016/j.ecoenv.2017.08.019.
[159] Anirudhan, T.S., Shainy, F., Christa, J., 2017. Synthesis and characterization of polyacrylic acid-grafted-carboxylic graphene/titanium nanotube composite for the effective removal of enrofloxacin from aqueous solutions: Adsorption and photocatalytic degradation studies. J. Hazard. Mater. 324, 117–130. https://doi.org/10.1016/j.jhazmat.2016.09.073.
[160] Chen, C., Xu, K., Ji, X., Miao, L., Jiang, J., 2014. Enhanced adsorption of acidic gases (CO2, NO2 and SO2) on light metal decorated graphene oxide. Phys. Chem. Chem. Phys. 16, 11031. https://doi.org/10.1039/c4cp00702f.
[161] Yang, K., Chen, B., Zhu, X., Xing, B., 2016. Aggregation, adsorption, and morphological transformation of graphene oxide in aqueous solutions containing different metal cations. Environ. Sci. Technol. 50, 11066–11075. https://doi.org/10.1021/acs.est.6b04235.
[162] Wang, X., Liang, Y., An, W., Hu, J., Zhu, Y., Cui, W., 2017. Removal of chromium (VI) by a self-regenerating and metal free g-C3N4/graphene hydrogel system via the synergy of adsorption and photo-catalysis under visible light. Appl. Catal., B 219, 53–62. https://doi.org/10.1016/j.apcatb.2017.07.008.
[163] Vu, H.C., Dwivedi, A.D., Le, T.T., Seo, S.-H., Kim, E.-J., Chang, Y.-S., 2017. Magnetite graphene oxide encapsulated in alginate beads for enhanced adsorption of Cr(VI) and As(V) from aqueous solutions: Role of crosslinking metal cations in pH control. Chem. Eng. J. 307, 220–229. https://doi.org/10.1016/j.cej.2016.08.058.
[164] Wu, Z., Sundén, B., 2016. Convective heat transfer performance of aggregate-laden nanofluids. Int. J. Heat Mass Transf. 93, 1107–1115. https://doi.org/10.1016/j.ijheatmasstransfer.2015.11.032.
[165] Robati, D., Mirza, B., Rajabi, M., Moradi, O., Tyagi, I., Agarwal, S., et al., 2016. Removal of hazardous dyes-BR 12 and methyl orange using graphene oxide as an adsorbent from aqueous phase. Chem. Eng. J. 284, 687–697. https://doi.org/10.1016/j.cej.2015.08.131.
[166] Qi, Y., Yang, M., Xu, W., He, S., Men, Y., 2017. Natural polysaccharides-modified graphene oxide for adsorption of organic dyes from aqueous solutions. J. Colloid Interface Sci. 486, 84–96. https://doi.org/10.1016/j.jcis.2016.09.058.
[167] Sitko, R., Musielak, M., Zawisza, B., Talik, E., Gagor, A., 2016. Graphene oxide/cellulose membranes in adsorption of divalent metal ions. RSC Adv. 6, 96595–96605. https://doi.org/10.1039/C6RA21432K.
[168] Yu, W., Zhan, S., Shen, Z., Zhou, Q., Yang, D., 2017. Efficient removal mechanism for antibiotic resistance genes from aquatic environments by graphene oxide nanosheet. Chem. Eng. J. 313, 836–846. https://doi.org/10.1016/j.cej.2016.10.107.
[169] Guo, Z., Huang, J., Xue, Z., Wang, X., 2016. Electrospun graphene oxide/carbon composite nanofibers with well-developed mesoporous structure and their adsorption performance for benzene and butanone. Chem. Eng. J. 306, 99–106. https://doi.org/10.1016/j.cej.2016.07.048.
[170] Xu, J., Wang, L., Zhu, Y., 2012. Decontamination of bisphenol A from aqueous solution by graphene adsorption. Langmuir 28, 8418–8425. https://doi.org/10.1021/la301476p.
[171] Fan, L., Luo, C., Sun, M., Qiu, H., Li, X., 2013. Synthesis of magnetic β-cyclodextrin–chitosan/graphene oxide as nanoadsorbent and its application in dye

adsorption and removal. Colloids Surf. B Biointerfaces 103, 601–607. https://doi.org/10.1016/j.colsurfb.2012.11.023.

[172] Che, G., Lakshmi, B.B., Martin, C.R., Fisher, E.R., Ruoff, R.S., 1998. Chemical vapor deposition based synthesis of carbon nanotubes and nanofibers using a template method. Chem. Mater. 10, 260–267. https://doi.org/10.1021/cm970412f.

[173] Lee, C.J., Park, J., Kim, J.M., Huh, Y., Lee, J.Y., No, K.S., 2000. Low-temperature growth of carbon nanotubes by thermal chemical vapor deposition using Pd, Cr, and Pt as co-catalyst. Chem. Phys. Lett. 327, 277–283. https://doi.org/10.1016/S0009-2614(00)00877-0.

[174] Peng, S., Li, L., Kong Yoong Lee, J., Tian, L., Srinivasan, M., Adams, S., et al., 2016. Electrospun carbon nanofibers and their hybrid composites as advanced materials for energy conversion and storage. Nano Energy 22, 361–395. https://doi.org/10.1016/j.nanoen.2016.02.001.

[175] Peng, S., Jin, G., Li, L., Li, K., Srinivasan, M., Ramakrishna, S., et al., 2016. Multifunctional electrospun nanofibres for advances in tissue regeneration, energy conversion and storage, and water treatment. Chem. Soc. Rev. 45, 1225–1241. https://doi.org/10.1039/C5CS00777A.

[176] Erdem, S., Hanbay, S., Blankson, M.A., 2017. Self-sensing damage assessment and image-based surface crack quantification of carbon nanofibre reinforced concrete. Constr. Build. Mater. 134, 520–529. https://doi.org/10.1016/j.conbuildmat.2016.12.197.

[177] Su, C.-I., Huang, Y.-X., Wong, J.-W., Lu, C.-H., Wang, C.-M., 2012. PAN-Based carbon nanofiber absorbents prepared using electrospinning. Fibers Polym. 13, 436–442. https://doi.org/10.1007/s12221-012-0436-x.

[178] Ruiz-Rosas, R., Bedia, J., Lallave, M., Loscertales, I.G., Barrero, A., Rodríguez-Mirasol, J., et al., 2010. The production of submicron diameter carbon fibers by the electrospinning of lignin. Carbon N Y 48, 696–705. https://doi.org/10.1016/j.carbon.2009.10.014.

[179] Nan, D., Liu, J., Ma, W., 2015. Electrospun phenolic resin-based carbon ultrafine fibers with abundant ultra-small micropores for CO2 adsorption. Chem. Eng. J. 276, 44–50. https://doi.org/10.1016/j.cej.2015.04.081.

[180] Chen, Y., Yue, M., Huang, Z.-H., Kang, F., 2014. Electrospun carbon nanofiber networks from phenolic resin for capacitive deionization. Chem. Eng. J. 252, 30–37. https://doi.org/10.1016/j.cej.2014.04.099.

[181] Bai, Y., Huang, Z.-H., Kang, F., 2014. Electrospun preparation of microporous carbon ultrafine fibers with tuned diameter, pore structure and hydrophobicity from phenolic resin. Carbon N Y 66, 705–712. https://doi.org/10.1016/j.carbon.2013.09.074.

[182] Cao, Y., Huang, J., Peng, X., Cao, D., Galaska, A., Qiu, S., et al., 2017. Poly (vinylidene fluoride) derived fluorine-doped magnetic carbon nanoadsorbents for enhanced chromium removal. Carbon N Y 115, 503–514. https://doi.org/10.1016/j.carbon.2017.01.033.

[183] Carabineiro, S.A., Pereira, M.F., Pereira, J.N., Caparros, C., Sencadas, V., Lanceros-Mendez, S., 2011. Effect of the carbon nanotube surface characteristics on the conductivity and dielectric constant of carbon nanotube/poly(vinylidene fluoride) composites. Nanoscale Res. Lett. 6, 302. https://doi.org/10.1186/1556-276X-6-302.

[184] Hong, S.-M., Choi, S.W., Kim, S.H., Lee, K.B., 2016. Porous carbon based on polyvinylidene fluoride: Enhancement of CO2 adsorption by physical activation. Carbon N Y 99, 354–360. https://doi.org/10.1016/j.carbon.2015.12.012.

[185] Jung, K.-H., Ferraris, J.P., 2012. Preparation and electrochemical properties of carbon nanofibers derived from polybenzimidazole/polyimide precursor blends. Carbon N Y 50, 5309–5315. https://doi.org/10.1016/j.carbon.2012.07.019.
[186] Jung, K.-H., Ferraris, J.P., 2016. Preparation of porous carbon nanofibers derived from PBI/PLLA for supercapacitor electrodes. Nanotechnology 27, 425708. https://doi.org/10.1088/0957-4484/27/42/425708.
[187] Le, T., Yang, Y., Huang, Z., Kang, F., 2015. Preparation of microporous carbon nanofibers from polyimide by using polyvinyl pyrrolidone as template and their capacitive performance. J. Power Sources 278, 683–692. https://doi.org/10.1016/j.jpowsour.2014.12.055.
[188] Xu, W., Feng, Y., Ding, Y., Jiang, S., Fang, H., Hou, H., 2015. Short electrospun carbon nanofiber reinforced polyimide composite with high dielectric permittivity. Mater. Lett. 161, 431–434. https://doi.org/10.1016/j.matlet.2015.09.014.
[189] Zhang, Y., Ou, H., Liu, H., Ke, Y., Zhang, W., Liao, G., et al., 2018. Polyimide-based carbon nanofibers: A versatile adsorbent for highly efficient removals of chlorophenols, dyes and antibiotics. Colloids Surf. A Physicochem. Eng. Asp. 537, 92–101. https://doi.org/10.1016/j.colsurfa.2017.10.014.
[190] Lai, C., Zhou, Z., Zhang, L., Wang, X., Zhou, Q., Zhao, Y., et al., 2014. Free-standing and mechanically flexible mats consisting of electrospun carbon nanofibers made from a natural product of alkali lignin as binder-free electrodes for high-performance supercapacitors. J. Power Sources 247, 134–141. https://doi.org/10.1016/j.jpowsour.2013.08.082.
[191] Wang, P., Xiang, R., Maruyama, S., 2018. Thermal conductivity of carbon nanotubes and assemblies. 43–122. https://doi.org/10.1016/bs.aiht.2018.07.004.
[192] Costa, P., Silva, J., Ansón-Casaos, A., Martinez, M.T., Abad, M.J., Viana, J., et al., 2014. Effect of carbon nanotube type and functionalization on the electrical, thermal, mechanical and electromechanical properties of carbon nanotube/styrene–butadiene–styrene composites for large strain sensor applications. Compos. B Eng. 61, 136–146. https://doi.org/10.1016/j.compositesb.2014.01.048.
[193] Fujii, M., Zhang, X., Xie, H., Ago, H., Takahashi, K., Ikuta, T., et al., 2005. Measuring the thermal conductivity of a single carbon nanotube. Phys. Rev. Lett. 95, 065502. https://doi.org/10.1103/PhysRevLett.95.065502.
[194] Wang, G., Zhang, J., Kuang, S., Zhou, J., Xing, W., Zhuo, S., 2015. Nitrogen-doped hierarchical porous carbon as an efficient electrode material for supercapacitors. Electrochim. Acta 153, 273–279. https://doi.org/10.1016/j.electacta.2014.12.006.
[195] Zussman, E., Chen, X., Ding, W., Calabri, L., Dikin, D.A., Quintana, J.P., et al., 2005. Mechanical and structural characterization of electrospun PAN-Derived carbon nanofibers. Carbon N Y 43, 2175–2185. https://doi.org/10.1016/j.carbon.2005.03.031.
[196] Bikshapathi, M., Mathur, G.N., Sharma, A., Verma, N., 2012. Surfactant-enhanced multiscale carbon webs including nanofibers and Ni-Nanoparticles for the removal of gaseous persistent organic pollutants. Ind. Eng. Chem. Res. 51, 2104–2112. https://doi.org/10.1021/ie200741e.
[197] Bikshapathi, M., Singh, S., Bhaduri, B., Mathur, G.N., Sharma, A., Verma, N., 2012. Fe-nanoparticles dispersed carbon micro and nanofibers: Surfactant-mediated preparation and application to the removal of gaseous VOCs. Colloids Surf. A Physicochem. Eng. Asp. 399, 46–55. https://doi.org/10.1016/j.colsurfa.2012.02.023.
[198] Díaz, E., Ordóñez, S., Vega, A., 2007. Adsorption of volatile organic compounds onto carbon nanotubes, carbon nanofibers, and high-surface-area graphites. J. Colloid Interface Sci. 305, 7–16. https://doi.org/10.1016/j.jcis.2006.09.036.

[199] Si, Y., Ren, T., Ding, B., Yu, J., Sun, G., 2012. Synthesis of mesoporous magnetic Fe3O4@carbon nanofibers utilizing in situ polymerized polybenzoxazine for water purification. J. Mater. Chem. 22, 4619. https://doi.org/10.1039/c2jm00036a.
[200] Lee, J.-W., Kang, H.-C., Shim, W.-G., Kim, C., Yang, K.-S., Moon, H., 2006. Heterogeneous adsorption of activated carbon nanofibers synthesized by electrospinning polyacrylonitrile solution. J. Nanosci. Nanotechnol. 6, 3577–3582. https://doi.org/10.1166/jnn.2006.060.
[201] Serp, P., 2008. Carbon nanotubes and nanofibers in catalysis. Carbon Materials for Catalysis. Wiley, pp. 309–372. https://doi.org/10.1002/9780470403709.ch9.
[202] Landis, E.C., Klein, K.L., Liao, A., Pop, E., Hensley, D.K., Melechko, A.V., et al., 2010. Covalent functionalization and electron-transfer properties of vertically aligned carbon nanofibers: The importance of edge-plane sites. Chem. Mater. 22, 2357–2366. https://doi.org/10.1021/cm9036132.
[203] Lee, K.J., Shiratori, N., Lee, G.H., Miyawaki, J., Mochida, I., Yoon, S.-H., et al., 2010. Activated carbon nanofiber produced from electrospun polyacrylonitrile nanofiber as a highly efficient formaldehyde adsorbent. Carbon N Y 48, 4248–4255. https://doi.org/10.1016/j.carbon.2010.07.034.
[204] Li, X., Chen, S., Fan, X., Quan, X., Tan, F., Zhang, Y., et al., 2015. Adsorption of ciprofloxacin, bisphenol and 2-chlorophenol on electrospun carbon nanofibers: In comparison with powder activated carbon. J. Colloid Interface Sci. 447, 120–127. https://doi.org/10.1016/j.jcis.2015.01.042.
[205] Liu, Q., Zhong, L.-B., Zhao, Q.-B., Frear, C., Zheng, Y.-M., 2015. Synthesis of Fe_3O_4/Polyacrylonitrile composite electrospun nanofiber mat for effective adsorption of tetracycline. ACS Appl. Mater. Interfaces 7, 14573–14583. https://doi.org/10.1021/acsami.5b04598.
[206] Bai, Y., Huang, Z.-H., Wang, M.-X., Kang, F., 2013. Adsorption of benzene and ethanol on activated carbon nanofibers prepared by electrospinning. Adsorption 19, 1035–1043. https://doi.org/10.1007/s10450-013-9524-5.
[207] Bai, Y., Huang, Z.-H., Kang, F., 2014. Surface oxidation of activated electrospun carbon nanofibers and their adsorption performance for benzene, butanone and ethanol. Colloids Surf. A Physicochem. Eng. Asp. 443, 66–71. https://doi.org/10.1016/j.colsurfa.2013.10.057.
[208] Li, Y.-K., Yang, T., Chen, M.-L., Wang, J.-H., 2021. Recent advances in nanomaterials for analysis of trace heavy metals. Crit. Rev. Anal. Chem. 51, 353–372. https://doi.org/10.1080/10408347.2020.1736505.
[209] Peng, H., Gao, P., Chu, G., Pan, B., Peng, J., Xing, B., 2017. Enhanced adsorption of Cu(II) and Cd(II) by phosphoric acid-modified biochars. Environ. Pollut. 229, 846–853. https://doi.org/10.1016/j.envpol.2017.07.004.
[210] Kwak, J.-H., Islam, M.S., Wang, S., Messele, S.A., Naeth, M.A., El-Din, M.G., et al., 2019. Biochar properties and lead(II) adsorption capacity depend on feedstock type, pyrolysis temperature, and steam activation. Chemosphere 231, 393–404. https://doi.org/10.1016/j.chemosphere.2019.05.128.
[211] Fan, S., Wang, Y., Wang, Z., Tang, J., Tang, J., Li, X., 2017. Removal of methylene blue from aqueous solution by sewage sludge-derived biochar: Adsorption kinetics, equilibrium, thermodynamics and mechanism. J. Environ. Chem. Eng. 5, 601–611. https://doi.org/10.1016/j.jece.2016.12.019.
[212] Li, C., Zhu, X., He, H., Fang, Y., Dong, H., Lu, J., et al., 2019. Adsorption of two antibiotics on biochar prepared in air-containing atmosphere: Influence of biochar porosity and molecular size of antibiotics. J. Mol. Liq. 274, 353–361. https://doi.org/10.1016/j.molliq.2018.10.142.

[213] Zhu, X., Chen, B., Zhu, L., Xing, B., 2017. Effects and mechanisms of biochar-microbe interactions in soil improvement and pollution remediation: A review. Environ. Pollut. 227, 98–115. https://doi.org/10.1016/j.envpol.2017.04.032.
[214] Wang, L., Wang, Y., Ma, F., Tankpa, V., Bai, S., Guo, X., et al., 2019. Mechanisms and reutilization of modified biochar used for removal of heavy metals from wastewater: A review. Sci. Total Environ. 668, 1298–1309. https://doi.org/10.1016/j.scitotenv.2019.03.011.
[215] Mandal, A., Singh, N., Purakayastha, T.J., 2017. Characterization of pesticide sorption behaviour of slow pyrolysis biochars as low cost adsorbent for atrazine and imidacloprid removal. Sci. Total Environ. 577, 376–385. https://doi.org/10.1016/j.scitotenv.2016.10.204.
[216] Ahmed, M.B., Zhou, J.L., Ngo, H.H., Guo, W., Johir, M.A.H., Sornalingam, K., 2017. Single and competitive sorption properties and mechanism of functionalized biochar for removing sulfonamide antibiotics from water. Chem. Eng. J. 311, 348–358. https://doi.org/10.1016/j.cej.2016.11.106.
[217] Park, J.-H., Ok, Y.S., Kim, S.-H., Cho, J.-S., Heo, J.-S., Delaune, R.D., et al., 2016. Competitive adsorption of heavy metals onto sesame straw biochar in aqueous solutions. Chemosphere 142, 77–83. https://doi.org/10.1016/j.chemosphere.2015.05.093.
[218] Zhu, N., Yan, T., Qiao, J., Cao, H., 2016. Adsorption of arsenic, phosphorus and chromium by bismuth impregnated biochar: Adsorption mechanism and depleted adsorbent utilization. Chemosphere 164, 32–40. https://doi.org/10.1016/j.chemosphere.2016.08.036.
[219] Cederlund, H., Börjesson, E., Lundberg, D., Stenström, J., 2016. Adsorption of pesticides with different chemical properties to a wood biochar treated with heat and iron. Water Air Soil Pollut. 227, 203. https://doi.org/10.1007/s11270-016-2894-z.
[220] Kelm, M.A.P., da Silva Júnior, M.J., de Barros Holanda, S.H., de Araujo, C.M.B., de Assis Filho, R.B., Freitas, E.J., et al., 2019. Removal of azo dye from water via adsorption on biochar produced by the gasification of wood wastes. Environ. Sci. Pollut. Control Ser. 26, 28558–28573. https://doi.org/10.1007/s11356-018-3833-x.
[221] Abd El-Monaem, E.M., Omer, A.M., El-Subruiti, G.M., Mohy-Eldin, M.S., Eltaweil, A.S., 2024. Zero-valent iron supported-lemon derived biochar for ultra-fast adsorption of methylene blue. Biomass Convers. Biorefin. 14, 1697–1709. https://doi.org/10.1007/s13399-022-02362-y.
[222] Iqbal, M.M., Imran, M., Hussain, T., Naeem, M.A., Al-Kahtani, A.A., Shah, G.M., et al., 2021. Effective sequestration of Congo red dye with ZnO/cotton stalks biochar nanocomposite: MODELING, reusability and stability. J. Saudi Chem. Soc. 25, 101176. https://doi.org/10.1016/j.jscs.2020.101176.
[223] Zhao, Y., Qamar, S.A., Qamar, M., Bilal, M., Iqbal, H.M.N., 2021. Sustainable remediation of hazardous environmental pollutants using biochar-based nanohybrid materials. J. Environ. Manage. 300, 113762. https://doi.org/10.1016/j.jenvman.2021.113762.
[224] Chausali, N., Saxena, J., Prasad, R., 2021. Nanobiochar and biochar based nanocomposites: Advances and applications. J. Agric. Food Res. 5, 100191. https://doi.org/10.1016/j.jafr.2021.100191.
[225] Ramanayaka, S., Vithanage, M., Alessi, D.S., Liu, W.-J., Jayasundera, A.C.A., Ok, Y.S., 2020. Nanobiochar: Production, properties, and multifunctional applications. Environ. Sci. Nano 7, 3279–3302. https://doi.org/10.1039/D0EN00486C.
[226] Mahmoud, M.E., Abdelfattah, A.M., Tharwat, R.M., Nabil, G.M., 2020. Adsorption of negatively charged food tartrazine and sunset yellow dyes onto positively charged triethylenetetramine biochar: Optimization, kinetics and thermodynamic study. J. Mol. Liq. 318, 114297. https://doi.org/10.1016/j.molliq.2020.114297.

[227] Omanović-Mikličanin, E., Badnjević, A., Kazlagić, A., Hajlovac, M., 2020. Nanocomposites: A brief review. Health Technol. 10, 51–59. https://doi.org/10.1007/s12553-019-00380-x.

[228] Almoisheer, N., Alseroury, F.A., Kumar, R., Almeelbi, T., Barakat, M.A., 2019. Synthesis of graphene oxide/silica/carbon nanotubes composite for removal of dyes from wastewater. Earth Syst. Environ. 3, 651–659. https://doi.org/10.1007/s41748-019-00109-w.

[229] Raeiszadeh, M., Hakimian, A., Shojaei, A., Molavi, H., 2018. Nanodiamond-filled chitosan as an efficient adsorbent for anionic dye removal from aqueous solutions. J. Environ. Chem. Eng. 6, 3283–3294. https://doi.org/10.1016/j.jece.2018.05.005.

[230] Sohni, S., Gul, K., Ahmad, F., Ahmad, I., Khan, A., Khan, N., et al., 2018. Highly efficient removal of acid red-17 and bromophenol blue dyes from industrial wastewater using graphene oxide functionalized magnetic chitosan composite. Polym. Compos. 39, 3317–3328. https://doi.org/10.1002/pc.24349.

[231] Zhang, L., Luo, H., Liu, P., Fang, W., Geng, J., 2016. A novel modified graphene oxide/chitosan composite used as an adsorbent for Cr(VI) in aqueous solutions. Int. J. Biol. Macromol. 87, 586–596. https://doi.org/10.1016/j.ijbiomac.2016.03.027.

[232] Wang, J., Zhuang, S., 2019. Covalent organic frameworks (COFs) for environmental applications. Coord. Chem. Rev. 400, 213046. https://doi.org/10.1016/j.ccr.2019.213046.

[233] Yaghi, O.M., O'Keeffe, M., Ockwig, N.W., Chae, H.K., Eddaoudi, M., Kim, J., 2003. Reticular synthesis and the design of new materials. Nature 423, 705–714. https://doi.org/10.1038/nature01650.

CHAPTER 4

Impact of carbonization temperatures on structural and morphological characteristics of sugar palm fiber-based biochar

Mohamad Faisal Mohamad Zauqi[1,2], Muhammad Syabbatul Al Hamdi Mustaffa AlBakri[1,2], Mohd Nurazzi Norizan[1,2,3], Nurul Fazita Mohammed Rawi[1,2,3], Mohamad Haafiz Mohamad Kassim[1,2,3], Muhammad Asyraf Muhammad Rizal[4,5], Ahmad Ilyas Rushdan[5] and Emin Bayraktar[6]

[1]Bioresource Technology Division, School of Industrial Technology, Universiti Sains Malaysia, Gelugor, Penang, Malaysia; [2]Green Biopolymer, Coatings & Packaging Cluster, School of Industrial Technology, Universiti Sains Malaysia, Gelugor, Penang, Malaysia; [3]Research Collaboration Center for Nanocellulose, BRIN - Andalas University, Padang, Indonesia; [4]Faculty of Mechanical Engineering, Universiti Teknologi Malaysia, Johor Bahru, Johor, Malaysia; [5]Centre for Advanced Composite Materials (CACM), Universiti Teknologi Malaysia, Johor Bahru, Johor, Malaysia; [6]ISAE-Supmeca, School of Mechanical and Manufacturing Engineering, Saint-Ouen, France

1. Introduction

Biochar is a carbonaceous material produced by the pyrolysis or carbonization of biomass in limited oxygen environments, and its uses have spanned into soil improvement, water purification, and energy storage. This interest in biochar production from biomass is a result of the increasing global demand for sustainable and eco-friendly solutions to waste management and resource utilization. One such promising feedstock is sugar palm fiber, which is a byproduct of the sugar palm tree, available in abundance in most tropical regions [1].

Attention to the valorization of biomass has been driven by the increasing demand for sustainable waste management approaches. Sugar palm fiber waste is a lignocellulosic by-product from the sugar palm tree, which is one of the most abundant but underutilized feedstocks. Such

Carbon-Based Materials and Composites for Water and Wastewater Treatment
ISBN 978-0-443-33681-2
https://doi.org/10.1016/B978-0-443-33681-2.00012-0

biomass conversion into biochar via carbonization would address the challenges in waste management and provide useful material in various applications, including environmental remediation, agriculture, and energy storage [2,3].

Carbonization is a pyrolytic reaction; therefore it is considered a complex process in which many reactions take place concurrently, such as dehydrogenation, condensation, hydrogen transfer, and isomerization [4,5]. The carbonization process has been considered vital in dictating the structural and morphological characteristics of biochar, to which functionality is directly related. Of particular importance is the temperature of carbonization, which dictates the level of organic component degradation, such as cellulose, hemicellulose, and lignin, and the development of porosity, surface area, and structural stability [6]. At the lower temperatures, most of the biochar retains a great deal of the original feedstock structure, including fibers, while higher temperatures lend aromaticity, volume increase in micropores, and an increase in adsorption capability.

Sugar palm fiber is considered a highly lignocellulosic material and therefore one of the most promising feedstocks for biochar production. Besides valorizing the waste into a functional material for different applications, using such waste would also reduce environmental problems caused by its disposal [7–9]. However, the effect of different carbonization temperatures on the morphological and structural properties of the resulting biochar from sugar palm fiber has not been fully investigated. Biochar produced from biomass waste might start at temperatures lower than 450°C and retains much of the original feedstock's fibrous structure, with only a limited formation of pores due to lignocellulosic elements' presence. Hemicellulose and cellulose start to decompose; thus slight changes occur in the structure [10,11]. Besides, proximate analysis reveals that volatile matter content decreases with the increase in temperature due to the release of organic compounds, while fixed carbon content increases, enhancing the stability and calorific value of biochar. Other changes in elemental composition include an increase in the carbon content with a simultaneous decrease in oxygen and hydrogen, hence more aromatic from the lignocellulosic compositions of sugar palm fiber. This study aims to investigate how different carbonization temperatures at 400°C, 450°C, 500°C, 550°C, and 600°C for 2 h of conditioning time impact the morphological characteristics through FESEM analysis, CHNS, and Raman for the structural characteristics and stability through zeta potential analysis of biochar from the SPF.

2. Methodology

2.1 Material

This study employed sugar palm fiber (Fig. 4.1) possessing a density of 1.28 g/cm^3 as the main material for the carbonization process. The fiber was obtained from Kampung Kuala Jempol, Negeri Sembilan, Malaysia.

2.2 Preparation of sugar palm fiber

Before it can be used, the sugar palm fiber (Fig. 4.2a) undergoes several preparation steps. First, the fiber is combed using steel wool hand combs to tease, align, and clean it, removing spikes and unwanted leaf material, as shown in Fig. 4.2. Afterward, the fiber is washed with distilled water to eliminate dirt and surface contaminants (Fig. 4.2b). It is then dried at 105°C for 24 h, followed by an additional 6 h of oven drying to further reduce its moisture content, as illustrated in Fig. 4.2c and d. Finally, the fiber is crushed using a grinding machine.

2.3 Carbonization process

Five portions of 10 g each crushed sugar palm fiber were measured out for carbonization in Fig. 4.3a, while one 10 g portion was left uncarbonized to serve as a control sample. The carbonization process was conducted to produce biochar. It was carried out in a furnace at several temperatures within 400°C, 450°C, 500°C, 550°C, 600°C for 2 h each sample, respectively. The process began with a preheating stage, which took approximately half an hour, followed by a carbonization step lasting an additional hour, as illustrated in Fig. 4.3b. After carbonization was completed, the samples were cooled at room temperature for 1 h Fig. 4.3c. The samples were then ground using a mortar and pestle for 5 min to produce fine biochar before putting it into a glass bottle.

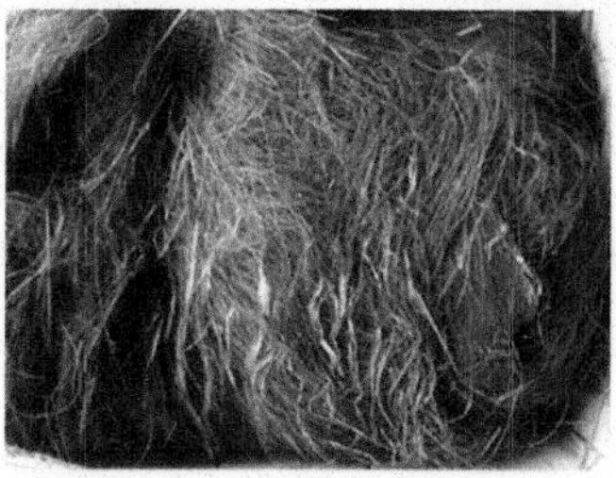

Figure 4.1 Sugar palm fiber.

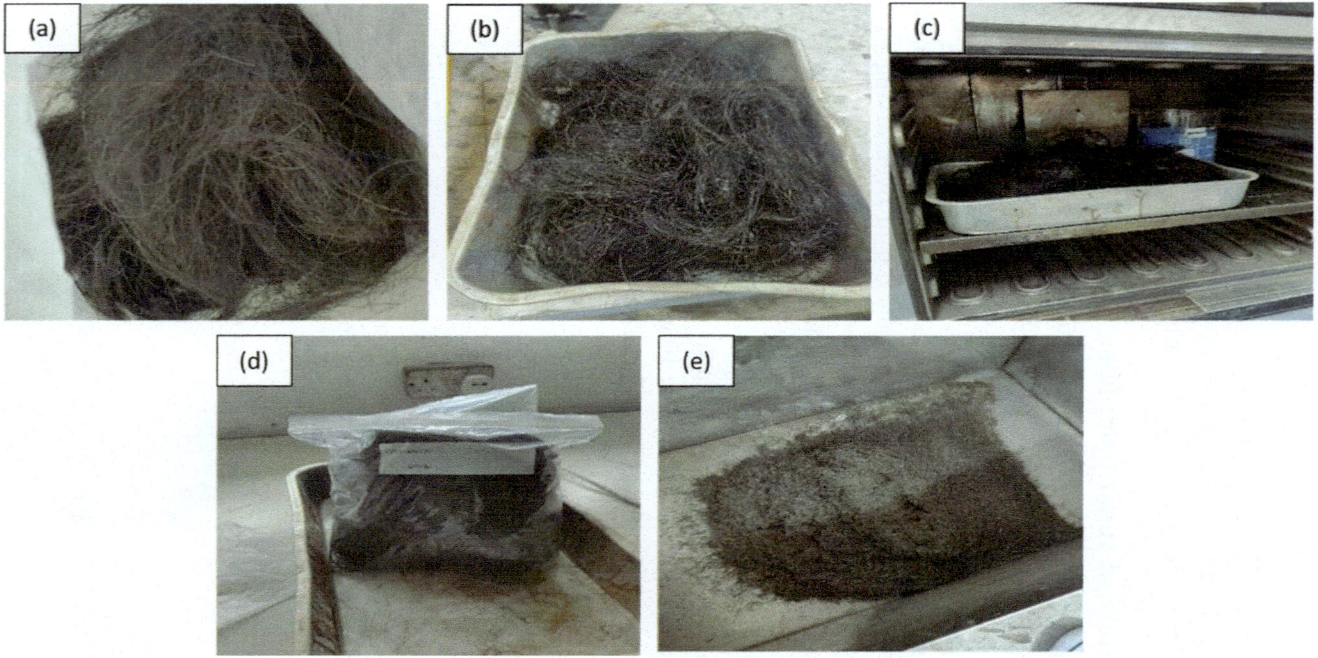

Figure 4.2 Step-by-step preparation of sugar palm fiber: (a) extracted sugar palm fiber, (b) dirt and dust removed from sugar palm fiber and clean the sugar palm fiber by using distilled water, (c) oven-dried at 105°C for 24 h (d) completely cleaned and dried sugar palm fiber, (e) crush the sugar palm fiber into small particles.

Figure 4.3 Preparation of carbonized sugar palm fiber: (a) preparation of 10 g crushed sugar palm fiber, (b) carbonize the sugar palm fiber using a furnace, (c) cool the sample at room temperature, (d) grind the sample using a mortar and pestle for 5 min before transferring it into a glass bottle, (e) grounded sample.

2.4 Characterizations

2.4.1 Morphological analysis

The morphological properties of carbonized samples were observed using FESEM to investigate their localized morphology. A Zeiss Supra PGT/HKL system was utilized. At an accelerating voltage of 1 kV and a pressure of typically 10−5–10−4 torr, FESEM was used to examine the effects of deicers on the morphology of the concrete, by collecting data from at least five randomly selected areas of the cement paste. The precaution was taken to examine only the cement paste, and the areas were randomly selected to avoid overlapping areas and aggregates. Morphology photographs were taken using an SE2 detector and at magnifications from 500 times to 6000 times. FESEM morphological examination at 500× magnification, and a minimum of five areas were analyzed for the given depth of each sample. Each of the areas analyzed was approximately 212.5 μm by 143.75 μm.

2.4.2 Zeta potential analysis

The zeta potential of the biochar was measured to determine its surface charge, which is critical for evaluating the stability of suspended biochar particles and their potential interactions in aqueous or soil environments. Measurements were conducted using a Particle Analyzer Litesizer DLS 500 (Anton Paar) at the Science and Engineering Research Center (SERC), Engineering Campus, Nibong Tebal. For sample preparation, approximately 1 mg (0.001 g) of biochar was accurately weighed and dispersed in 10 mL of distilled water to prepare a 0.01% (w/v) suspension. Distilled water was used to minimize interference from extraneous ions or impurities that could influence the zeta potential measurement. The suspension was subjected to ultrasonication for 5 min to ensure uniform dispersion by breaking up any particle agglomerates. Ultrasonication is essential for preventing inaccuracies caused by uneven particle distribution. After sonication, 1 mL of the homogenized suspension was carefully transferred into a specialized cuvette compatible with the Litesizer DLS 500. The instrument measures particle motion under an applied electric field using dynamic light scattering (DLS) principles to determine the zeta potential, reflecting the surface charge characteristics of the biochar particles.

2.4.3 CHNS analysis

CHNS analysis is a popular method for the determination of the elemental composition of carbon (C), hydrogen (H), nitrogen (N), and sulfur (S) in biochar. The analysis is required to assess the quality and potential application of biochar as a soil fertilizer, carbon sequestrant, and cleaner pollutant. It begins with sample preparation, where the biochar is dried, powdered to a fine state, and accurately weighed, usually in very small quantities of 1–5 mg. The sample is placed in a tin capsule and into a CHNS elemental analyzer. In the analyzer, the sample undergoes combustion at high temperature (around 950°C–1100°C) under pure oxygen. When the elements are combusted, they transform to their respective gases—carbon to CO_2, hydrogen to H_2O, nitrogen to N_2 or NOx, and sulfur to SO_2. The gases are then carried by an inert gas through a reduction column and separated by gas chromatography. The analysis is done with a thermal conductivity detector or an infrared detector, and the elemental concentrations are approximated by determining the amount of gas produced. Results are provided in weight percentages of single elements, and oxygen content is usually approximated from 100% minus carbon, hydrogen, nitrogen, sulfur, and ash contents. This analysis provides valuable information regarding the chemical structure and stability of biochar, which could be very dissimilar for various feedstocks and pyrolysis conditions used for its production.

2.4.4 Raman analysis

A handheld Raman spectrometer model Bravo by Bruker was used to record the Raman spectra shown in this study at room temperature. This instrument is appropriate for analyzing carbonaceous materials because it has dual laser excitation at 852 and 785 nm, which offers a wide spectral range from 170 to 3200 cm^{-1}. The powdered biochar sample was examined. To guarantee precise spectral readings, the spectrometer was calibrated using standardized graphite (1580 cm^{-1} peak) before analysis. The Bravo spectrometer's probe and the biochar sample came into direct contact during the measurement. The user started the spectral acquisition process, which usually took 10–30 s to gather enough data for analysis.

3. Results and discussions

3.1 Zeta potential analysis

As tabulated in Table 4.1, the influence of the carbonization temperature on two prominent physicochemical attributes of biochar: zeta potential (mV) and electrical conductivity (mS/cm) on a two-axis graph. The zeta potential, represented as square markers and against the left y-axis, is strongly dependent on temperature. With rising carbonization temperature from 400°C to 550°C, zeta potential becomes increasingly negative with an absolute maximum value of approximately −28 mV at 550°C. This kind of trend can be a marker for enhanced surface charge of biochar after thermal treatment, which is likely attributed to the degradation of oxygen-

Table 4.1 Zeta potential and conductivity value for sample at different carbonization temperature.

Carbonization temperature (°C)	Zeta potential (mV)	Conductivity value (mS/cm)
400	−23.73	0.051
450	−24.47	0.041
500	−26.17	0.047
550	−28.93	0.037
600	−27.83	0.056

functional groups such as carboxyl (-COOH) and hydroxyl (-OH). These functional groups dissociate in water to yield negative surface charges. The thermal breakdown of hemicellulose, cellulose, and some of the lignin at 400–550°C leads to the formation of more aromatized, thermostable carbonaceous networks that are more abundant in acidic functional moieties, hence increasing the surface acidity and electrostatic repulsion of the biochar.

There is a sharp deviation from this trend at 600°C, where the zeta potential becomes less negative by approximately −11 mV. This reversal may be due to additional degradation or restructuring of surface functional groups at excessively high temperatures. At additional carbonization beyond 550°C, removal of acidic groups and formation of a more compressed, graphitic structure carbon may reduce the surface density of charges. Moreover, increased thermal severity will lead to biochar particle fusing or sintering, reducing the surface area exposed and altering electrokinetic properties. The phenomenon highlights the importance of setting an optimal carbonization temperature range wherein the surface functionalization is intact without compromising structural stability, particularly in water treatment, adsorption of pollutants, or colloidal stabilization applications, wherein the charge on the surface is a crucial parameter.

On the other hand, the electrical conductivity, represented by Table 4.1, shows a more erratic pattern over the temperature range of 0.038–0.057 mS/cm. The highest conductivity is at 600°C and suggests that higher temperatures are conducive to the formation of conductive networks of carbon via increased aromaticity and structural ordering. This enhanced conductivity is due to the removal of nonconductive organic fractions and the exposure of graphitized domains that allow electron transport. Even at 550°C and despite the very negative zeta potential, the conductivity remains relatively low, indicating that conductivity is more dependent on the degree of carbon matrix organization than on surface functional groups. This highlights the different mechanistic origins of zeta potential and conductivity: the first being regulated through surface chemistry and the second through bulk structural carbonization of carbonaceous material. Overall, the findings highlight the multifaceted influence of pyrolysis temperature in regulating biochar properties toward target end-use functionalities. The optimum carbonization temperature for the manufacture of biochar from the trends noted in the graph is 550°C, particularly when the application has a main emphasis on surface charge attributes such as adsorption and ion exchange. At this point, 550°C, the biochar exhibits the maximum value of zeta potential (approximately −28 mV), indicating a high concentration of dissociable surface functional groups such as carboxyl and hydroxyl. These groups are especially valuable in enhancing the electrostatic attraction of the biochar surface to other pollutants, and hence are especially valuable in applications such as adsorption of heavy metals, soil remediation, and treatment of wastewater. The increased negativity in zeta potential at 550°C also indicates increased colloidal stability in aqueous systems, valuable in terms of conductivity and interaction in environmental matrices.

Despite such maximum electrical conductivity of 600°C, with attendant implication of increased graphitic ordering and electron mobility, this is accompanied by a significant loss in surface charge (lesser zeta potential negativity at about −11 mV). Such transition means loss of valuable oxygen-containing surface groups through further thermal decomposition and condensation of carbon structure. Therefore although 600°C may be desirable for use where electrical performance is at the forefront, such as with energy storage or electrochemical devices, it is not suitable for applications centered on surface interaction. 550°C, however, has optimal surface chemistry and structural integrity best balanced and is therefore best for multifunctional biochar generation for use in environmental and agronomic applications.

3.2 FESEM analysis

The FESEM micrographs in Fig. 4.4 illustrate the morphological evolution of a biochar material under pyrolysis treatment at increasing temperatures ranging from 400°C to 600°C. All in Fig. 4.4 were recorded at 500× magnification and illustrate the major microstructural evolution associated with pyrolysis-induced devolatilization, carbonization, and pore development.

At 400°C (Fig. 4.4a), the material surface exhibits partially degraded cell walls with apparent micropores and a layered, delaminated structure, indicating the onset of thermal decomposition. At 450°C (Fig. 4.4b), the structure becomes denser with rough surfaces and partially collapsed pores, indicating devolatilization in progress and additional carbon condensation. In contrast, the 500°C-treated sample (Fig. 4.4c) shows exaggerated pore formation and fragmentation with readily visible porous channels and expanded intercellular spaces, indicating an optimum degree for the

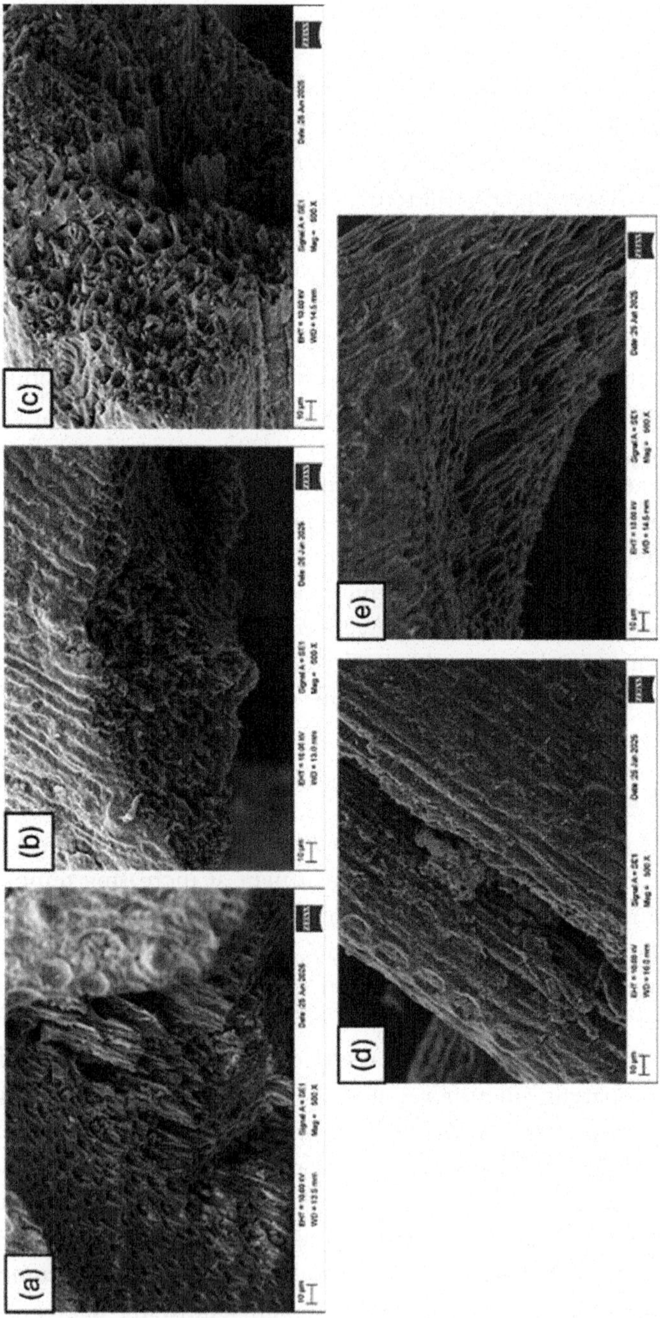

Figure 4.4 Carbonized sugar palm fiber under 500× magnification at different temperatures: (a) 400°C, (b) 450°C, (c) 500°C, (d) 550°C, and (e) 600°C.

formation of porosity. Upon further rise in temperature to 550°C (Fig. 4.4d), the surface shows further coarsening and distortion of the pore network, possibly due to enhanced carbon reorganization or structural shrinkage. At 600°C (Fig. 4.4e), the material exhibits a more organized lamellar structure with less surface roughness, which is a sign of thermal stabilization and carbon matrix densification.

These results demonstrate a strong correlation between structural morphology and carbonization temperature, which is critical for the design of material properties such as surface area, porosity, and mechanical integrity for applications in energy storage, adsorption, or catalysis [12]. Subsequently, biochars were used as fillers in Epoxy resin to create a composite. The optimum temperature for the maximization of structure development is observed to be at or near 500°C, where pore development is most active without experiencing collapse or densification to any significant degree.

3.3 CHNS analysis

The CHNS elemental analysis presented in Table 4.2 illustrates the compositional changes of sugar palm fiber subjected to carbonization at temperatures ranging from 400°C to 600°C. This analysis provides critical insights into the thermal decomposition behavior and chemical stability of the resulting biochar.

The carbon content (C%) shows a marked increase with temperature, from 44.490% in the raw material to 75.850% at 600°C. This steady rise is indicative of progressive carbon enrichment through the elimination of volatile components and the decomposition of cellulose, hemicellulose,

Table 4.2 Composition of CHNS in raw sugar palm fiber and biochar of sugar palm fiber at different temperatures.

Carbonization temperatures (°C)	Compositions			
	C (%)	H (%)	N (%)	S (%)
Raw sugar palm fiber	44.490	5.541	0.250	0.084
400	65.610	3.527	0.750	0.058
450	69.660	3.317	0.630	0.023
500	70.670	2.903	0.620	0.007
550	74.760	2.628	0.600	0.001
600	75.850	2.377	0.560	0.004

and lignin. The increase in carbon content is a key indicator of enhanced aromaticity and thermal stability, making the biochar more suitable for long-term applications such as carbon sequestration, soil amendment, or adsorbent materials.

Simultaneously, the hydrogen content (H%) decreases significantly from 5.541% in the raw fiber to 2.377% at 600°C. This reduction is primarily due to the loss of hydroxyl and aliphatic groups through dehydration and devolatilization reactions. The decreasing H/C ratio, inferred from this trend, suggests an increase in aromatic condensation, which is characteristic of high-temperature biochar with improved structural integrity and lower reactivity.

The nitrogen content (N%) displays an initial increase at 400°C (0.75%), which may be attributed to the partial retention or concentration of nitrogenous compounds such as proteins and amino acids. However, as temperature increases beyond this point, nitrogen content declines steadily to 0.56% at 600°C. This behavior reflects the progressive volatilization of nitrogen-containing species, potentially forming ammonia, hydrogen cyanide, or nitrogen oxides, which are released during pyrolysis. Sulfur content (S%) also exhibits a consistent reduction with temperature, decreasing from 0.084% in the raw fiber to as low as 0.004% at 600°C. The substantial decline in sulfur suggests efficient removal of sulfurous compounds during thermal decomposition, which is beneficial in minimizing emissions of sulfur-containing gases and improving the environmental compatibility of the produced biochar.

Overall, the results of the CHNS analysis confirm that increasing the carbonization temperature enhances the carbon content and stability of sugar palm fiber-based biochar, while simultaneously reducing undesirable elements such as hydrogen, nitrogen, and sulfur. These transformations underscore the importance of temperature optimization in tailoring biochar properties for specific environmental or industrial applications. Among the studied temperatures, 600°C appears most effective in producing high-carbon, low-contaminant biochar with potential for superior performance in energy storage, catalysis, or environmental remediation [13].

3.4 Raman analysis

The Raman spectra of carbonized sugar palm fiber demonstrated progressive structural transformation with increasing carbonization temperature (Fig. 4.5). The evolution of the characteristic D-band (~1350 cm^{-1}) and G-band (~1580 cm^{-1}) revealed significant changes in the graphitic ordering of the material. The IG/ID ratio, a key indicator of graphitization degree, increased from 0.7 (400°C) to 1.3 (600°C), suggesting enhanced carbon crystallinity at higher temperatures. This trend aligns with observations by Malini et al. [14] for similar biomass-derived carbons, though our values were systematically 10%–15% lower, likely due to differences in the initial lignin content (15%) compared to their study (20%–25%) and slower heating rates employed in our process.

The prominent D-band intensity across all samples indicates the persistence of structural defects and disordered carbon, which is characteristic of biomass-derived graphitic materials. Notably, the 2D-band (~2700 cm^{-1}) emerged weakly at 550°C and became more defined at 600°C (Fig. 4.5), suggesting the initial formation of few-layer graphene-like structures. This observation partially supports the findings of Lua et al. [15] but extends their work by demonstrating that such features can develop at relatively low temperatures (≥550°C) in palm-based fibers, which provides an advantage for energy-efficient carbonization.

The practical implications of these results are twofold. First, the temperature-dependent IG/ID evolution allows precise tuning of electrical conductivity for applications in flexible electronics. Second, the retained structural defects at all temperatures may benefit catalytic applications by

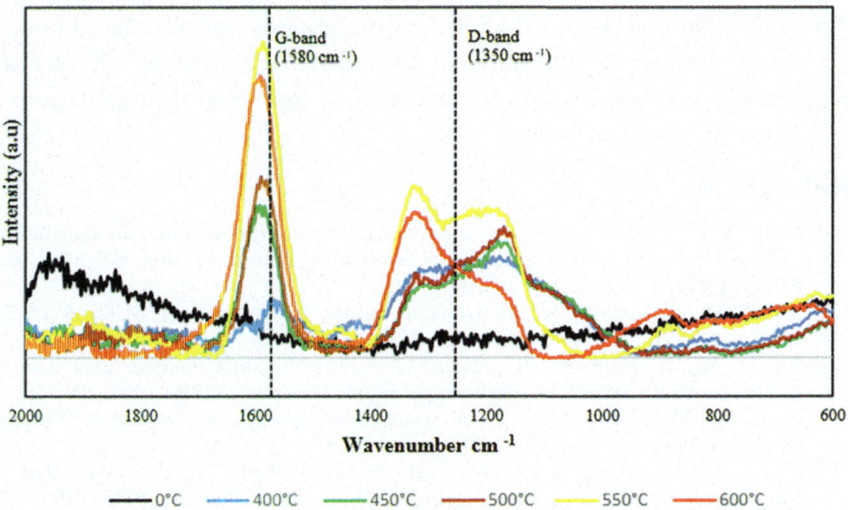

Figure 4.5 Raman spectra of 0°C, 400°C, 450°C, 500°C, 550°C, and 600°C.

providing active sites. However, the comparatively low IG/ID ratios ($\leq$1.3) suggest that additional activation steps may be required to achieve performance competitive with synthetic graphitic materials in high-end applications.

4. Conclusions

In conclusion, this study has demonstrated that carbonization temperature plays a crucial role in shaping the morphological and structural properties of biochar derived from sugar palm fiber, a lignocellulosic biomass. Biochar samples produced at temperatures ranging from 400°C to 600°C revealed significant transformations in surface structure, carbon content, and stability. Higher temperatures, particularly at 550°C, enhanced the development of porous structures, increased fixed carbon content, and reduced volatile matter. These changes are indicative of improved thermal stability, surface reactivity, and potential adsorption capabilities. Elemental analysis confirmed a clear increase in carbon concentration, while zeta potential results reflected greater colloidal stability, making the biochar more suitable for applications such as water purification and soil amendment. Therefore sugar palm fiber waste shows strong potential as a sustainable feedstock for producing high-performance biochar, supporting both environmental remediation and resource valorization. Graphitic structures increased with increasing temperature (>500°C), but biochar with a balance of carbon content around 40% and residual functionality was obtained in the optimal 400–450°C range. These findings unambiguously specify the chemical development of sugar palm fiber biochar during carbonization and identify temperature as a dominant parameter to tune its chemical properties.

References

[1] Mulla, M.H., et al., 2025. Surface and interfaces effects of concentrations and alkaline treatment durations on sugar palm fiber as structural reinforcement in polymer composites. J. Nat. Fibers 22 (1), 2527277.

[2] Bartoli, M., et al., 2020. A review of non-soil biochar applications. Materials 13 (2), 261.

[3] Bartoli, M., Giorcelli, M., Tagliaferro, A., 2023. A comprehensive overview on biochar-based materials for catalytic applications. Catalysts 13 (10), 1336.

[4] Nasir, A.H.M.P., et al., 2025. Plastic waste carbonization-a review. Part II. Thetmal properties. Polimery 70 (6), 367–377.

[5] Nasir, A.H.M.P., et al., 2025. Plastic waste carbonization-a review. Part I. Morphology and characteristics of carbonization products. Polimery 70 (5), 299–312.

[6] Manimaran, M., et al., 2025. Tunable structural, thermal and colloidal properties of anionic cellulose nanocrystals from oil palm biomass/metal oxide nanoparticles hybrid nanocomposites. Int. J. Biol. Macromol. 322, 146530.
[7] Manimaran, M., et al., 2025. Surfaces and interfaces analysis on different carboxymethylation reaction time of anionic cellulose nanoparticles derived from oil palm biomass. Nanotechnol. Rev. 14 (1), 20250152.
[8] Mulla, M.H., et al., 2024. Dynamic mechanical properties of sugar palm fiber-reinforced polymer composites: A review. Polym. Bull. 81 (9), 7737–7789.
[9] Asyraf, M., et al., 2023. Lignocellulosic sugar palm fibre-reinforced thermoplastic composites: Mechanical, thermal and dynamic mechanical properties. Fibers Polym. 24 (8), 2625–2639.
[10] Asyraf, M., et al., 2023. Thermal properties of oil palm lignocellulosic fibre reinforced polymer composites: A comprehensive review on thermogravimetry analysis. Cellulose 30 (5), 2753–2790.
[11] Ilyas, R., Nurazzi, N., Norrrahim, M., 2022. Fiber-reinforced Polymer Nanocomposites. MDPI, p. 3045.
[12] Giorcelli, M., et al., 2019. Biochar as a cheap and environmental friendly filler able to improve polymer mechanical properties. Biomass Bioenergy 120, 219–223.
[13] Giorcelli, M., Bartoli, M., 2019. Development of coffee biochar filler for the production of electrical conductive reinforced plastic. Polymers 11 (12), 1916.
[14] Malini, K., Selvakumar, D., Kumar, N., 2022. Activated carbon from biomass: preparation, factors improving basicity and surface properties for enhanced CO_2 capture capacity – a review. J. CO_2 Util. 67, 102318. https://doi.org/10.1016/j.jcou.2022.102318.
[15] Lua, A.C., Yang, T., Guo, J., 2004. Effects of pyrolysis conditions on the properties of activated carbons prepared from pistachio-nut shells. J. Anal. Appl. Pyrolysis 72 (2), 279–287.

CHAPTER 5

Carbon-based photocatalyst materials and composites in water and wastewater treatment

N.B. Mohamed[1], S.F. Mohd Din[1], M.H. Mohamed Noor[1], Y.M.N.S. Ismail[1,2], Abu Hassan Nordin[3], C.A. Luqman[4,5] and Ahmad Ilyas Rushdan[1,6]

[1]Faculty of Chemical and Energy Engineering, Universiti Teknologi Malaysia (UTM), Skudai, Johor, Malaysia; [2]Department of Environment Pulau Pinang, Aras Bawah, Zon B, Wisma Persekutuan, Seberang Perai Utara, Pulau Pinang, Malaysia; [3]Faculty of Applied Sciences, Universiti Teknologi MARA (UiTM) Arau Campus, Perlis, Malaysia; [4]Laboratory of Biocomposite, Institute of Tropical Forestry and Forest Products (INTROP), Universiti Putra Malaysia, Serdang, Selangor, Malaysia; [5]Advanced Engineering Materials and Composites Research Centre (AEMC), Department of Mechanical and Manufacturing Engineering, Faculty of Engineering, Universiti Putra Malaysia, UPM Serdang, Selangor Darul Ehsan, Malaysia; [6]Centre for Advanced Composite Materials (CACM), Universiti Teknologi Malaysia, Johor Bahru, Johor, Malaysia

1. Introduction

Water pollution is one of the most serious environmental problems linked to industrial growth and rapid urbanization. Traditional treatment systems such as filtration, coagulation, and adsorption often fail to remove persistent pollutants like dyes, pharmaceuticals, and heavy metals. In recent years, advanced oxidation processes (AOPs) have gained attention because they can mineralize hazardous compounds into harmless forms. Among these, photocatalysis is an attractive option because it uses sunlight to generate highly reactive radicals that can destroy pollutants under mild conditions [1,2].

Carbon-based materials including graphene, carbon nanotubes, biochar, and graphitic carbon nitride are often combined with semiconductor photocatalysts to enhance efficiency and stability. Their good conductivity, large surface area, and chemical flexibility help promote charge separation and pollutant adsorption, creating a combined adsorption–photocatalysis effect [3,4]. These materials are now considered key components of advanced water treatment technologies.

Carbon-Based Materials and Composites for Water and Wastewater Treatment
ISBN 978-0-443-33681-2
https://doi.org/10.1016/B978-0-443-33681-2.00004-1

2. Fundamentals of carbon-based photocatalysis (balqis)

Photocatalysis involves light-induced excitation in semiconductors, which triggers oxidation and reduction reactions that degrade pollutants [4]. The addition of carbon improves semiconductor photocatalysts by increasing light absorption, surface activity, and charge movement [5].

2.1 Mechanism of photocatalytic reactions

When exposed to light, electrons in the valence band of a semiconductor move to the conduction band, leaving holes behind. The excited electrons reduce oxygen to form superoxide radicals ($O_2\bullet^-$), while the holes oxidize water to form hydroxyl radicals (•OH) [3,4]. These radicals break down organic compounds into carbon dioxide and water. However, charge carriers tend to recombine quickly, which limits efficiency. Carbon helps slow this recombination by providing a conductive pathway for electrons, extending their lifetime [5]. Materials such as graphene and carbon nanotubes are particularly effective because their structure allows fast electron movement and better use of visible light.

2.2 Functional role of carbon in composite photocatalysts

Carbon plays multiple roles in composite photocatalysts. Its porous structure increases the surface area available for pollutant adsorption and improves contact with active sites [6,7].

Surface groups such as hydroxyl, carboxyl, and carbonyl aid electron transfer and help bind nanoparticles. Carbon also keeps semiconductor particles well-dispersed and prevents clumping. In some cases, it enhances visible-light absorption, allowing the photocatalyst to work more effectively under solar light [8]. Carbon materials are also durable, resisting chemical attack and maintaining stability through repeated use [9,10].

2.3 Influence of structural and electronic properties on photocatalytic efficiency

The structure and electronic features of carbon strongly affect photocatalytic performance. Highly graphitic carbons conduct electricity well and reduce charge loss, while amorphous carbons and biochar contain many reactive groups that boost adsorption [11–13]. Doping with nitrogen, sulfur, or phosphorus changes electronic density and narrows the bandgap, improving visible-light absorption [14]. Combining carbon with semiconductors like TiO_2 or ZnO creates heterojunctions that direct charge movement and suppress recombination [15]. Well-designed pore structures help light and reactants reach inner surfaces efficiently [16,17]. The right balance between conductivity, porosity, and surface chemistry is essential for optimum performance.

3. Types of carbon-based photocatalyst materials

Carbon-based composite photocatalysts offer major benefits for water decontamination and disinfection, highlighting their adsorption capacity for pollutant removal. Carbon-based nanocomposites material (CNMs) can first gather trace pollutants and microorganisms on their surface and then oxidize them in situ via photocatalytic reactions [18]. Quantum dots and nanotubes are carbon-based nanostructures with good catalytic effectiveness.

3.1 Carbon quantum dots (CQDs)

CQDs are quasi-spherical carbon less than 10 nm in size [19]. CQDs in nanocomposites are donors and acceptors for e^- transfer and increase the contact area [20]. Various authors have described a higher photocatalytic activity for the CQD/TiO_2 composite than for pure TiO_2, attributed to the production of more electrons and e^-/h^+ pairs under light irradiation and the CQD/TiO_2 nanocomposite is a good photocatalyst [21].

Pan, Zhou [22] found that the SnO_2/CQD composite has excellent photocatalytic properties under visible light irradiation (Fig. 5.1). Up-conversion luminescence results from simultaneous multiphoton absorption that has been related to graphite nitrogen and oxygen-containing functional groups on the surface of the CQDs [23]. Figure 5.1a presents the SEM image of the SnO_2 nanotube surface, which shows a granular appearance that may favor adsorption of contaminant molecules onto the nanotubes and improve the contact area between them. Figure 5.1b depicts the XRD spectra of SnO_2 nanotubes at different concentrations of CQDs. No diffraction peaks are observed other than those corresponding to diffractions of typical crystal planes of tetragonal SnO_2, indicating the highly crystalline SnO_2 structure of the nanotube compounds. Figure 5.1c–5.1e exhibit TEM images that show the attachment of CQDs to the surface of SnO_2 nanotubes. The lattice spacings in the HRTEM images are consistent with the (002) and (110) planes of CQDs and SnO_2, respectively. Figure 5.1f reveals the photocatalytic degradation under visible light

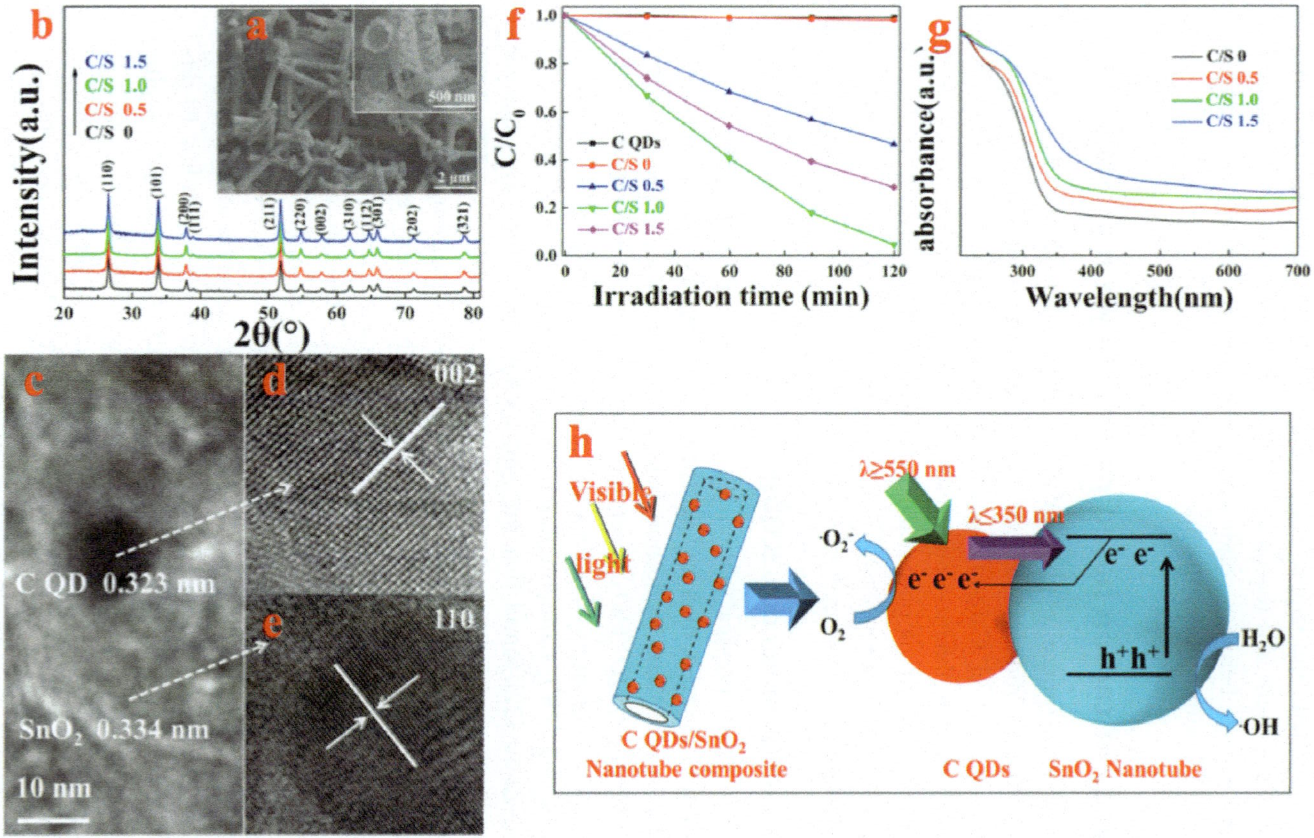

Figure 5.1 The structure of CQD-modified SnO2 nanotube composites: (a) Scanning electron microscopy (SEM) image of CQD-modified SnO2 nanotubes; (b) X-ray diffractometer (XRD) image of CQD-modified SnO2 nanotubes, labeling composites coupled with a CQD solution of 1.5 mg mL− 1 as C/S 1.5, etc.; (c) Transmission electron microscopy (TEM) image of CQD-modified SnO2 nanotube; (d) High-resolution TEM (HRTEM) image of CQD; (e) HRTEM of SnO2; (f) photocatalytic degradation of different CQD-modified SnO2 nanotubes under visible light irradiation; (g) UV–visible absorption spectra of the samples; (h) Schematic illustration of the visible light photocatalytic

irradiation of CQD-modified SnO_2 nanotubes, showing the resulting improvement in their photocatalytic activity. Figure 5.1g shows the intensification of absorption at higher CQD concentrations, in agreement with the increase in photocatalysis under visible light. Figure 5.1h provides a schematic representation of the proposed system, in which CQDs/SnO_2 nanotube composites are excited under visible light irradiation and generate photogenerated e^-/h^+ pairs, improving the photocatalytic activity along with the transfer by the CQDs of photogenerated electrons.

3.2 Carbon nanotubes (CNTs)

CNTs are characterized by sp^2 hybridization between C–C bonds. The two main types are multiwalled CNTs (MWCNTs) and single-walled CNTs (SWCNTs), classified by the number of rolled layers that form their structure. CNTs can increase the photocatalytic activity of nanocomposites in three possible ways: (i) avoiding photogenerated charge recombination in the metal oxide by functioning as an e^- reservoir [24]; (ii) acting as a photosensitizer to produce positive holes and electrons that can form reactive radical species [25]; and (iii) generating new bonds and energy levels through the formation of C–O–Metal bonds [26].

Figure 5.2a–5.2c shows SEM micrographs of pure ZnO, Ag@ZnO and Ag@ZnO/MWCNT. Addition of MWCNTs improves the distribution of smaller particles on the surface of MWCNTs. Figure 5.2d displays the specific surface areas of Ag@ZnO and Ag@ZnO/MWCNT. The loading of nanoadsorbents with MWCNTs enlarges the specific surface area of the nanocomposite which correlates with an increase in its sorption capacity during adsorption. Furthermore, the surface properties of CNTs can be functionalized by introducing functional groups such as ketone, alcohol, phenol and acid moieties, or nitrogen, among others [27]. These modifications can improve the adsorption of specific compounds, adjust specific metal-support interactions and change their chemical behavior (e.g., hydrophilicity, pH, e^- donor/acceptor properties, etc.) [26,28].

4. Mechanistic insights and photodegradation performance

Reactive oxygen species such as hydroxyl radicals, superoxide radicals, and singlet oxygen are the main agents in pollutant degradation. Efficient photocatalysis depends on how well electrons and holes separate and move.

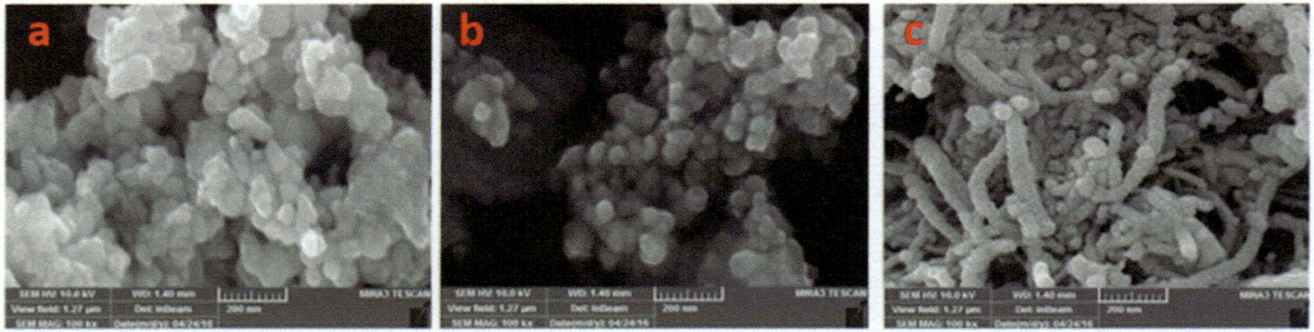

d

Nano-adsorbent	BET ($m^2\ g^{-1}$)	Total pore volume ($cm^3\ g^{-1}$)	Average pore size (nm)
Ag@ZnO	20.89	0.21	4.66
Ag@ZnO/MWCNT	175.50	0.423	2.56

Figure 5.2 SEM images of (a) ZnO, (b) silver-doped ZnO Ag@ZnO, and (c) Ag@ZnO/MWCNT; (d) BET surface area and structural properties of Ag@ZnO and Ag@ZnO/MWCNT nanocomposite. *(Adapted from Ref (Ahmadi Azqhandi et al., 2018).*

Defect engineering and heterojunctions improve this process by reducing recombination [29–31].

Degradation of organic pollutants occurs in several stages, including dealkylation, ring opening, and eventual mineralization to carbon dioxide and water [32]. Factors like pH, light strength, and catalyst dose affect reaction speed [33]. Inorganic ions such as calcium and magnesium, or organic matter like humic substances, may interfere with reactive species or block active sites, lowering efficiency [34–36].

Carbon doping changes the electronic structure of semiconductors, creating mid-gap states that absorb visible light and improve charge mobility [29,37]. Oxygen vacancies formed during doping also help trap electrons and generate reactive species [38].

Reaction conditions must be optimized to avoid problems such as dopant excess or recombination [39–43]. Combining photocatalysis with processes like ozonation or Fenton oxidation can improve degradation [40]. Immobilizing catalysts on membranes enhances recovery and durability [41].

5. Challenges, limitations, and future perspectives

5.1 Challenges in carbon-based photocatalysis

Despite many promising developments in carbon-based photocatalysts for water and wastewater treatment, several key challenges remain. One major issue is the fast recombination of photogenerated electron-hole pairs, which significantly reduces the effectiveness of the photocatalytic process. This recombination slows down redox reactions that are essential for breaking down pollutants, which lowers the overall treatment efficiency [44]. Many carbon materials, like pristine graphene and carbon nanotubes (CNTs), also have limited reactivity under visible light because of their wide band gaps or poor absorption in that spectrum.

Another challenge is surface fouling. Over time, organic and inorganic substances in wastewater can accumulate on the surface of the photocatalyst, blocking active sites and reducing performance. The durability of these materials is also a concern. Some composites can degrade structurally or chemically, especially under harsh conditions or extended light exposure [44].

Additionally, producing high-performance carbon-based photocatalysts often involves complex methods and expensive materials like graphene oxide or multi-walled CNTs, which makes it hard to scale up production [45]. These problems make it difficult to move from lab-scale success to real-world applications.

5.2 Limitations of current applications

Even though carbon-based photocatalysts have shown strong potential in removing pollutants, there are still several real-world limitations. Most studies are done in labs using synthetic wastewater, which doesn't fully represent the complex mixtures found in actual wastewater streams. Also, these studies usually use batch systems, whereas industries need continuous flow systems that are rarely explored in research [45].

Another issue is recovery and reuse. Carbon-based nanomaterials are difficult to separate from treated water because of their small size. This makes them hard to recycle and poses a risk of secondary pollution [45]. There's also no standard way to evaluate how well different photocatalysts perform, making it hard to compare results across studies. Concerns about the potential health and environmental risks of carbon nanomaterials are also growing. Some studies suggest that these materials might be toxic to humans or aquatic life, so detailed safety and life-cycle studies are necessary before scaling up their use [34].

5.3 Future research directions

To deal with current challenges, future work should focus on developing hybrid photocatalysts that combine the strengths of different materials. For example, combining carbon with metal oxides like TiO2 or ZnO can improve light absorption and reduce charge recombination [46]. Doping carbon with elements like nitrogen, boron, or sulfur can also improve performance under visible light. Researchers should also explore more sustainable ways to make carbon photocatalysts, such as using biomass or agricultural waste. Materials like biochar and activated carbon have shown good results in both adsorption and photocatalytic degradation of pollutants [47].

Advanced production methods, including 3D printing, electrospinning, and atomic layer deposition, could help create films or membranes with more surface area and better light absorption. Combining photocatalysts with other treatment methods, such as membranes or electrochemical processes, might lead to better performance overall [46]. Artificial intelligence (AI) and machine learning can also help by predicting how new materials might perform based on their structure, which could make research more efficient.

5.4 Outlook for practical implementation

For carbon-based photocatalysts to be used widely in wastewater treatment, several technical and non-technical issues need to be addressed. Clear guidelines and regulations on the safe use and disposal of nanomaterials will help build public and industrial trust. Using cheaper materials and simpler manufacturing methods can also help lower production costs.

These systems could be especially useful in rural or off-grid areas, where solar-powered, low-maintenance water treatment systems are needed. This fits well with global goals for sustainable development and access to clean water. Stronger collaboration between universities, industry, and government agencies is needed to move these technologies from the lab to the field. Pilot studies and long-term testing in real environments will be key to proving that these systems work and are safe.

6. Conclusion

Carbon-based photocatalysts provide a versatile and effective platform for advanced water and wastewater treatment. Their unique structural and electronic properties enhance light harvesting, charge-carrier separation, and the formation of reactive oxygen species responsible for pollutant degradation. Integration with semiconductor materials such as TiO_2, ZnO, or $BiVO_4$ enables improved performance under visible light while maintaining strong chemical stability and recyclability. The combination of adsorption capacity and photocatalytic activity makes these materials highly adaptable for removing diverse organic and inorganic contaminants.

Despite their potential, several challenges remain, including limited scalability, difficulties in catalyst recovery, and inconsistent performance in real wastewater systems. Addressing these issues requires continued innovation in synthesis strategies, reactor design, and material optimization. Future directions include developing biomass-derived carbon composites, exploring Z-scheme heterojunctions, and applying data-driven modeling to accelerate material discovery. With sustained interdisciplinary efforts, carbon-based photocatalysts are poised to play a major role in realizing next-generation, solar-driven purification technologies that contribute to global water sustainability goals.

Acknowledgment

The authors acknowledge Universiti Teknologi Malaysia and the Ministry of Higher Education Malaysia for their financial support under the Fundamental Research Grant Scheme (FRGS) under grant number R.J130000.7809.5F767.

References

[1] Durodola, S.S., Akeremale, O.K., Ore, O.T., Bayode, A.A., Badamasi, H., Olusola, J. A., 2024. A review on nanomaterial as photocatalysts for degradation of organic pollutants. J. Fluoresc. 34 (2), 501–514. https://doi.org/10.1007/s10895-023-03332-x.
[2] Gołąbiewska, A., Kobylański, M.P., Zaleska-Medynska, A., 2018. Fundamentals of metal oxide-based photocatalysis. In: Metal oxide-based photocatalysis: Fundamentals and prospects for application. https://doi.org/10.1016/B978-0-12-811634-0.00002-0.
[3] Wang, H., Duan, Y., Fei, G., Yan, T., Kang, Y.M., Dionysiou, D.D., 2023. Design, synthesis and modification of 2D nanomaterials-based photocatalysts for pollutant degradation and photodegradation experiments from lab-scale to grand-scale. Chem. Eng. J. 477. https://doi.org/10.1016/j.cej.2023.147219.
[4] Wang, M., Cai, Y., Zhou, B., Yuan, R., Chen, Z., Chen, H., 2022. Removal of PFASs from water by carbon-based composite photocatalysis with adsorption and catalytic properties: A review. Sci. Total Environ. 836. https://doi.org/10.1016/j.scitotenv.2022.155652.
[5] Liu, B., Wu, H., Parkin, I.P., 2020. New insights into the fundamental principle of semiconductor photocatalysis. ACS Omega 5 (2), 14847–14856. https://doi.org/10.1021/acsomega.0c02145.
[6] Liu, X., Gu, S., Zhang, X., Li, X., Zhao, Y., Li, W., 2021. The production discipline and mechanism of hydroxyl radical by investigating the Ln_2O_3-Bi_2MoO_6 heterojunction photocatalysts. J. Alloys Compd. 864. https://doi.org/10.1016/j.jallcom.2021.158894.
[7] Sohail, M., Rauf, S., Irfan, M., Hayat, A., Alghamdi, M.M., El-Zahhar, A.A., Ghernaout, D., Al-Hadeethi, Y., Lv, W., 2024. Recent developments, advances and strategies in heterogeneous photocatalysts for water splitting. Nanoscale Adv. 6 (5), 1286–1330. https://doi.org/10.1039/d3na00442b.
[8] Ding, Y., Li, Z., Zhang, S., Yang, G., Zheng, R., Wang, C., 2025. Visible-light driven photocatalytic water body purification on 2D carbon materials. J. Environ. Manag. 392. https://doi.org/10.1016/j.jenvman.2025.126774.
[9] Jabbar, Z.H., Graimed, B.H., Ammar, S.H., Sabit, D.A., Najim, A.A., Radeef, A.Y., Taher, A.G., 2024. The latest progress in the design and application of semiconductor photocatalysis systems for degradation of environmental pollutants in wastewater: Mechanism insight and theoretical calculations. Mater. Sci. Semicond. Process. 173. https://doi.org/10.1016/j.mssp.2024.108153.
[10] Usman, M., Humayun, M., Shah, S.S., Ullah, H., Tahir, A.A., Khan, A., Ullah, H., 2021. Bismuth-graphene nanohybrids: Synthesis, reaction mechanisms, and photocatalytic applications—a review. Energies 14 (8). https://doi.org/10.3390/en14082281.
[11] Farooq, N., Qureshi, N., Hussain, S., Qureshi, A.M., Sattar, Z., Manavalan, R.K., Ullah, N., Aslam, M.K., 2025. Recent advances, synthetic and adaptive approaches of graphitic carbon nitride based nanocomposites towards sustainable applications: A review. Int. J. Hydrogen Energy 180. https://doi.org/10.1016/j.ijhydene.2025.151763.

[12] Maarisetty, D., Baral, S.S., 2020. Defect engineering in photocatalysis: Formation, chemistry, optoelectronics, and interface studies. J. Mater. Chem. A. https://doi.org/10.1039/d0ta04297h.

[13] Mennani, M., Ait Benhamou, A., Mekkaoui, A.A., El Bachraoui, F., El Achaby, M., Moubarik, A., Kassab, Z., 2024. Probing the evolution in catalytic graphitization of biomass-based materials for enduring energetic applications. J. Mater. Chem. A. https://doi.org/10.1039/d3ta07449h.

[14] Wu, X., Su, Y., Wang, Y., Amina, K., Zhu, P., Wang, P., Wei, G., 2024. TiO_2 nanotube arrays-based photoelectrocatalyst: Tri-doping engineering and carbon coating engineering boosting visible activity, and stable hydrogen evolution. J. Colloid Interface Sci. 568, 247–257. https://doi.org/10.1016/j.jcis.2023.12.034.

[15] Benko, A., Duch, J., Gajewska, M., Marzec, M., Bernasik, A., Nocuń, M., Piskorz, W., Kotarba, A., 2021. Covalently bonded surface functional groups on carbon nanotubes: From molecular modeling to practical applications. Nanoscale 13 (22), 10152–10166. https://doi.org/10.1039/d0nr09057c.

[16] He, X., Yang, Y., Li, Y., Chen, J., Yang, S., Liu, R., Xu, Z., 2022. Effects of structure and surface properties on the performance of ZnO towards photocatalytic degradation of methylene blue. Appl. Surf. Sci. 599. https://doi.org/10.1016/j.apsusc.2022.153898.

[17] Mohtar, S.S., Aziz, F., Ismail, A.F., Sambudi, N.S., Abdullah, H., Rosli, A.N., Ohtani, B., 2021. Impact of doping and additive applications on photocatalyst textural properties in removing organic pollutants: A review. Catalysts 11 (10). https://doi.org/10.3390/catal11101160.

[18] Qamar, M.A., Shahid, S., Javed, M., Sher, M., Iqbal, S., Bahadur, A., Li, D., 2021. Fabricated novel g-C3N4/Mn doped ZnO nanocomposite as highly active photocatalyst for the disinfection of pathogens and degradation of the organic pollutants from wastewater under sunlight radiations. Colloids Surf. A Physicochem. Eng. Asp. 611. https://doi.org/10.1016/j.colsurfa.2020.125863.

[19] Liu, W., Li, C., Ren, Y., Sun, X., Pan, W., Li, Y., Wang, J., Wang, W., 2016. Carbon dots: Surface engineering and applications. J. Mater. Chem. B 4 (53). https://doi.org/10.1039/c6tb00976j.

[20] Gao, W., Zhang, S., Wang, G., Cui, J., Lu, Y., Rong, X., Gao, C., 2022. A review on mechanism, applications and influencing factors of carbon quantum dots based photocatalysis. Ceram. Int. 48 (24). https://doi.org/10.1016/j.ceramint.2022.10.116.

[21] Sendão, R.M.S., Esteves da Silva, J.C.G., Pinto da Silva, L., 2022. Photocatalytic removal of pharmaceutical water pollutants by TiO_2 – Carbon dots nanocomposites: A review. Chemosphere 301. https://doi.org/10.1016/j.chemosphere.2022.134731.

[22] Pan, J., Zhou, Y., Cao, J., Sheng, Y., Wu, Y., Cui, C., Li, C., Feng, B., 2016. Fabrication of carbon quantum dots modified granular SnO_2 nanotubes for visible light photocatalysis. Mater. Lett. 170, 187–191. https://doi.org/10.1016/j.matlet.2016.02.033.

[23] Li, M., Ma, C., Wang, G., Zhang, X., Dong, X., Ma, H., 2021. Controlling the upconversion photoluminescence property of carbon quantum dots (CQDs) by modifying its surface functional groups for enhanced photocatalytic performance of CQDs/$BiVO_4$ under a broad-spectrum irradiation. Res. Chem. Intermed. 47 (8), 3469–3485. https://doi.org/10.1007/s11164-021-04459-x.

[24] Yadav, G., Ahmaruzzaman, M., 2023. Recent development of novel nanocomposites for photocatalysis mediated remediation of phenolic derivatives: A comprehensive review. J. Ind. Eng. Chem. 127, 18–35. https://doi.org/10.1016/j.jiec.2023.07.029.

[25] Ma, G., Zhu, Y., Zhang, Z., Li, L., 2014. Preparation and characterization of multi-walled carbon nanotube/TiO_2 composites: Decontamination organic pollutant in water. Appl. Surf. Sci. 313, 817–822. https://doi.org/10.1016/j.apsusc.2014.06.079.

[26] Arora, B., Attri, P., 2020. Carbon nanotubes (CNTs): A potential nanomaterial for water purification. J. Compos. Sci. 4 (3). https://doi.org/10.3390/jcs4030135.
[27] Nurazzi, N.M., Asyraf, M.R.M., Khalina, A., Abdullah, N., Sabaruddin, F.A., Kamarudin, S.H., Ahmad, S., Mahat, A.M., Lee, C.L., Aisyah, H.A., Norrrahim, M. N.F., Ilyas, R.A., Harussani, M.M., Ishak, M.R., Sapuan, S.M., 2021. Fabrication, functionalization, and application of carbon nanotube-reinforced polymer composite: An overview. Polymers 13 (7). https://doi.org/10.3390/polym13071047.
[28] Kharlamova, M.V., Paukov, M., Burdanova, M.G., 2022. Nanotube functionalization: Investigation, methods and demonstrated applications. Materials 15 (15). https://doi.org/10.3390/ma15155386.
[29] Kurban, M., Polat, C., Serpedin, E., Kurban, H., 2024. Enhancing the electronic properties of TiO_2 nanoparticles through carbon doping: An integrated DFTB and computer vision approach. Comput. Mater. Sci. 244. https://doi.org/10.1016/j.commatsci.2024.113248.
[30] Tripathy, S.R., Baral, S.S., 2025. Defect engineering in semiconductor photocatalysts: Enhancing photocatalytic activity for green energy production. Adv. Energy Sustain. Res. 6 (11). https://doi.org/10.1002/aesr.202500110.
[31] Zhang, P., Peng, C., Li, H., Huang, J., Wang, Y., Yu, Y., Ding, S., Liu, S., Zhao, Y., 2022. Wavelength-dependent generation of reactive species in the photodegradation process over pure and C-doped Nb2O5. Separ. Purif. Technol. 286, 120406.
[32] Huang, Y., Wang, H., Huang, K., Huang, D., Yin, S., Guo, Q., 2020. Degradation kinetics and mechanism of 3-Chlorobenzoic acid in anoxic water environment using graphene/TiO2 as photocatalyst. Environ. Technol. 41 (17), 2165–2179. https://doi.org/10.1080/09593330.2018.1556741.
[33] Bekele, T., Alamnie, G., 2025. The photocatalytic degradation of organic pollutants-a comprehensive overview. Results Chem. 18. https://doi.org/10.1016/j.rechem.2025.102758.
[34] He, J., Kumar, A., Khan, M., Lo, I.M.C., 2021. Critical review of photocatalytic disinfection of bacteria: From noble metals- and carbon nanomaterials-TiO_2 composites to challenges of water characteristics and strategic solutions. Sci. Total Environ. 758. https://doi.org/10.1016/j.scitotenv.2020.143953.
[35] Ngulube, K.F., Abdelhaleem, A., Osman, A.I., Peng, L., Nasr, M., 2024. Advancing sustainable water treatment strategies: Harnessing magnetite-based photocatalysts and techno-economic analysis for enhanced wastewater management in the context of SDGs. Environ. Sci. Pollut. Res. 1–37. https://doi.org/10.1007/s11356-024-32680-9.
[36] Xu, H.-T., Fu, X.-H., Feng, W.-H., Wang, T., 2023. Photo-degradation mechanism and pathway for tetracycline in simulated seawater under irradiation of visible light. Huanjing Kexue/Environ. Sci. 44 (6), 3260–3269. https://doi.org/10.13227/j.hjkx.202206237.
[37] Xie, M., Deng, Q., Yang, T., Wang, C., Liu, J., Wen, F., 2024. One-step thermal oxidation synthesis of carbon-doped TiO_2 with enhanced photocatalytic activity using CO_2 as the carbon source. J. Photochem. Photobiol. Chem. 456. https://doi.org/10.1016/j.jphotochem.2024.115822.
[38] Kawakami, R., Matsumoto, T., Yanagiya, S.-I., Shirai, A., Nakano, Y., Niibe, M., 2025. Enhanced photocatalytic activity of Anatase/rutile-mixed phase titanium dioxide nanoparticles annealed with polyethylene glycol at low temperatures in aluminum foil-covered combustion boats. Physica Status Solidi (A) Appl. Mater. Sci. 222 (2). https://doi.org/10.1002/pssa.202400478.
[39] Issaka, E., Danso-Boateng, E., Baffoe, J., 2024. Harnessing the power of heterogeneous photocatalytic process for sustainable pharmaceutical contaminant remediation in water environments. Desalination Water Treat. 319. https://doi.org/10.1016/j.dwt.2024.100574.

[40] Liu, X., Liu, M., Dai, Y., Cui, J., Jamil, A., Liu, W., Li, J., Wang, J., 2024. Construction of self-cleaning g-C3N4/Bi2MoO6/PVDF membrane and coupling with photo-fenton-like reaction for sustainable removal of antibiotics in wastewater. J. Environ. Manag. 366. https://doi.org/10.1016/j.jenvman.2024.121928.
[41] Moslehyani, A., Ismail, A.F., Othman, M.H.D., Matsuura, T., 2015. Design and performance study of hybrid photocatalytic reactor-PVDF/MWCNT nanocomposite membrane system for treatment of petroleum refinery wastewater. Desalination 363, 99–111. https://doi.org/10.1016/j.desal.2015.01.044.
[42] Qu, J., Du, Y., Ji, P., Li, Z., Jiang, N., Sun, X., Xue, L., Li, H., Sun, G., 2021. Fe, Cu co-doped BiOBr with improved photocatalytic ability of pollutants degradation. J. Alloys Compd. 881. https://doi.org/10.1016/j.jallcom.2021.160391.
[43] Sui, X., Xu, T., Wang, G., Lu, W., 2025. Preparation of graphene oxide quantum dot-spherical carbon nitride catalyst and its photocatalytic degradation of sulfamethoxazole. Huanjing Kexue Xuebao/Acta Sci. Circumstantiae 45 (7), 285–295. https://doi.org/10.13671/j.hjkxxb.2025.0057.
[44] Tripti, T., Singh, P., Rani, N., Kumar, S., Kumar, K., Kumar, P., 2024. Carbon dots as potential candidate for photocatalytic treatment of dye wastewater. Environ. Sci. Pollut. Res. 31 (5), 6738–6765. https://doi.org/10.1007/s11356-023-31437-0.
[45] Zhao, X., Liu, Y., Chen, Y., Su, X., Feng, Y., Lu, C., Ma, Y., Lu, G., Ma, M., 2023. Research progress of novel magnetic two-dimensional carbon composites in photocatalytic degradation of pollutants: A review. Environ. Sci. Pollut. Res. 30 (27), 69774–69795. https://doi.org/10.1007/s11356-023-27234-4.
[46] Scaria, J., Karim, A.V., Divyapriya, G., Nidheesh, P.V., Suresh Kumar, M., 2019. Carbon-supported semiconductor nanoparticles as effective photocatalysts for water and wastewater treatment. In: Nano-materials as photocatalysts for degradation of environmental pollutants: Challenges and possibilities, pp. 245–278. https://doi.org/10.1016/B978-0-12-818598-8.00013-4.
[47] Rasouli, K., Rasouli, J., Mohtaram, M.S., Sabbaghi, S., Kamyab, H., Moradi, H., Chelliapan, S., 2023. Biomass-derived activated carbon nanocomposites for cleaner production: A review on aspects of photocatalytic pollutant degradation. J. Clean. Prod. 419. https://doi.org/10.1016/j.jclepro.2023.138181.

CHAPTER 6

Carbon-based electrode materials and composites in water and wastewater treatment

Balkis Hazmi[1], Muhammad Aliyu[2, 3] and Umer Rashid[2, 3, 4]
[1]Institute of Nanoscience and Nanotechnology (ION2), Universiti Putra Malaysia, Serdang, Selangor, Malaysia; [2]Department of Chemical Technology, Faculty of Science, Chulalongkorn University, Bangkok, Thailand; [3]Center of Excellence in Catalysis for Bioenergy and Renewable Chemicals (CBRC), Faculty of Science, Chulalongkorn University, Bangkok, Thailand; [4]Faculty of Applied Sciences, UCSI University, Cheras, Kuala Lumpur, Malaysia

1. Introduction

The presence of pesticide runoff, heavy metals, pharmaceutical effluent, organic pollutants, and microplastics in waterbodies is becoming increasingly concerning. These contaminations have a substantial negative impact on surface and groundwater quality and pose a risk to aquatic ecosystems and human health. Consequently, the increasing complexity and higher pollutant concentrations in wastewater worsen the water treatment process because current wastewater treatment technologies, which rely on conventional physical, chemical, and biochemical degradation methods, are inadequate for achieving cost-effective and efficient treatment, leaving trace micropollutants in treated water. Nevertheless, the ongoing research and development led to the innovation of advanced water treatment technologies that adopted active ecological material, such as carbon-based electrodes in electrochemical, physical adsorption, carbon membrane filtration, and advanced oxidation.

Electrochemistry has attracted significant attention among water treatment technologies due to its straightforward principles, ease of operation, environmental sustainability, and its capability to enhance the pollutant's biochemical properties that are difficult to decompose. Electrocoagulation, electrochemical oxidation, and advanced oxidation techniques are notable for their efficiency in reducing harmful

Carbon-Based Materials and Composites for Water and Wastewater Treatment
ISBN 978-0-443-33681-2
https://doi.org/10.1016/B978-0-443-33681-2.01006-1

contaminants without relying heavily on traditional chemical additives, thus minimizing environmental impacts associated with conventional treatment methods [1,2]. Electrochemical oxidation (EO) is a versatile wastewater treatment process that can be carried out via direct or indirect oxidation methods. Indirect oxidation minimizes electrode fouling and improves process efficiency [3]. This approach leverages electrochemically generated oxidants, such as hydrogen peroxide and Fenton's reagent, to effectively degrade organic contaminants [4].

Carbon-based electrodes, including graphite, carbon nanotubes (CNTs), and activated carbon, are vital to electrochemical treatment processes. Renowned for their excellent conductivity, high surface area, and outstanding electrocatalytic properties, these materials play a crucial role in optimizing pollutant degradation, such as fluorite ions, nitrate reduction, and organic pollutant removal, significantly improving the overall efficiency of treatment systems [5,6]. The efficiency of electrochemical treatment methods for wastewater is often evaluated based on several performance metrics, including removal rates of pollutants, Faradaic efficiency (FE), and selectivity toward target contaminants. The removal rates of contaminants, including heavy metals and organic pollutants, are key performance indicators in electrochemical treatment systems. Studies have shown that integrating electrochemical processes with nanocarbonaceous membranes can achieve high removal efficiencies for toxic substances such as arsenic and antimony. For example, graphene-modified Fe_3O_4@N-GO composite electrodes demonstrated efficient removal of organophosphorus pesticides from wastewater within 40 min, attributed to their ability to promote the generation of $\cdot OH$ radicals [3].

Besides, FE measures the effectiveness of an electrochemical reaction in converting electrical energy into chemical energy, providing insight into the process's selectivity [4]. For example, experiments have shown that FE for nitrate reduction can reach up to 78% under optimal nitrate concentrations where active anodes promote partial oxidation, while nonactive anodes are more effective for complete oxidation of organic pollutants [7]. By optimizing electrode design and operational parameters, selectivity toward specific contaminants can be enhanced, leading to more efficient and targeted treatment outcomes.

Despite the benefits, electrochemical treatment technologies face challenges, including issues related to electrode passivation and the release of harmful ions during operation. Addressing these concerns is vital for the sustainable implementation of these systems in real-world applications. In conclusion, carbon electrodes' electrochemical treatment techniques represent a significant advancement in wastewater treatment. They are a sustainable substitute for conventional treatment techniques because of their effectiveness in breaking down contaminants with less chemical usage, which emphasizes the necessity of ongoing study and development in this field.

2. Properties of electrode derived from carbon materials

An ideal electrode material should possess strong adsorption capabilities, excellent electrical conductivity, and high catalytic activity to efficiently promote pollutant degradation. The commonly utilized electrodes are derived from carbon-based and metal-based materials because of their compatibility with the nature of water/wastewater and the type of treatment process. Carbon-based electrodes have garnered interest due to their structural stability, affordability, and ease of doping and modification, in contrast to metal-based electrodes, which are more prone to secondary pollution, although they are active toward pollutant degradation. Thus, carbon materials have become essential in a wide range of electrocatalytic applications.

2.1 Electrical conductivity and surface textural

Carbon is a naturally abundant element that can be easily prepared and synthesized from biomass, waste materials, and polymers. In addition, carbon can exist in both amorphous and crystalline forms, such as activated carbon, graphite, diamond, CNTs, and graphene. Those carbon-based materials are considered excellent adsorbents due to their large surface area per volume. For example, electrolysis using a porous $Ni_3(HITP)_2$/graphene-based composite electrode exhibited a surface area of 236.64 m^2g^{-1} effectively degraded phenol in treated coal effluent, attributed to the electrode's efficient adsorption capacity and high conductivity [8]. In addition, modifying the particle size distribution can optimize and maximize the carbon-based material's specific surface area. Better particle dispersion throughout the material's active surface is made possible by an increase in specific surface area when particle size is reduced to the nanoscale. Good conductivity is a key characteristic of carbon electrodes in electrochemistry, as it facilitates efficient electron transfer necessary for degradation processes. Lamellar planes of sp^2 hybridized carbons in a hexagonal aromatic configuration structure of a typical carbon electrode. This structure promotes a high degree of electron delocalization and weak van der Waals forces, which result in good electrical conductivity. Wang et al. investigated the effect of pyrolysis temperatures (400°C, 600°C, 800°C, and 1000°C) on the production of graphitic biochar electrodes, where biochar pyrolyzed at 1000°C resulted in the most uniform crystalline structure, characterized by an intense signal at $2\theta = 25.7$ corresponding to the 002-crystal plane. This electrode demonstrated a cadmium (Cd) wastewater treatment efficiency of 88.96% with low power consumption [9].

2.2 Chemical stability and reactivity

Carbon-based electrodes are exceptionally suitable for a range of electrochemical applications due to their remarkable chemical stability and reactivity. Graphite and activated carbon are examples of carbon materials that exhibit noncorrosive behavior, even in severe conditions such as saline, acidic, or alkaline solution ensuring their longevity and reliability in electrochemical systems [10]. Carbon electrodes exhibit thermal and oxidative stability, enabling them to perform electrolysis effectively at high temperatures and in oxidative environments. Carbon electrode modification is often done prior to create a new sensor with desired, frequently predefined properties, therefore improving their catalytic efficiency, adsorption capability, and overall performance in applications such as pollutant removal and energy storage [11]. A flow electrode material derived from cellulose loaded with zero-valent (ZVI@C-2:1) was synthesized for treating uranium-containing wastewater. It demonstrated exceptional performance, achieving 97.75% uranium removal efficiency and was reused up to 36 electrochemical cycles of UO_2^{2+} electrosorption, significantly outperforming nondoped carbon electrodes [12].

2.3 Structural and morphological properties

Electrosorption and catalysis require the effective adsorption capacity and mass transfer of reactants and products, which is rendered feasible by the presence of well-defined porous structures, such as microporous or mesoporous. According to the International Union of Pure and Applied Chemistry (IUPAC), pore structures are classified into three types: micropores, mesopores, and macropores, with sizes of <2 nm, 2–50 nm, and >50 nm, respectively. Ge et al. studied the synergistic effects of functional groups and microporosity in carbon electrodes derived from ZIF-8 for the electroreduction removal of nitrate ions in wastewater. Functionalizing ZIF-8-C with organic silicon quaternary ammonium salt (ZIF-8-C-Q-2) resulted in a surface area of 213.43 m^2/g and micropores ranging from 0.5 to 2 nm. This microporous structure exhibited a 4.55-fold higher adsorption affinity for NO_3^- compared to activated carbon [13].

2.4 Adsorption capabilities

Adsorption is another property criterion that is sometimes beneficial but is typically seen as a negative influence on analytical response, especially in systems relying on diffusion-controlled mass transfer. In addition, the adsorption of nonelectroactive impurities or reaction products can deactivate the electrode surface, necessitating a regeneration process. Thus, activated carbon electrodes are widely utilized for the adsorption of organic and inorganic materials [14], with the selection of the appropriate type of carbon-based electrode being a crucial factor for specific applications.

3. Type of carbon-based electrodes

Carbonaceous electrodes have been made of a variety of carbon-based materials, including activated carbon, CNTs, graphene, and carbon aerogels. The technological developments in the creation of various kinds of electrode structures for (waste)water electrolysis are reviewed and discussed in this section.

3.1 Activated carbon electrodes

Activated carbon offers several advantages, including good conductivity, a large surface area, and low cost. It remains stable across a wide pH range and allows liquid in reactors to flow smoothly. As a result, it can be used directly or modified for use as a particle electrode, making it the most widely utilized particle electrode today. Duan et al. developed microporous activated carbon electrodes to study the simultaneous removal of inorganic salts (NaCl) and organic pollutants (phenol) from a solution mixture. The findings revealed that phenol was removed via adsorption and direct EO, while inorganic ions were eliminated through chlorine-mediated oxidation, highlighting the importance of ion electrosorption [15]. Brackish water desalination electrochemical tests were conducted using honeycomb-structured porous nanoflakes of cork-activated carbon

(CAC). The material's high microporosity (2.43 nm) and large surface area (1960 m^2/g) enabled efficient adsorption, with results showing that CAC-3, with an alkaline-carbon ratio of 3:1, exhibited high desalination capacity and charging efficiency, making it ideal for capacitive deionization (CDI) electrodes [16]. Quang et al. successfully developed a 3D Electro-Fenton system using pellet-activated carbon (PAC) particle electrodes to treat a mixture of pharmaceutical and steel industrial wastewater. The system achieved a total organic carbon (TOC) removal efficiency of 94.1% within 180 min under optimal operating conditions [17].

3.2 Carbon nanotube electrode

The efficiency of carbon nanotubes (CNTs) in adsorbing and degrading pollutants from effluent is largely because of their unique features, such as an enormous surface area in the form of cylindrical nanostructure, which provides ample sites for interaction with various impurities [18]. A Ti/SnO_2-Sb-CNT electrode was synthesized using pulse electro-deposition, showing better performance than a non-CNT electrode by posing a larger surface area and smaller lattice size, allowing for more active sites and better pollutant removal. It showed higher oxygen evolution potential (2.23 V), improving efficiency and reducing competing reactions. At lower energy consumption (121.15 kWh) and higher mineralization efficiency (11.95%), the Ti/SnO_2-Sb-CNT electrode achieved 80.12% chemical oxygen demand (COD) removal and 46.01% TOC removal, outperforming the non-CNT electrode. In combination with nano-filtration, the Ti/SnO_2-Sb-CNT electrode showed better flux and rejection rates compared to the nonmodified electrode [19]. Shao et al. highlight the impact of CNT loading on the performance of modified anode electrodes (MAEs) in wastewater treatment. For Azo dye Acid red G (ARG) degradation, Ti/PbO_2-CNT(0.04 g) excelled, while Ti/PbO_2-CNT(0.01 g) was effective under lower current. In phenol treatment, Ti/PbO_2-CNT(0.01 g) outperformed other configurations, effectively removing both phenol and its intermediates. For petrochemical wastewater, both CNT-modified electrodes showed improved efficiency, with Ti/PbO_2-CNT(0.01 g) better at COD removal and minimizing intermediate accumulation. Ti/PbO_2-CNT(0.01 g) effectively generated hydroxyl radicals, while Ti/PbO_2-CNT(0.04 g) focused on breaking specific chemical bonds [20].

3.3 Graphene and graphene oxide

Graphene is a single layer of carbon atoms organized in a hexagonal lattice. It is perfect for increasing the efficiency of electrochemical processes because of its remarkable mechanical strength, electrical conductivity, and high surface area [21]. However, graphene oxide (GO) is a graphene derivative that has been functionalized by adding oxygen groups such as hydroxy and epoxy using the Modified Hummers Method [22,23]. These functional groups enhance GO's dispersion in polar solvents and enable tunable electrical properties, expanding its use in various applications, including water purification. A study by Mousset et al. demonstrated the application of pristine graphene materials (monolayer, multilayer, and foam) as cathodes in the electro-Fenton process for wastewater treatment. Gfoam significantly achieving the highest H_2O_2 yield (4.25 mg-H_2O_2 L^{-1} cm^{-3}), phenol degradation rate (78% in 3 h), and mineralization yield (50% in 8 h) attributed to Gfoam's high purity, low charge-transfer resistance (1.6 Ω), large electroactive surface area (55.017 cm^2), and 3D porous structure, demonstrating its superior performance compared to conventional carbon-based materials and its potential in efficient wastewater treatment [24].

For an electrochemical deionization application, a bismuth/reduced graphene oxide (Bi/rGO) nanocomposite electrode was developed, enhancing ion selectivity for water purification. rGO fully coated the Bi nanoparticles, which made up about 79.56% of the material. A high Cl^- removal capacity of 62.59 mg g^{-1} demonstrated the Bi/rGO electrode's selective ion removal, which prioritized chlorine (Cl^-) above fluoride (F^-). The system also demonstrated potential for selective ionic species removal in water treatment by exhibiting good regeneration performance, retaining steady and reversible salt absorption/desorption across 10 cycles [25]. The conductive polyurethane/polypyrrole/graphene (CPU/PPy/Gr) particle electrode was fabricated via in situ oxidative polymerization and effectively degraded levofloxacin (LEV) in a 3D electrode reactor. The electrode achieved over 90% LEV removal under optimal conditions (pH 7, 6 V, 2 L/min aeration, 20 mg/L LEV, 7 mM Na_2SO_4) with a low energy consumption of 20.12 kWh/g LEV. Toxicity assessments indicated that while some intermediate products were more toxic than LEV, the overall toxicity decreased during the degradation process [26].

4. Carbon electrode synthesis, activation, modification, and fabrication techniques

This section comprehensively examines the diverse synthesis, activation, modification, and fabrication procedures used to develop carbon-based electrodes essential for electrochemical water and wastewater treatment [27]. These techniques are crucial for customizing carbon materials' physical, chemical, and electrochemical properties, enhancing their suitability for pollutant adsorption, ion exchange, and catalytic degradation applications [28]. Comprehending these processes is crucial for designing electrodes that improve performance, longevity, and efficiency.

4.1 Thermal and thermochemical properties of biomass for carbon-based electrode

The synthesis of carbon electrodes involves a range of physical and chemical methods, each tailored to control the carbon material's structure, surface area, conductivity, and porosity, which are crucial for its performance in water treatment applications [29]. These methods are designed to optimize the properties of carbon electrodes for applications such as pollutant adsorption, electrosorption, and catalytic degradation.

4.1.1 High-temperature pyrolysis

High-temperature pyrolysis is a process that involves the thermal decomposition of organic precursors (such as biomass, polymers, or carbon-rich compounds) at high temperatures in an inert atmosphere (e.g., nitrogen or argon) [30,31]. Pyrolysis produces carbon materials with a wide range of pore structures and high surface areas, which are ideal for adsorption applications. This process is widely used to produce activated carbon, which features micro- and mesoporous structures that maximize adsorption sites for pollutants in water treatment systems [32]. Thermal activation follows the pyrolysis stage, where the carbonized material is further heated with activating agents such as steam or carbon dioxide [33]. This step enhances the porosity and surface area, producing carbon electrodes with abundant micropores that improve adsorption capacity. Thermal activation is particularly effective for producing electrodes with high adsorption properties and is suitable for CDI and electrosorption processes [34].

4.1.2 *Low-temperature hydrothermal carbonization*

Hydrothermal carbonization (HTC) is a sustainable and eco-friendly method for converting biomass into hydrochar, a carbon-rich material with excellent structural and chemical properties [35]. This process involves treating biomass in water at subcritical temperatures, typically in a sealed reactor, reducing greenhouse gas emissions and environmental impacts [36]. The hydrochar produced has desirable properties such as high carbon content, tunable surface functionalities, and a well-developed pore structure, making it an ideal precursor for further modification. HTC generates minimal waste and can use various feedstocks, making it a sustainable waste valorization solution [37]. Hydrochar's versatility allows it to remove heavy metals, dyes, and organic pollutants from water. It also catalyzes advanced oxidation processes, making it a crucial technology in eco-friendly materials for environmental remediation and sustainable development [35].

4.2 Activation process

Activation techniques improve carbon materials' structural, physical, and chemical properties, rendering them more efficient for environmental and electrochemical applications [32]. These techniques aim to enhance surface area, porosity, and functional group density, consequently augmenting the efficacy of carbon electrodes in adsorption, catalysis, and energy storage systems [29]. Activation methods are categorized into physical and chemical activation, each providing distinct benefits according to the specific application.

4.2.1 *Physical activation*

This procedure exposes carbonized biomass to oxidizing agents, such as steam or carbon dioxide, at temperatures ranging from 700°C to 1000°C [38]. Physical activation emphasizes opening closed pores and expanding existing ones, augmenting the material's surface area and pore volume. The resultant porous carbon composites are especially efficient for applications necessitating elevated adsorption capacity, including CDI and electrosorption [39]. Physical activation is straightforward, environmentally friendly, and does not use chemicals. It is well suited for synthesizing electrodes with a well-balanced microporous and mesoporous structure, which increases ion transport and pollutant adsorption [40].

4.2.2 Chemical activation

Chemical activation involves the treatment of biomass with activating agents such as potassium hydroxide (KOH), zinc chloride ($ZnCl_2$), or phosphoric acid (H_3PO_4) before or during the carbonization process [29]. This procedure enables exact regulation of the material's porosity, surface chemistry, and functional group distribution.

Basic activating agents, such as potassium hydroxide (KOH) and potassium carbonate (K_2CO_3), are essential for the synthesis of carbon compounds with highly porous structures [28]. These chemicals promote the chemical reactions essential for establishing a refined pore structure, markedly improving the specific surface area and microporosity of the material. KOH is especially distinguished for its propensity to create interconnected pore networks, yielding carbon composites with enhanced structural integrity and elevated adsorption capacity [39]. KOH-activated carbons are well suited for applications necessitating effective ion or molecule transport, including CDI systems and sophisticated water treatment procedures. The potent alkaline properties of KOH enhance its efficacy in facilitating pore formation, rendering it a favored option for the synthesis of high-performance carbon electrodes [41].

Acidic activating agents, including phosphoric acid (H_3PO_4), are crucial for improving the oxygen content and surface chemistry of carbon compounds [32]. Incorporating oxygenated functional groups enhances the material's hydrophilicity, hence strengthening its interaction with aquatic environments and increasing its affinity for negatively charged pollutants, including anions and organic contaminants [42]. H_3PO_4 facilitates the formation of a stable porous structure while maintaining the integrity of the carbon framework, hence providing long-term durability and reusability [43]. Carbons treated with H_3PO_4 demonstrate remarkable catalytic capabilities, rendering them very efficient in advanced oxidation processes (AOPs) that promote the destruction of intricate organic contaminants. The twin roles of adsorption and catalysis render H_3PO_4-activated carbons essential for water purification and environmental remediation applications [29]. Chemical activation enables extensive customization, allowing for the fabrication of carbon electrodes tailored to specific requirements. The choice of activating agent, temperature, and biomass feedstock significantly influences the material's final properties [32]. For example, carbons activated with KOH exhibit enhanced conductivity and are ideal for energy storage applications, whereas H_3PO_4-activated carbons excel in adsorbing dyes and heavy metals from wastewater [44].

Integrating physical and chemical activation methods with pyrolysis or HTC can further enhance these materials. Such synergistic approaches facilitate the design of multifunctional carbon electrodes with superior adsorption capacity, conductivity, and reactivity. These electrodes are crucial for tackling global issues in water treatment, renewable energy, and environmental sustainability [45].

4.3 Modification techniques

4.3.1 Surface functionalization

Surface functionalization involves introducing certain chemical groups onto the surface of carbon materials, thereby establishing sites that selectively interact with contaminants [42]. This is accomplished using several chemical treatments, including acid treatment, ozone oxidation, and plasma treatment, each of which introduces certain functional groups, such as hydroxyl (-OH), carboxyl (-COOH), and carbonyl (-C=O) groups.

Acid treatment. Acid treatment is a prevalent technique for functionalizing carbon materials to improve their surface characteristics and reactivity [46]. The carbon material undergoes treatment with powerful acids, such as sulfuric acid (H_2SO_4) or nitric acid (HNO_3), which oxidize the surface and introduce various oxygen-containing functional groups, including carboxyl (–COOH), hydroxyl (–OH), and carbonyl (–C=O) groups [30]. These oxygenated groups enhance the material's hydrophilicity, facilitating improved interaction and contact with aqueous contaminants [47]. The improved wettability promotes the adsorption of various pollutants, including organic compounds, heavy metals, and dyes. The enhanced surface chemistry enhances the material's catalytic efficacy, rendering acid-treated carbon materials proficient in AOPs and electrochemical applications [30]. This technique is commonly utilized to customize the surface properties of carbon electrodes for water treatment, environmental remediation, and energy storage applications.

Ozone oxidation. Ozone oxidation is an effective and relatively gentle surface modification method that adds oxygen functions to carbon materials while preserving their structural integrity. In this process, ozone (O_3), a potent oxidizing agent, interacts with the carbon surface, producing oxygen-rich functional groups such as epoxides and hydroxyls [48]. These functions enhance the wettability of the carbon material, improving its interaction with aqueous contaminants. Ozone treatment's gentle characteristics maintain the carbon material's porosity architecture and mechanical integrity while enhancing its catalytic properties [49]. Ozone-treated carbon electrodes are very effective in applications necessitating close interaction with aqueous systems, including adsorption, EO, and catalytic degradation [27,28,43]. Moreover, ozone oxidation can be regulated to alter the surface, rendering it a versatile instrument for enhancing carbon materials for water treatment and environmental applications [30].

Plasma treatment. Plasma methods permit meticulous regulation of surface chemistry by the selective and homogeneous introduction of functional groups. Plasma-treated carbon materials are frequently employed for specialized adsorption applications, necessitating enhanced surface contact with contaminants [42]. Treatments that incorporate functional groups enhance the water compatibility of carbon electrodes, augmenting their affinity for contaminants and improving adsorption capacity. Functionalized carbon materials enhance the hydrophilicity and binding capacity of the electrode surface, hence improving the capture of organic molecules, heavy metals, and ions, making them suitable for applications such as CDI and electrosorption [50].

4.3.2 Doping with heteroatoms

Doping involves introducing heteroatoms such as nitrogen (N), sulfur (S), or phosphorus (P) into the carbon framework. This process alters the material's electronic configuration, introduces active sites, and enhances its catalytic, adsorption, and electrochemical properties. By modifying the electronic density and surface chemistry, doped carbon materials are tailored for high performance in water treatment and environmental applications [51].

Nitrogen doping. Nitrogen atoms integrate into the carbon matrix, introducing electron-donating sites and improving the material's conductivity and reactivity. This enhances redox reactions, making nitrogen-doped carbons particularly effective in EO processes. Because of their enhanced electrocatalytic activity, these materials excel in degrading organic pollutants, removing heavy metals, and facilitating faster and more efficient pollutant degradation [52].

Sulfur doping. Sulfur atoms generate active sites that increase reactivity and selectivity toward specific pollutants. Sulfur-doped carbons are especially effective for adsorbing heavy metals such as lead and cadmium, as sulfur's affinity for metal ions strengthens the interaction. This doping method is advantageous in processes requiring selective adsorption and catalytic functions, making it ideal for water treatment and specific catalytic applications [53].

Phosphorus doping. Phosphorus-doped carbons possess unique surface properties, enhanced stability, and improved chemical resistance. These materials are suitable for AOPs and excel in harsh environments, including acidic or high-voltage conditions often encountered in wastewater treatment. Phosphorus doping enhances the durability of carbon electrodes, enabling long-term application in challenging conditions [54]. Doped carbon materials demonstrate improved electronic conductivity, chemical stability, and interaction with pollutants, leading to higher removal efficiency and selectivity. Carbon electrodes can be precisely customized to meet the unique demands of modern water and wastewater treatment technologies through functionalization and doping [55].

Overall, surface functionalization and doping techniques greatly enhance the versatility and functionality of carbon-based electrodes. These modifications allow for fine-tuning of the electrode's surface properties, essential for achieving optimal performance in different water treatment applications, such as pollutant degradation, ion removal, and contaminant adsorption [27]. Through functionalization and doping, carbon electrodes become more adaptable, efficient, and resilient, addressing the specific demands of water and wastewater treatment systems [28].

4.3.3 Metal impregnation

Metal impregnation involves the incorporation of metal oxides or nanoparticles, such as Fe_2O_3, MnO_2, or NiO, into a carbon matrix to improve its catalytic efficacy. This procedure may occur either before or after carbonization, with precarbonization involving the incorporation of metal salts into raw biomass and postcarbonization involving the application of metal precursors to activated carbon surfaces [56]. Both approaches enhance the catalytic activity of carbon materials, especially in redox reactions, hence improving the degradation of pollutants in AOPs. Metal-impregnated carbon compounds are essential in EO systems, including electro-Fenton reactions and photocatalysis, for treating persistent organic pollutants (POPs) and facilitating environmental restoration [27].

Metal oxides, such as titanium dioxide (TiO_2) or manganese oxide (MnO_2), can be coated onto the carbon surface to improve corrosion resistance, especially in electrolytic conditions. These oxides also enhance the electrode's ability to withstand high voltages and prolonged oxidation reactions, reducing degradation over time [57]. These coatings improve corrosion resistance and make the electrodes more hydrophilic, increasing wettability and improving the interaction between the electrode and contaminants in water, further enhancing the adsorption and electrochemical reaction rates [56].

4.3.4 Formation of composite

Composite production involves the integration of biomass-derived carbon with advanced materials such as graphene, CNTs, or conductive polymers to create a multifunctional material [58]. This method utilizes various synthesis processes, including chemical vapor deposition (CVD) [59], hydrothermal synthesis [60], and in situ polymerization [58], to produce a cohesive and structurally robust composite. The choice of procedure depends on the necessary attributes and the compatibility of the biomass carbon with the advanced material [27]. CVD is commonly employed to produce graphene layers directly on the carbon substrate. Conversely, hydrothermal synthesis is utilized to distribute CNTs or conductive polymers within the carbon matrix consistently. The resulting composite leverages the intrinsic properties of both components, combining the sustainable and porous qualities of biomass carbon with the enhanced conductivity, strength, and surface functionality of contemporary materials. Combining these elements produces synergistic properties that significantly enhance the composite's performance [57].

Notable improvements include enhanced mechanical stability, electrical conductivity, and tailored surface chemistry, making the material adaptable to various environmental conditions. These attributes allow the composite to interact effectively with diverse pollutants, making it particularly adept at addressing complex environmental challenges [42]. Moreover, the composites are designed to improve durability and reactivity, ensuring extended stability in challenging working environments. Integrating carbon materials with reinforcing agents, such as polymers or ceramic nanoparticles, enhances stiffness and load-bearing capability [28]. Carbon-polymer composites demonstrate enhanced flexibility and strength, enabling them to withstand physical pressures that usually compromise unreinforced carbon materials. These structural alterations are essential when electrodes encounter large currents or mechanical loads, as they preserve the electrode's integrity and reduce performance degradation [32]. Consequently, reinforced carbon electrodes exhibit enhanced reliability and durability, diminishing maintenance needs and prolonging the operational lifespan of water treatment systems.

Biomass-carbon composites are especially advantageous in AOPs and CDI. AOPs' superior electrical conductivity and chemical stability facilitate the effective production of reactive species, such as hydroxyl radicals, essential for degrading organic contaminants [47]. In CDI, their improved electrical characteristics enable the extraction of ions from saline or polluted water, positioning them as a viable solution for desalination and water purification [28]. The versatility and multifunctionality of biomass-carbon composites render them a crucial element in advancing sustainable technologies.

4.4 Fabrication of carbon-based and composite electrodes

Fabricating carbon-based and composite electrodes for water and wastewater treatment is crucial for enhancing electrochemical systems for pollutant degradation and water purification [61]. Carbon derived from biomass, noted for its inherent porosity, high surface area, and customizable surface chemistry, is particularly suitable for these applications. The manufacturing process transforms carbon materials into functional electrode structures tailored to meet the specifications of water and wastewater treatment systems [62]. The primary fabrication techniques are discussed in the following sections.

4.4.1 Electrode casting

Electrode casting is a prevalent method for producing carbon-based electrodes, especially for electrochemical applications in water and wastewater treatment. The procedure entails combining biomass-derived carbon with binders, such as polytetrafluorethylene (PTFE) or Nafion, to formulate a slurry, which is subsequently applied to a conductive substrate such as graphite, stainless steel, or titanium [30]. Upon drying, the coating establishes a stable electrode layer characterized by homogeneous adherence and effective electrical conductivity. The resultant electrodes display smooth, homogeneous surfaces and exceptional compatibility with electrochemical systems, rendering them highly efficient in facilitating redox processes. This technique is frequently utilized in electrochemical reactors for wastewater treatment [38]. These encompassing systems employ AOPs, electro-Fenton reactions, and CDI, wherein the electrodes are essential for the degradation of POPs and for recovering valuable resources from wastewater [63].

4.4.2 Pressing and sintering

Pressing and sintering is a production method wherein biomass-derived carbon powders are crushed into electrode forms using mechanical or hydraulic presses, followed by high-temperature sintering in a controlled environment to improve mechanical stability and cohesiveness [58]. The resultant electrodes are dense, durable, and able to endure extreme circumstances, including corrosive environments and elevated operational temperatures, rendering them suitable for industrial-scale water and wastewater treatment. This technique is frequently utilized in extensive electrochemical systems to extract heavy metals, desalinate, and treat high-strength industrial effluents, where durability and longevity are essential [42].

4.4.3 3D printing and additive manufacturing

3D printing and additive manufacturing are sophisticated techniques for fabricating electrodes with complex and customized shapes, employing biomass-derived carbon powders as precursors in conjunction with suitable binders. This approach enables the creation of intricate designs that enhance surface area, improve mass transfer, and increase accessibility to active regions, resulting in more efficient electrochemical processes [64]. The versatility of 3D printing enables the fabrication of electrodes tailored to specific treatment needs, making them highly effective in advanced water treatment technologies, such as electrochemical reactors for organic pollutant removal, electrosorption for desalination, and integrated systems for simultaneous pollutant degradation and resource recovery [27]. Moreover, the precision of 3D printing enables the fabrication of multifunctional electrodes for hybrid systems, incorporating catalytic and adsorptive properties to enhance performance [42].

5. Applications of carbon-based and composite electrodes in water and wastewater treatment

Carbon-based electrodes, particularly those derived from biomass or integrated with advanced materials to form composites, are increasingly valued in water and wastewater treatment for their outstanding physicochemical properties [13]. These electrodes possess a high surface area, which promotes pollutant adsorption and facilitates electrochemical reactions, enhancing their efficiency in contaminant removal processes. Their tunable porosity allows precise pore size and distribution control, optimizing mass transport and access to active sites [22]. With superior electrical conductivity, they enable efficient electron transfer, a critical factor in electrochemical systems. Their excellent chemical and thermal stability ensures durability across diverse modern treatment methods. These electrodes address environmental challenges, ion separation processes, and resource recovery initiatives [17]. The sustainability and renewability of biomass-derived carbon further increase its appeal, aligning with global efforts to create eco-friendly and efficient water and wastewater management solutions [25].

5.1 Electrochemical oxidation processes

EO is a highly efficient and eco-friendly water and wastewater treatment method, with carbon-based electrodes playing a pivotal role [48]. These electrodes facilitate the generation of reactive oxygen species (ROS), such as hydroxyl radicals, which are essential for breaking down POPs, such as pharmaceuticals, pesticides, and synthetic dyes [46]. Processes such as anodic oxidation rely on water oxidation at the electrode surface to produce hydroxyl radicals, which nonselectively degrade contaminants. In contrast, the electro-Fenton process generates hydrogen peroxide at the cathode, producing ROS in the presence of catalysts such as ferrous ions [43]. Carbon-based electrodes are favored for their high surface area, electrical conductivity, and chemical stability, allowing efficient and complete pollutant degradation. They also offer advantages such as environmental compatibility, cost-effectiveness, and adaptability to various EO processes, including advanced oxidation and hybrid systems [47]. However, EO requires significant energy input for ROS generation, and challenges such as electrode fouling, by-product formation, and limited long-term durability may affect efficiency. While laboratory and pilot-scale applications have proven effective, scaling up EO systems for industrial use requires careful optimization of operational parameters and electrode materials [28]. Despite these limitations, carbon-based electrodes remain a promising solution for addressing complex environmental challenges, making EO a cornerstone of sustainable water and wastewater management.

5.1.1 Mechanism of electrochemical oxidation

Anodic oxidation. In anodic oxidation, the carbon-based electrode functions as the anode. When an electrical current is applied, water molecules are oxidized on the electrode surface, generating hydroxyl radicals (·OH) and other ROS. These species subsequently interact with organic contaminants, decomposing their molecular structures [32].

For example:

$$H_2O \rightarrow HO^{\cdot} + H^+ + e^- \tag{6.1}$$

$$HO^{\cdot} + R \rightarrow CO_2 + H_2O \tag{6.2}$$

Electro-Fenton process. In the electro-Fenton reaction, carbon-based electrodes produce hydrogen peroxide (H_2O_2) in situ, primarily at the cathode, by reducing dissolved oxygen in the aqueous medium.

$$O_2 + 2H^+ + 2e^- \rightarrow H_2O_2 \tag{6.3}$$

In the presence of a ferrous ion catalyst (Fe^{2+}), hydrogen peroxide decomposes into hydroxyl radicals:

$$Fe^{2+} + H_2O_2 \rightarrow Fe^{3+} + OH^- + HO^{\cdot} \tag{6.4}$$

These radicals oxidize organic pollutants, breaking them into smaller, less harmful compounds or fully mineralizing them.

5.1.2 Advantages of electrochemical oxidation

- **High efficiency in pollutant degradation**: Carbon-based electrodes facilitate the generation of highly reactive species, completely breaking down complex and recalcitrant organic pollutants into harmless by-products such as CO_2 and H_2O.
- **Environmental compatibility**: EO does not require external chemical oxidants, reducing the risk of secondary pollution. The process relies on electricity and in situ ROS generation, making it environmentally sustainable.
- **Versatility**: Carbon-based electrodes are adaptable to various EO processes, including anodic oxidation, electro-Fenton reactions, and hybrid AOP.
- **Chemical and thermal stability**: Carbon-based electrodes exhibit excellent resistance to corrosion and degradation under harsh operational conditions, including acidic environments and elevated temperatures.
- **Cost-effectiveness**: Compared to noble metals such as platinum, carbon-based electrodes are less expensive and offer comparable performance in many EO applications. Biomass-derived electrodes further enhance economic and environmental benefits.

5.1.3 Limitations of electrochemical oxidation

- **Energy requirements**: The generation of ROS, particularly hydroxyl radicals, requires significant energy input. The cost and sustainability of EO depend heavily on the availability of low-cost, renewable energy sources.
- **Electrode fouling and scaling**: Carbon-based electrodes can experience fouling due to the accumulation of organic matter, biofilms, or scaling from salts in wastewater. This reduces their efficiency over time and requires regular maintenance.
- **Formation of by-products**: While EO effectively degrades pollutants, it can produce intermediate by-products, including halogenated compounds, when chloride ions are present in the water. These by-products may require additional treatment to ensure safety.
- **Limited long-term durability**: Despite their stability, carbon-based electrodes can degrade after prolonged use in highly aggressive environments, necessitating periodic replacement.

EO using carbon-based and composite electrodes is a promising water and wastewater treatment technology, offering an efficient and environmentally friendly solution for tackling POP. The ability to generate ROS in situ and the adaptability of these electrodes to various processes make them indispensable in advanced treatment systems [57]. However, addressing limitations such as energy consumption, electrode fouling, and scalability is crucial to enhancing this technology's practicality and economic feasibility for widespread adoption [42].

5.2 Capacitive deionization

Capacitive deionization (CDI) is an emerging water treatment technology that leverages carbon-based and composite electrodes to remove ions from water through an electrostatic adsorption mechanism [65]. This technology is particularly suitable for desalination, water softening, and recycling industrial wastewater, offering an energy-efficient and environmentally friendly alternative to conventional processes such as reverse osmosis. Carbon-based and composite electrodes are critical to CDI systems due to their exceptional properties, including high surface area, tunable porosity, superior electrical conductivity, and chemical stability, enabling efficient ion removal and regeneration [33].

5.2.1 Mechanism of capacitive deionization

CDI operates on the principle of electrosorption, where ions are removed from water as they pass between oppositely charged electrodes.

Ion adsorption. When a low voltage is applied between a pair of electrodes, cations (e.g., Na^+ Ca^{2+}) migrate toward the negatively charged electrode (cathode), while anions (e.g., Cl^-, SO_4^{2-}) move to the positively charged electrode (anode). These ions are electrostatically adsorbed onto the electrode surfaces, reducing the water's ionic content.

Ion desorption and regeneration. Once the electrodes become saturated with ions, the applied voltage is reversed or removed, releasing the adsorbed ions into a brine stream. This regenerates the electrodes, making CDI a cyclic and reusable process.

Role of carbon-based and composite electrodes. Carbon-based and composite materials, including activated carbon, graphene, CNTs, and biomass-derived carbon, are commonly used for CDI electrodes due to their high surface area, which increases ion adsorption capacity and excellent conductivity, which ensures efficient charge transfer [66]. Functionalizing carbon electrodes with dopants (e.g., nitrogen) or coatings (e.g., polymers) can further enhance their performance by improving ion selectivity or resistance to fouling [65].

5.2.2 Advantages of CDI

Energy efficiency. CDI operates at low voltages (typically below 2V), significantly lowering energy consumption compared to pressure-driven processes like reverse osmosis.

Eco-friendliness. The process does not require chemical regenerants, reducing the environmental impact and minimizing secondary pollution.

Cost-effectiveness. Using inexpensive and renewable carbon-based materials, particularly biomass-derived electrodes, reduces overall system costs while promoting sustainability.

High recovery rate. CDI systems can achieve high water recovery rates by minimizing the volume of waste brine produced during regeneration.

Scalability and modularity. The modular design of CDI systems makes them scalable for applications ranging from small-scale household water purification to industrial-scale wastewater treatment.

Versatility. CDI is effective for treating brackish water, softening hard water, and recovering valuable ions (e.g., lithium and potassium) from industrial effluents, making it suitable for various applications.

5.2.3 Limitations of CDI

Fouling and scaling. Over time, electrodes can become fouled by organic matter, biofilms, or scaling from dissolved salts, reducing efficiency and requiring regular cleaning or replacement.

Selective ion removal. CDI is more effective for removing monovalent ions (e.g., Na^+ and Cl^-) than multivalent ions (e.g., Mg^{2+} and SO_4^{2-}), which may limit its applicability for certain water sources.

Energy demand for high-salinity water. While energy-efficient for low- to medium-salinity water, CDI becomes less practical for desalinating highly saline water, such as seawater, because of limited adsorption capacity.

Electrode degradation. Prolonged operation, especially under harsh conditions, can degrade carbon electrodes, reducing their conductivity and structural integrity over time.

Capital costs. Although operational costs are low, the initial investment in high-quality CDI systems and advanced electrode materials can be significant, potentially limiting adoption in low-resource settings.

By-product management. The brine produced during regeneration contains concentrated ions, which, depending on the contaminants present, may require additional treatment or disposal.

CDI represents a promising solution for water desalination, softening, and resource recovery. It leverages the unique properties of carbon-based electrodes to achieve efficient and sustainable ion removal [28]. While the technology offers significant energy efficiency, environmental impact, and versatility, addressing challenges such as electrode fouling, degradation, and limitations in treating high-salinity water is essential to enhancing its effectiveness and scalability. With ongoing electrode materials and system design advancements, CDI holds immense potential as a cornerstone of future water treatment technologies [42].

5.3 Photoelectrocatalytic degradation

Photoelectrocatalytic (PEC) degradation is a hybrid water treatment method that integrates photocatalysis with electrocatalysis to decompose complex organic contaminants. This procedure is enabled by altered carbon-based electrodes combined with photoactive substances, like titanium dioxide (TiO_2), bismuth oxyhalides, or graphitic carbon nitride [31]. Combining photoactivity and electrocatalytic functionality improves pollutant degradation efficiency and reduces energy consumption, especially under visible light exposure.

5.3.1 Mechanism of photoelectrocatalytic degradation

5.3.1.1 Photocatalytic mechanism

Upon exposure to light, particularly visible light, the photoactive material combined with the carbon electrode experiences photon absorption, wherein photons possessing energy equivalent to or above the material's bandgap promote electrons from the valence band to the conduction band [27]. This excitation produces electron-hole pairs, wherein:

Electrons (e^-) in the conduction band participate in reduction reactions.

Holes (h^+) in the valence band participate in oxidation reactions.

These charge carriers facilitate the formation of ROS, such as hydroxyl radicals (·OH) and superoxide anions ($O_2^-\cdot$), which nonselectively degrade organic pollutants into harmless end products such as CO_2 and H_2O.

5.3.1.2 Electrocatalytic mechanism

A voltage is applied across the carbon-based electrode, accelerating electron flow and enhancing the separation of electron-hole pairs [32]. This reduces recombination losses, improving the generation of ROS.

5.3.1.3 Synergistic action

The integration of photocatalysis and electrocatalysis enhances the overall degradation process:

Improved Light Utilization: Modified carbon electrodes coated with visible-light-responsive photocatalysts maximize the absorption of solar or artificial light.

Charge Separation: The electric field induced by the applied voltage efficiently separates photogenerated charge carriers, prolonging their lifetimes.

Enhanced ROS Formation: Electrocatalytic activity ensures the continuous supply of electrons and holes for sustained ROS production, boosting degradation efficiency.

5.4 Role of carbon-based and composite electrodes

Carbon-based and composite materials are robust, conductive, and chemically stable platforms for immobilizing photocatalysts. Their large surface area facilitates pollutant adsorption, enhancing contact with active sites. Surface modifications, such as doping with nitrogen or metals, can improve photoelectrocatalytic (PEC) performance by increasing conductivity and catalytic activity [29].

5.4.1 Advantages of photoelectrocatalytic degradation

Enhanced pollutant degradation. Compared to standalone photocatalysis or electrocatalysis, PEC systems achieve faster and more complete degradation of complex organic pollutants, including pharmaceuticals, dyes, pesticides, and endocrine-disrupting chemicals.

Visible light utilization. The integration of visible-light-responsive photocatalysts enables effective operation under sunlight, reducing energy demands and operational costs.

Minimized energy consumption. PEC systems use light energy to facilitate reactions, necessitating lower applied voltages than traditional electrocatalytic methods, which yields substantial energy savings.

Reduced recombination losses. The applied electric field enhances the separation of photogenerated charge carriers, minimizing recombination losses and improving overall system efficiency.

Versatile application. PEC systems can degrade various organic pollutants, making them suitable for industrial wastewater treatment, municipal effluent polishing, and advanced water purification systems.

Environmental compatibility. EC systems operate without chemical additives, producing fewer secondary pollutants and reducing environmental impact.

5.4.2 Limitations of photoelectrocatalytic degradation

Electrode fouling and deactivation. Over time, organic matter, biofilms, or reaction by-products may accumulate on the electrode surface, reducing activity and necessitating regular cleaning or regeneration.

Material stability. Prolonged operation, particularly under aggressive wastewater conditions, can degrade photoactive materials or carbon electrodes, compromising system durability.

Limited scalability. While effective at laboratory and pilot scales, PEC systems face challenges in scaling up for industrial applications due to the complexity of maintaining uniform light distribution and electrical parameters.

Energy and light requirements. Although PEC systems minimize energy consumption compared to conventional methods, they still require external energy inputs and reliable light sources, which may limit their use in resource-constrained settings.

Complex reactor design. The integration of light sources, electrodes, and flow systems in PEC reactors adds to the design complexity, increasing initial capital costs.

Selective pollutant removal. PEC systems may struggle with certain nonorganic pollutants, such as heavy metals, which require complementary treatment methods for complete remediation.

Dependence on catalyst performance. The efficiency of PEC degradation heavily depends on the properties of the photoactive material, including its bandgap, stability, and resistance to photochemical corrosion. Materials with suboptimal properties may limit performance.

PEC degradation is an efficient and sustainable method for addressing complex organic contaminants, utilizing the advantages of both photocatalysis and electrocatalysis. Although scalability, material stability, and system complexity require attention, continuous improvements in electrode materials, reactor configurations, and operating methodologies present considerable potential for the extensive implementation of PEC systems in sophisticated water and wastewater treatment [27].

6. Conclusion

Using carbon-based and composite electrodes to treat water and wastewater has surfaced as a viable and sustainable approach to tackle worldwide water quality issues. Carbon materials' versatility, cost-effectiveness, and efficiency make them appealing for diverse electrochemical water treatment methods, such as PEC, CDI, and advanced oxidation processes (AOPs). Carbon-based electrodes, including activated carbon, graphene, and CNTs, provide a high surface area, superior electrical conductivity, and chemical stability, crucial for efficient pollutant removal and sustained operational stability in water treatment systems.

6.1 Summary of utilization of carbon-based electrodes in sustainable water treatment

Carbon-based and composite electrodes have demonstrated significant efficacy in eliminating various impurities, such as heavy metals, organic pollutants, salts, and microbes. Their capacity to engage in diverse electrochemical reactions, including oxidation and reduction, renders them suitable for various water treatment applications. Furthermore, carbon composites, including carbon-metal oxide hybrids and nitrogen-doped carbon materials, augment their efficacy by enhancing pollutant degradation rates, ion exchange efficiency, and selectivity for particular pollutants. The ecological sustainability of carbon compounds reinforces their function in sustainable water treatment. Employing renewable and plentiful resources such as biomass and carbon-based electrodes diminishes the environmental impact of water treatment technology. Moreover, their capacity for minimal waste production and the regeneration and reuse of resources is consistent with the ideas of a circular economy and sustainable resource utilization.

6.2 Future trends of carbon-based and carbon composite electrodes in water and wastewater treatment

As research continues to advance, the future of carbon-based electrodes in water treatment looks promising, with several key trends expected to shape their development:

Enhanced performance through nanoengineering and surface modifications. The development of advanced nanostructured carbon materials and surface functionalization techniques will further improve the efficiency of carbon-based electrodes. Nano-structuring and doping with heteroatoms (e.g., nitrogen, sulfur, and phosphorus) will enhance their catalytic activity, stability, and specificity, enabling more targeted removal of specific contaminants such as pharmaceuticals, pesticides, and industrial chemicals.

Integration with hybrid systems. Future trends may involve integrating carbon-based electrodes into hybrid water treatment systems and combining them with other advanced technologies like photocatalysis, UV irradiation, or biological treatment. These hybrid systems can offer synergistic effects, improving overall treatment efficiency and enabling the removal of a broader range of pollutants in a single treatment step.

Improved durability and regeneration capabilities. Developing carbon-based electrodes with enhanced durability and self-cleaning or regeneration capabilities will be key to extending their lifespan and reducing maintenance costs. Research into protective coatings, composite materials, and regenerative electrochemical processes will help address fouling and improve the long-term performance of carbon electrodes.

Scalability for large-scale applications. As the demand for clean water increases globally, carbon-based electrodes must be optimized for large-scale industrial and municipal water treatment systems.

Advancements in electrode manufacturing and scalability will enable these materials to be produced at a lower cost while maintaining high performance, making them suitable for widespread implementation in various regions and settings.

Circular economy and sustainability. The ongoing development of sustainable materials and waste-minimizing technologies will drive the adoption of carbon-based electrodes in water treatment. By utilizing waste-derived carbon materials, such as biochar or activated carbon from agricultural residues, the carbon electrode industry can align with circular economy principles, reducing water purification processes' environmental impact and cost.

In conclusion, carbon-based electrodes represent a transformative technology in water treatment, offering both environmental and economic advantages. As advancements in material science, surface engineering, and process optimization continue, carbon-based electrodes will play a pivotal role in the next generation of sustainable water and wastewater treatment technologies.

References

[1] Fan, X., Wei, G., Quan, X., 2023. Carbon nanomaterial-based membranes for water and wastewater treatment under electrochemical assistance. Environ. Sci. Nano 10, 11–40. https://doi.org/10.1039/D2EN00545J.

[2] Y.H. Teow, K.C. Ho, S.R. Wickramasinghe, M.G. Jebur, Z.H. Chang, Integrated process technology for recycling and re-use of industrial and municipal wastewater: A review, Integrated Environmental Technologies for Wastewater Treatment and Sustainable Development, Elsevier, 222, pp. 49–80.

[3] Chai, D., Chen, Y., Jiang, Z., Li, Z., Zhang, X., Chen, X., et al., 2024. Insight on magnetic nitrogen-doped graphene oxide composite electrode constructing heterogeneous electro-Fenton system for treating organophosphorus pesticide wastewater. Sep. Purif. Technol. 345. https://doi.org/10.1016/j.seppur.2024.127419.

[4] Son, J.Y., Choe, S., Jang, Y.J., Kim, H., 2024. Waste paper-derived porous carbon via microwave-assisted activation for energy storage and water purification. Chemosphere 355. https://doi.org/10.1016/j.chemosphere.2024.141798.

[5] Bagbi, Y., Yomgam, P., Libang, E., Boruah, B., Kaur, J., Jayanthi, S., et al., 2024. Waste bamboo-derived magnetically separable bamboo-activated carbon: From characterization to effective remediation of fluoride (F−) ions from water. RSC Adv. 14, 24952–24968. https://doi.org/10.1039/d4ra03752a.

[6] Goglio, A., Carrara, A., Elboghdady, H.G.E., Cucina, M., Clagnan, E., Soggia, G., et al., 2025. The performance of biochar waste-derived electrodes in different bioelectrochemical applications. J. Power Sources 625. https://doi.org/10.1016/j.jpowsour.2024.235623.

[7] Lee, J., Antonini, G., Al-Omari, A., Muller, C., Mathew, J., Bell, K., et al., 2024. Electrochemical methods for nutrient removal in wastewater: A review of advanced

electrode materials, processes, and applications. Sustainability 16, 9764. https://doi.org/10.3390/su16229764.
[8] Gao, S., Liu, W., Wang, M., Zhao, Z., Liu, X., 2024. A three-dimensional in-situ self-electrolysis system based on Ni3(HITP)2/graphene-based composite aerogel particle electrodes for efficient deep removal of phenol from coking wastewater. Appl. Catal., B 340. https://doi.org/10.1016/j.apcatb.2023.123276.
[9] Wang, Z., Zhou, T., Dai, J., Zhao, W., Tan, Z., 2024. A biochar electrode based on conductive straw for Cd removal and recovery from wastewater by DC electrochemical deposition. Biomass Bioenergy 189. https://doi.org/10.1016/j.biombioe.2024.107347.
[10] Zhang, B., Kong, L., Yan, X., Zhang, H., Wang, Z., Xia, S., et al., 2025. Recent progress in graphitic carbon nitride-based catalysts for water treatment: Contaminant elimination, disinfection and membrane applications. Sep. Purif. Technol. 354. https://doi.org/10.1016/j.seppur.2024.129420.
[11] Ma, Q., Yu, Y., Sindoro, M., Fane, A.G., Wang, R., Zhang, H., 2017. Carbon-based functional materials derived from waste for water remediation and energy storage. Adv. Mater. 29. https://doi.org/10.1002/adma.201605361.
[12] Zhao, E., lv, H., Fan, J., Zhang, H., Wu, S., Yu, Y., et al., 2025. Iron-carbon composite as flow electrode active material for highly efficient electrosorption of uranium from radioactive wastewater. Sep. Purif. Technol. 354. https://doi.org/10.1016/j.seppur.2024.128405.
[13] Ge, Y., Wang, Y., Chen, Y., Xu, X., Liu, Z., Yin, Z., et al., 2024. Electrically driven enhancement on selective adsorption of nitrate by microporous carbon from wastewater: Synergism of functional groups and micropores. Chem. Eng. J. 495. https://doi.org/10.1016/j.cej.2024.153462.
[14] Khee, Y.L., Kiew, P.L., Chung, Y.T., et al., 2023. Valorizing papaya seed waste for wastewater treatment: a review. Int. J. Environ. Sci. Technol. 20, 2327–2346. https://doi.org/10.1007/s13762-022-04178-9.
[15] Duan, F., Li, Y., Cao, H., Wang, Y., Crittenden, J.C., Zhang, Y., 2015. Activated carbon electrodes: Electrochemical oxidation coupled with desalination for wastewater treatment. Chemosphere 125, 205–211. https://doi.org/10.1016/J.CHEMOSPHERE.2014.12.065.
[16] Hu, S., Li, B., Tan, C., Sun, K., Xu, W., Wang, A., et al., 2024. Promising honeycombed cork activated carbon for high desalination performance brackish water treatment. Ind. Crops Prod. 222. https://doi.org/10.1016/j.indcrop.2024.119887.
[17] Phan Quang, H.H., Nguyen, T.P., Duc Nguyen, D.D., Ngoc Bao, L.T., Nguyen, D. C., Nguyen, V.H., 2022. Advanced electro-Fenton degradation of a mixture of pharmaceutical and steel industrial wastewater by pallet-activated-carbon using three-dimensional electrode reactor. Chemosphere 297, 134074. https://doi.org/10.1016/J.CHEMOSPHERE.2022.134074.
[18] Dubey, R., Dutta, D., Sarkar, A., Chattopadhyay, P., 2021. Functionalized carbon nanotubes: Synthesis, properties and applications in water purification, drug delivery, and material and biomedical sciences. Nanoscale Adv. 3, 5722–5744. https://doi.org/10.1039/D1NA00293G.
[19] Zhang, L., Xu, L., He, J., Zhang, J., 2014. Preparation of Ti/SnO2-Sb electrodes modified by carbon nanotube for anodic oxidation of dye wastewater and combination with nanofiltration. Electrochim. Acta 117, 192–201. https://doi.org/10.1016/j.electacta.2013.11.117.
[20] Shao, D., Li, W., Wang, Z., Yang, C., Xu, H., Yan, W., et al., 2022. Variable activity and selectivity for electrochemical oxidation wastewater treatment using a magnetically assembled electrode based on Ti/PbO2 and carbon nanotubes. Sep. Purif. Technol. 301, 122008. https://doi.org/10.1016/j.seppur.2022.122008.

[21] Tewari, C., Pathak, M., Tatrari, G., Kumar, S., Dhali, S., Saha, B., et al., 2024. Waste plastics derived reduced graphene oxide-based nanocomposite with Fe3O4 for water purification and supercapacitor applications. J. Ind. Eng. Chem. 130, 346–356. https://doi.org/10.1016/j.jiec.2023.09.038.
[22] Li, C., Yang, J., Zhang, L., Li, S., Yuan, Y., Xiao, X., et al., 2021. Carbon-based membrane materials and applications in water and wastewater treatment: A review. Environ. Chem. Lett. 19, 1457–1475. https://doi.org/10.1007/s10311-020-01112-8.
[23] Duan, X., Ren, D., Wang, S., Zhang, M., Sun, Y., Sun, S., et al., 2023. Carbon materials in electrocatalytic oxidation systems for the treatment of organic pollutants in wastewater: A review. Carbon Resour. Convers. 6, 262–273. https://doi.org/10.1016/j.crcon.2023.03.006.
[24] Mousset, E., Wang, Z., Hammaker, J., Lefebvre, O., 2016. Physico-chemical properties of pristine graphene and its performance as electrode material for electro-Fenton treatment of wastewater. Electrochim. Acta 214, 217–230. https://doi.org/10.1016/j.electacta.2016.08.002.
[25] Min, X., Zhu, M., He, Y., Wang, Y., Deng, H., Wang, S., et al., 2020. Selective removal of Cl− and F− from complex solution via electrochemistry deionization with bismuth/reduced graphene oxide composite electrode. Chemosphere 251, 126319. https://doi.org/10.1016/j.chemosphere.2020.126319.
[26] Guo, C., Liu, H., Wang, C., Zhao, J., Zhao, W., Lu, N., et al., 2020. Electrochemical removal of levofloxacin using conductive graphene/polyurethane particle electrodes in a three-dimensional reactor. Environ. Pollut. 260, 114101. https://doi.org/10.1016/j.envpol.2020.114101.
[27] Liu, Y.H., Kuo, Y.S., Liu, W.C., Chou, W.L., 2021. Photoelectrocatalytic activity of perovskite YFeO3/carbon fiber composite electrode under visible light irradiation for organic wastewater treatment. J. Taiwan Inst. Chem. Eng. 128. https://doi.org/10.1016/j.jtice.2021.08.029.
[28] Cen, B., Yang, R., Li, K., Lv, C., Liang, B., 2021. Large capacity and rapid rate of ion removal from synthetic municipal wastewater via CDI using chitosan-based nitrogen-doped porous carbon electrode. Process Saf. Environ. Prot. 147, 857–865. https://doi.org/10.1016/j.psep.2021.01.011.
[29] Wilson, M.E., Rukh, M.G., Ashraf, M.A., 2021. The role of nanotechnology, based on carbon nanotubes in water and wastewater treatment. Desalination Water Treat. 242, 12–21. https://doi.org/10.5004/dwt.2021.27568.
[30] Liu, J., Dang, Q., Wang, L., Wang, D., Tang, L., 2024. Applications of flexible electrochemical electrodes in wastewater treatment: A review. Chin. Chem. Lett. 35. https://doi.org/10.1016/j.cclet.2023.109277.
[31] Huang, X.H., Yang, X.K., Gui, L., Liu, S.G., Wang, K., Rong, H.W., et al., 2024. Carbon electrodes for the electrocatalytic synthesis of hydrogen peroxide: A review. Xinxing Tan Cailiao/New Carbon Mater. 39, 254–270. https://doi.org/10.1016/S1872-5805(24)60846-2.
[32] Ramírez, Á., Muñoz-Morales, M., López-Fernández, E., Fernández-Morales, F.J., Llanos, J., 2024. Advancing circular economy: Critical insights into waste biomass derived carbon electrodes for (bio)electrochemical water treatment. Curr. Opin. Electrochem. 46. https://doi.org/10.1016/j.coelec.2024.101492.
[33] Sayed, E.T., Olabi, A.G., Shehata, N., Al Radi, M., Majdy Muhaisen, O., Rodriguez, C., et al., 2023. Application of bio-based electrodes in emerging capacitive deionization technology for desalination and wastewater treatment. Ain Shams Eng. J. 14. https://doi.org/10.1016/j.asej.2022.102030.
[34] Liu, Y.H., Guo, H.W., Zeng, F.Y., 2023. Green fabrication of nanostructured Ni (OH)2/Ni/Carbon felt electrodes with water-containing deep eutectic solvent for

enhanced water electrolysis performance. J. Power Sources 570. https://doi.org/10.1016/j.jpowsour.2023.233043.
[35] Aliyu, M., Rashid, U., Ghani, W.A.W.A.K., Salleh, MA. bin M., Hazmi, B., Shitu, I. G., et al., 2024. Heterogeneous hydrochar-based catalysts for biodiesel production. In: Energy, Environment, and Sustainability. Springer, pp. 323–359. https://doi.org/10.1007/978-981-97-2523-6_14.
[36] Aliyu, M., Moser, B.R., Alharthi, F.A., Rashid, U., 2024. Efficient production of biodiesel from palm fatty acid distillate using a novel hydrochar-based solid acid catalyst derived from palm leaf waste. Process Saf. Environ. Prot. 187, 1126–1139. https://doi.org/10.1016/j.psep.2024.05.040.
[37] Aliyu, M., Rashid, U., Ghani, W.A.W.A.K., Salleh, M.A.M., Hazmi, B., Alharthi, F. A., et al., 2024. Waste palm leaves-based hydrochar support catalytic esterification of palm fatty acid distillate. Bioenergy Res. 17. https://doi.org/10.1007/s12155-024-10789-6.
[38] Wang, N., Quan, H., Tan, B., Zhang, Q., Zhang, Z.X., Chen, D., 2024. Selective electrosorption of heavy metal ions from wastewater with S-doped hierarchical porous carbon derived from waste Camellia oleifera shell. Desalination 592. https://doi.org/10.1016/j.desal.2024.118086.
[39] Liu, C., Xu, H., Huo, B., Wang, J., Wang, Z., Chen, X., et al., 2023. Research progress of PVDF based piezoelectric polymer composites in water pollution remediation. J. Water Process Eng. 55. https://doi.org/10.1016/j.jwpe.2023.104181.
[40] Jangid, N.K., Kaliraman, S., Singh, A., Srivastava, A., Srivastava, M., Jadoun, S., et al., 2024. Removal of fluoride ions from water using MgO-based materials with special emphasis on MgO/PPy nanocomposites: A review. J. Mol. Liq. 399. https://doi.org/10.1016/j.molliq.2024.124473.
[41] Tong, T., Zhang, M., Chen, W., Huo, X., Xu, F., Yan, H., et al., 2024. Recent advances in carbon-based material/semiconductor composite photoelectrocatalysts: Synthesis, improvement strategy, and organic pollutant removal. Coord. Chem. Rev. 500. https://doi.org/10.1016/j.ccr.2023.215498.
[42] Ghaffarian Khorram, A., Fallah, N., Nasernejad, B., Afsham, N., Esmaelzadeh, M., Vatanpour, V., 2023. Electrochemical-based processes for produced water and oily wastewater treatment: A review. Chemosphere 338. https://doi.org/10.1016/j.chemosphere.2023.139565.
[43] Piao, M., Zhang, J., Du, H., Du, H., Teng, H., 2023. Preparation of Pb/biochar composite as particle electrode for three-dimensional electrochemical oxidation of chemical oxygen demand from chemical wastewater. Desalination Water Treat. 316, 317–326. https://doi.org/10.5004/dwt.2023.30198.
[44] Vasconcelos, V.M., Santos, G.O.S., Eguiluz, K.I.B., Salazar-Banda, G.R., de Fatima Gimenez, I., 2022. Recent advances on modified reticulated vitreous carbon for water and wastewater treatment – A mini-review. Chemosphere 286. https://doi.org/10.1016/j.chemosphere.2021.131573.
[45] Men, L., Chen, C., Liu, A., Yu, S., Zhou, J., Xie, Y., et al., 2022. N-doped porous carbon-based capacitive deionization electrode materials loaded with activated carbon fiber for water desalination applications. J. Environ. Chem. Eng. 10. https://doi.org/10.1016/j.jece.2022.107943.
[46] Rodríguez-Narváez, O.M., Picos, A.R., Bravo-Yumi, N., Pacheco-Alvarez, M., Martínez-Huitle, C.A., Peralta-Hernández, J.M., 2021. Electrochemical oxidation technology to treat textile wastewaters. Curr. Opin. Electrochem. 29. https://doi.org/10.1016/j.coelec.2021.100806.
[47] Li, W., Song, G., Sun, J., Zhou, M., 2024. Electrochemical advanced oxidation processes towards carbon neutral wastewater treatment: A review. Chem. Eng. J. 480. https://doi.org/10.1016/j.cej.2023.148044.

[48] Zhang, H., Li, S., Zhang, C., Ren, X., Zhou, M., 2024. A critical review of ozone-based electrochemical advanced oxidation processes for water treatment: Fundamentals, stability evaluation, and application. Chemosphere 365. https://doi.org/10.1016/j.chemosphere.2024.143330.

[49] Sarlaki, E., Kianmehr, M.H., Ghorbani, M., Kermani, A.M., Asefpour Vakilian, K., Angelidaki, I., et al., 2023. Highly humified nitrogen-functionalized lignite activated by urea pretreatment and ozone plasma oxidation. Chem. Eng. J. 456. https://doi.org/10.1016/j.cej.2022.140978.

[50] Cao, W., Zeng, C., Guo, X., Liu, Q., Zhang, X., Mameda, N., 2020. Enhanced electrochemical degradation of 2,4-dichlorophenol with the assist of hydrochar. Chemosphere 260. https://doi.org/10.1016/j.chemosphere.2020.127643.

[51] Liu, T., Li, Y., Peng, N., Lang, Q., Xia, Y., Gai, C., et al., 2017. Heteroatoms doped porous carbon derived from hydrothermally treated sewage sludge: Structural characterization and environmental application. J. Environ. Manag. 197, 151–158. https://doi.org/10.1016/j.jenvman.2017.03.082.

[52] Shen, Y., Zhang, N., 2020. A facile synthesis of nitrogen-doped porous carbons from lignocellulose and protein wastes for VOCs sorption. Environ. Res. 189. https://doi.org/10.1016/j.envres.2020.109956.

[53] Yang, X., Wang, B., Song, X., Yang, F., Cheng, F., 2022. Co-hydrothermal carbonization of sewage sludge and coal slime with sulfuric acid for N, S doped hydrochar. J. Clean. Prod. 354. https://doi.org/10.1016/j.jclepro.2022.131615.

[54] Pan, B., Feng, M., McDonald, T.J., Manoli, K., Wang, C., Huang, C.H., et al., 2020. Enhanced ferrate(VI) oxidation of micropollutants in water by carbonaceous materials: Elucidating surface functionality. Chem. Eng. J. 398. https://doi.org/10.1016/j.cej.2020.125607.

[55] Zeghioud, H., Fryda, L., Djelal, H., Assadi, A., Kane, A., 2022. A comprehensive review of biochar in removal of organic pollutants from wastewater: Characterization, toxicity, activation/functionalization and influencing treatment factors. J. Water Process Eng. 47. https://doi.org/10.1016/j.jwpe.2022.102801.

[56] Wang, B., Liao, Y., Wang, T., 2024. Metal oxide particle electrodes for degradation of high concentration phenol wastewater via electrocatalytic advanced oxidation. Chin. J. Chem. Eng. 75, 204–213. https://doi.org/10.1016/j.cjche.2024.09.006.

[57] Rahmani, A., Shabanloo, A., Shabanloo, N., Torkshavand, Z., Dargahi, A., Ansari, A., 2023. The integration of PbO2-based EAOPs with other advanced oxidation processes for improved treatment of water and wastewater. Curr. Opin. Electrochem. 37. https://doi.org/10.1016/j.coelec.2022.101204.

[58] Tripathy, D.B., 2024. MXenes polymer composites/nanocomposites: High-strength multifunctional composite materials for water remediation and desalination. Desalination 586. https://doi.org/10.1016/j.desal.2024.117781.

[59] Kamel, M.S.A., Oelgemöller, M., Jacob, M.V., 2024. Chemical vapor deposition-grown graphene transparent conducting electrode for organic photovoltaics: Advances towards scalable transfer-free synthesis. Renew. Sustain. Energy Rev. 203. https://doi.org/10.1016/j.rser.2024.114740.

[60] Alatalo, S.M., Qiu, K., Preuss, K., Marinovic, A., Sevilla, M., Sillanpää, M., et al., 2016. Soy protein directed hydrothermal synthesis of porous carbon aerogels for electrocatalytic oxygen reduction. Carbon NY 96, 622–630. https://doi.org/10.1016/j.carbon.2015.09.108.

[61] Fereidooni, L., Morais, A.R.C., Shiflett, M.B., 2024. Application of pistachio shell waste in composites, nanocomposites, and carbon electrode fabrication: A review. Resour. Conserv. Recycl. 203. https://doi.org/10.1016/j.resconrec.2023.107403.

[62] Liu, Z., Wang, Z., Tang, S., Liu, Z., 2021. Fabrication, characterization and sorption properties of activated biochar from livestock manure via three different approaches. Resour. Conserv. Recycl. 168. https://doi.org/10.1016/j.resconrec.2020.105254.

[63] Kumar, R., Awino, E., Njeri, D.W., Basu, A., Chattaraj, S., Nayak, J., et al., 2024. Advancing pharmaceutical wastewater treatment: A comprehensive review on application of catalytic membrane reactor-based hybrid approaches. J. Water Process Eng. 58. https://doi.org/10.1016/j.jwpe.2024.104838.

[64] Yan, B., Dai, Y., Xin, L., Li, M., Zhang, H., Long, H., et al., 2024. Research progress in the degradation of printing and dyeing wastewater using chitosan based composite photocatalytic materials. Int. J. Biol. Macromol. 263. https://doi.org/10.1016/j.ijbiomac.2024.130082.

[65] Younes, H., Lou, D., Mao, M., Rahman, M.M., AlNahyan, M., Younis, H., et al., 2024. A review on capacitive deionization: Recent advances in Prussian blue analogues and carbon materials based electrodes. Hybrid Adv. 6, 100191. https://doi.org/10.1016/j.hybadv.2024.100191.

[66] Sivasubramanian, P.D., Kumar, M., Kirankumar, V.S., Samuel, M.S., Dong, C Di, Chang, J.H., 2023. Capacitive deionization and electrosorption techniques with different electrodes for wastewater treatment applications. Desalination 559. https://doi.org/10.1016/j.desal.2023.116652.

CHAPTER 7

Carbon-based Fenton-catalyst materials and composites in water and wastewater treatment

Anees Ameera Fauzi, N. Chitraningrum and I. Budiman
Research Center for Biomass and Bioproducts, National Research and Innovation Agency (BRIN), Cibinong, Indonesia

1. Introduction

As industry and society continue to advance, a growing number of organic pollutants, including antibiotics, pesticides, and dyes, are being released into aquatic habitats. This poses a threat to both ecological security and human health [1]. The Fenton process is one of the well-known advanced oxidation processes (AOPs), which is commonly employed for the purification of wastewater and the breakdown of organic contaminants in water [2]. This process involves the formation of active oxygen from peroxides and iron ions. Active oxygen species can oxidize organic and inorganic compounds that exist in contaminated water and soils. Conventionally, the Fenton reaction is performed by the decomposition of hydrogen peroxide (H_2O_2) via ferrous iron Fe^{2+} [3]. The Fenton oxidation reaction is feasible and could enhance the capacity for oxidizing and breaking down recalcitrant organic compounds in effluent. However, it encounters challenges such as the inadequate complete conversion of the organics into minerals, the generation of excessive iron sludge leading to secondary pollution, and high concentration of H_2O_2 utilization. Currently, scientists utilize iron oxides in various forms, including goethite, lepidocrocite, hematite, and magnetite, as potential heterogeneous catalysts for diverse applications such as bioseparation, drug delivery, and wastewater treatment [4,5]. Moreover, diverse categories of magnetic materials and iron oxides (Fe_2O_3 and γ-Fe_2O_3) are optimal selections for the investigation of catalytic reactions owing to their magnetic characteristics

Carbon-Based Materials and Composites for Water and Wastewater Treatment
ISBN 978-0-443-33681-2
https://doi.org/10.1016/B978-0-443-33681-2.00013-2

and minimal toxicity. Previous studies indicate that magnetite is favored because its octahedral sites contain both Fe^{2+} and Fe^{3+}, enabling the rapid production of hydroxyl radicals. Nevertheless, it is unstable in the aqueous phase owing to its elevated surface energy and inclination to attain more stable phases. Various sophisticated approaches have been devised to address these issues, extending upon the traditional Fenton process. These include transition to Fenton-like (photo-Fenton (PF), electro-Fenton (EF), microwave-Fenton (MW-Fenton), and ultrasonic-Fenton (US-Fenton)) technologies and the addition of support materials. Incorporating a stable support material that is not influenced by pH may address this problem and enhance the effectiveness of the Fenton reaction by inhibiting the leaching of iron oxide into water.

Carbon-based materials have been proven as effective supports to couple with Fe-based catalysts to enhance performance due to the synergistic effect between these two components. It is indispensable in wastewater treatment procedures as an adsorbent and support material. Carbon-based materials with a large specific surface area (SSA) and highly opened pore structure can serve as an adequate dispersion of Fe-based catalysts on their surface and favor the interfacial contact of the reactants to the catalytically active centers. Some carbon-based materials themselves can function as an activator of H_2O_2 to yield hydroxyl radicals (·OH). Thus, there is only a small percentage of Fe needed, which consequently reduces the sludge containing Fe. One proposed mechanism for the interaction of carbon-based materials, for example, activated carbon with H_2O_2, is that the surface hydroxyl group is substituted by a hydroperoxyl group after binding with H_2O_2, which is then restored to the initial form by another H_2O_2 molecule and generates a ·OH radical. Another general mechanism proposed that the decomposition of H_2O_2 on the carbon-based materials is analogous to the Haber–Weiss mechanism involving both reduced (carbon) and oxidized (carbon+) states of catalysts. Additionally, the introduction of metals also plays a significant role in the conversion of H_2O_2 into active ·OH. Li and his colleagues demonstrated that the reactivity and stability of the manganese–iron disulfide/multiwalled carbon nanotubes (Mn–FeS_2/MWCNTs) may be enhanced by electron transfer of the Fe–S–C bond, facilitated by the presence of MWCNTs [6]. Sulfur species were essential in speeding up the Fe(II) regeneration, and the addition of Mn lowered the band gap of Mn–FeS_2, which further increased the creation of superoxide radical anion (O_2^-) due to the charge difference in the Fe–S–Mn connection. This enabled the continuous activation of

H_2O_2 and degradation of pollutants. Hence, this chapter centers on incorporating carbon-based materials with Fenton-like catalysts. This section attempts to provide a brief overview of the latest developments in the study of carbon-based Fenton catalysts for the treatment of wastewater. The information is derived from a comprehensive analysis of the current literature, commencing with an evaluation of carbon-based Fenton catalysts employed in the reactions. Additionally, the chapter provides a thorough examination of different carbon-based Fenton catalysts, including an analysis of their effectiveness, the key operational factors that impact their performance, and the mechanism for elimination of organic pollutants from wastewater.

2. Overview of Fenton catalysis

Fenton catalysis is a significant type of AOP used for treating wastewater. It involves the production of ·OH by reacting Fe(II) with H_2O_2 [7]. The Fenton process can be classified into homogeneous and heterogeneous systems depending on the condition of the catalyst, as depicted in Fig. 7.1.

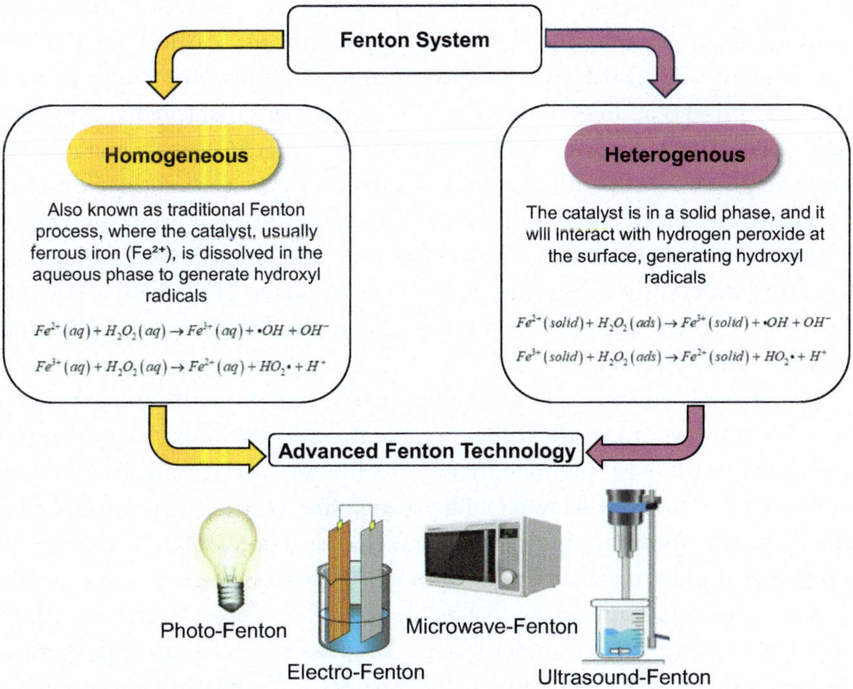

Figure 7.1 Differences between homogeneous and heterogeneous Fenton system and its advanced technologies.

Homogeneous Fenton is usually attributed to the conventional Fenton process, in which the catalyst, typically ferrous iron (Fe^{2+}), is completely dissolved in the liquid phase. Nevertheless, the conventional homogeneous Fenton reaction is still restricted by drawbacks, including a narrow pH range, a high rate of iron leaching, and the formation of iron sludge. Lately, there has been a growing interest in the heterogeneous Fenton process because of its remarkable effectiveness and ability to be recovered in the treatment of complex wastewater matrices. In the heterogeneous Fenton process, the catalyst exists in a solid state and undergoes interaction with H_2O_2 at its surface, resulting in the production of ·OH. However, the effectiveness of the heterogeneous Fenton reaction is limited due to the slow Fe^{3+}/Fe^{2+} cycle and excessive consumption of H_2O_2 [8]. Therefore, it is essential to investigate a highly effective heterogeneous Fenton method for the fast degradation of recalcitrant pollutants.

Furthermore, the selection between Fenton and Fenton-like methods should be determined by the specific requirements of the treatment, budgetary constraints, and the characteristics of the contaminants. The Fenton oxidation reaction is easily operated and effectively oxidizes and breaks down recalcitrant organic compounds in wastewater. However, it has certain drawbacks, including high expenses for operation due to the consumption of large amounts of H_2O_2, significant ion leaching, which can lead to additional contamination, and a limited pH range (usually 2.5–3.5) [9]. To address these issues, there are many Fenton-like strategies that have been devised [10]. Fenton-like reagents are generated by replacing Fe^{2+} with either a heterogeneous catalyst (such as Fe_3O_4 or CuO/Al) or a homogeneous catalyst (such as Fe^{3+} or Cu^{2+}) in the Fenton reagent. It is crucial to acknowledge that the Fenton-like technology operates differently from the Fenton technology [11]. Heterogeneous Fenton-like processes provide substantial benefits for advanced applications and specific solutions.

In general, advanced Fenton/Fenton-like refers to the integration of the Fenton system with other conventional methods, as previously mentioned. EF is a highly efficient technique for eliminating organic contaminants from wastewater. The in situ free radicals were produced at the electrode upon the application of an electrical potential to the electrode surface [12]. There are different types of EF processes, each using specific catalysts and mechanisms as shown in Fig. 7.2. These variations offer a spectrum of choices for implementing EF processes, customizing them to address individual water treatment parameters and contaminant properties,

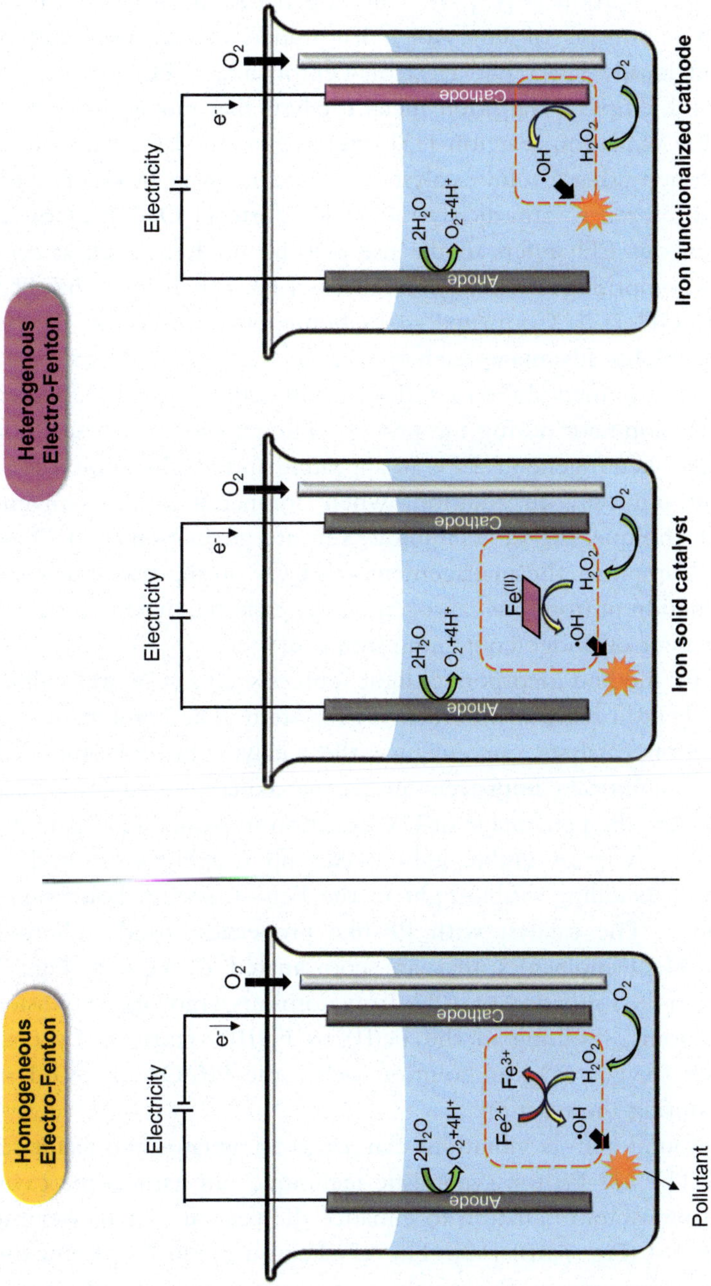

Figure 7.2 Degradation process by the electro-Fenton method.

with the shared objective of breaking down organic pollutants into less detrimental byproducts [13]. In a homogeneous electrochemical system, the catalyst consists of Fe^{2+} ions that undergo a reaction with H_2O_2 produced locally [14]. The electron-Fenton-like (EFL) process, on the other hand, employs transition metal catalysts that interact with H_2O_2 to trigger the oxidation reaction [15]. Heterogeneous EF reduction can be accomplished using a solid catalyst that contains iron. In this process, the dissolved oxygen is cathodically reduced to form H_2O_2 [16]. Conversely, heterogeneous EFL refers to the use of solid catalysts, such as synthetic materials comprising any transition metals or Fe, which react with H_2O_2 to produce ·OH [17]. Compared to its homogeneous version, the heterogeneous EFL has advantages such as easier recovery of solid catalyst, a softer reaction environment due to a wider working pH range, and hindered iron sludge development during the process. The presence of these characteristics makes heterogeneous EFL an appealing and feasible option for water treatment purposes. In addition, when compared to the conventional Fenton technique, the EF technology's in situ generation of H_2O_2 lowers costs and simplifies the management of H_2O_2 during transit and storage. The reduction of iron ions can occur at the cathode, making it suitable for the treatment of sludge containing iron [18].

The PF method incorporates light sources such as UV and visible light into the Fenton reaction to create a PF system. The combination of UV with Fenton catalysts can enhance the catalytic breakdown of H_2O_2, leading to significant improvements in the oxidation of Fenton reagents. Nevertheless, the percentage of UV radiation in natural light is rather low (about 4%), whereas visible light makes up roughly 43% of the total. Therefore, including visible light in the Fenton system holds significant importance. The catalyst with PF-like properties absorbs photons and oxidizes H_2O molecules that are coordinated to Fe(III) (Fig. 7.3A), resulting in the formation of ·OH. Iron atoms undergo charge transfer with oxygen atoms, resulting in the Fe(II) to Fe(III) oxidation by dissolved molecular oxygen (O_2) in solution or by the HOO· and ·OH species formed during the reaction.

Ultrasound and its vibrations can act as an agitator and facilitate mass transfer. The US-Fenton synergistic technique can harness the cavitation effect of ultrasound radiation to enhance the pace at which contaminants are degraded (Fig. 7.3B) [19]. The US-Fenton method operates through two main components: the pyrolysis of contaminants caused by the local high pressure and high temperature induced by the cavitation of ultrasonic

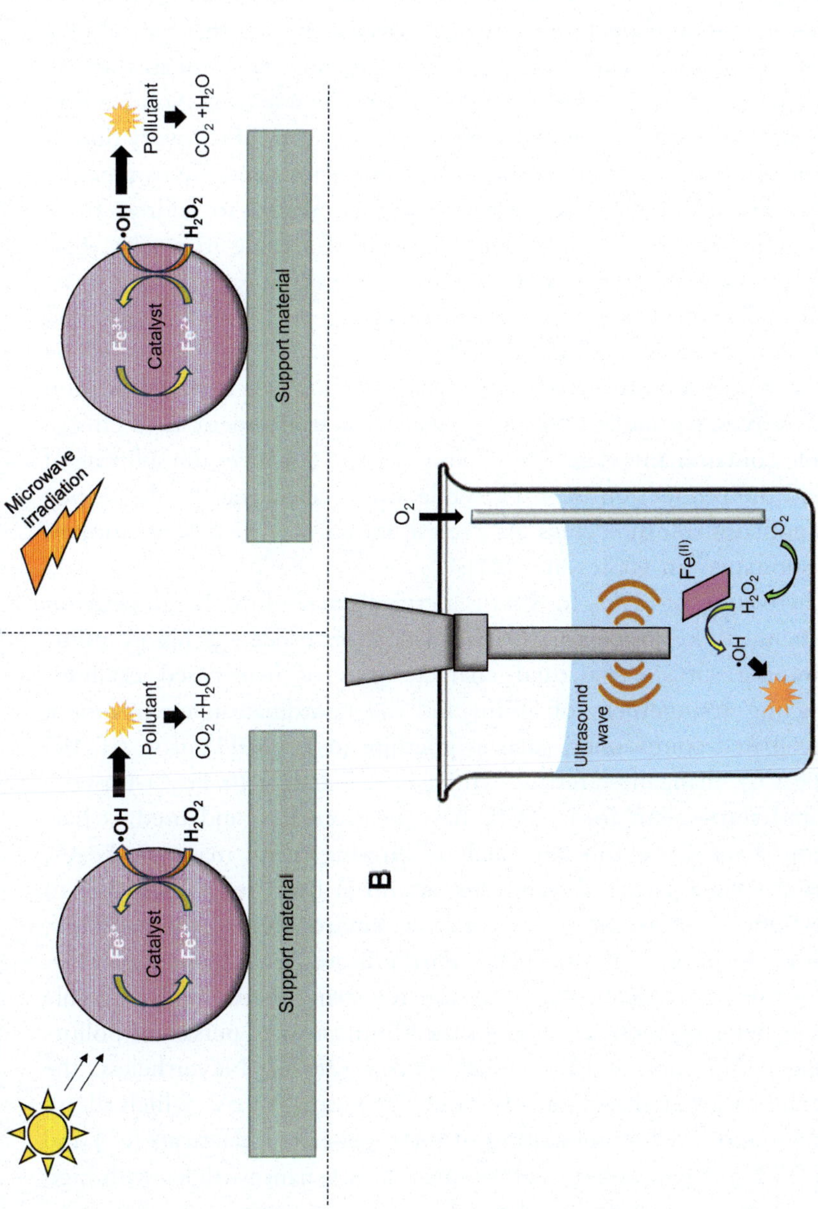

Figure 7.3 The general mechanism of (A) photo-Fenton, (B) ultrasonic-Fenton, and (C) microwave-Fenton systems.

waves. The ·OH created in the high-temperature and high-pressure environment possesses a powerful oxidation potential, which leads to the oxidation of contaminants [20]. However, despite the fact that typical US-Fenton systems are capable of degrading organic contaminants in an effective manner, recycling Fe^{2+} presents a considerable issue. Soluble iron salts exhibit high solubility and pose challenges in terms of recovery during the reaction process, leading to the loss of metal ions and environmental pollution. The US-Fenton-like technique was created to address these limitations by utilizing a solid catalyst instead of soluble iron salts. This approach provides the benefits of a catalyst that can be reused and can be applied to different types of wastewaters. Likewise, the US-Fenton method and the MW-Fenton method work by a comparable mechanism. The MW-Fenton method utilizes the synergistic effects of free radical oxidation and microwaves, particularly the thermal and electromagnetic field effects, to degrade contaminants (Fig. 7.3C). This method enhances the splitting of H_2O_2 and the production of ·OH, while also accelerating the decomposition of contaminant molecules. As a result, it catalyzes the deterioration of organic substances in wastewater [21].

The activation of H_2O_2 to ·OH is a critical stage in the heterogeneous Fenton/Fenton-like process [22]. This activation process is highly influenced by the surface structural characteristics of iron-based catalysts, including the arrangement of atoms and the coordination environment [23]. Iron-based compounds, such as goethite (α-FeOOH), iron (II, III) oxide (Fe_3O_4), manganese ferrite ($MnFe_2O_4$), hematite (α-Fe_2O_3), pyrite (FeS_2), and zero-valent iron (ZVI), have been created and used as heterogeneous Fenton/Fenton-like catalysts. Among them, magnetic Fe_3O_4 nanoparticles have garnered significant attention from researchers due to their cost-effectiveness, ease of separation, environmental compatibility, and favorable photoresponse [24]. Furthermore, Fe_3O_4 nanoparticles exhibit intrinsic peroxidase-like behavior, rendering them highly valuable as efficient heterogeneous catalysts in the elimination of persistent pollutants using AOPs such as PF and EF systems [25,26]. Nevertheless, the Fe_3O_4 exhibits a remarkably narrow band gap of only 0.1 eV, which allows for efficient and quick combination of photogenerated electrons (e^-) and holes (h^+) [27]. Agglomeration of the small Fe_3O_4 nanoparticles with high surface energy is unavoidable throughout the preparation and reaction [28]. Hence, it is imperative to discover an appropriate substrate for Fe_3O_4 nanoparticles in order to address the issue of their tendency to aggregate and their subpar photoelectrochemical characteristics. Iron-based catalysts, namely those loaded on carbon, easily release iron ions during the reaction process, leading to the formation of significant quantities of iron sludge, which hinders their practical use [29]. This shortcoming could probably be solved by enhancing the properties of the support materials.

In addition, pyrite (FeS_2), which is the most prevalent sulfide mineral found on Earth, has shown great potential as a catalyst for the Fenton reaction in the degradation of various pollutants. This is due to its significant amount of Fe(II), and the presence of low-coordinated sulfide species that can serve as electron donors for the continuous reduction of Fe(III). A prior investigation showcased the efficacy of this process in degrading dyes, carbamazepine, 2,4-dichlorophenol, and diethyl phthalate. Nevertheless, the process of aggregation and surface dissolving resulted in a reduced number of active sites being exposed and restricted the in situ cycling of Fe (III) to Fe(II), thereby impeding its performance. Hence, it is imperative and arduous to design a highly dispersed and stable heterogeneous catalyst that can effectively remove pollutants for a prolonged period, while requiring less expense for preparation, in the current heterogeneous catalytic process.

3. Carbon-based Fenton catalyst and composite strategy

In order to surmount the aforementioned challenges, carbon materials have garnered significant interest as a promising choice for cathode material and metal support because of their exceptional optical features, chemical stability, mechanical characteristics, and electrical conductivity [30,31]. Carbon-based nanomaterials consist of carbon elements combined with other nanomaterials, and their structural units vary in size from 1 to 100 nm. To date, there are numerous carbon-based materials that have been developed to be incorporated with Fe and other metals in the Fenton process, as shown in Fig. 7.4, including porous carbon (PC) [32], carbon aerogel (CA) [31], activated carbon (AC) [33,34], carbon nanotube (CNT) [6,35], graphene oxide (GO) [36–38], graphitic carbon nitride (g-C_3N_4) [39], fullerenes [40,41], carbon dots (CDs) [42] and carbon nanofiber (CNF) [43].

PC can be classified into two main categories, including carbon foams that have a specific pore structure, and AC that is made up of PC with additional chemically active surface groups. PC are frequently used in the production of carbon-based catalysts because of their exceptional electronic

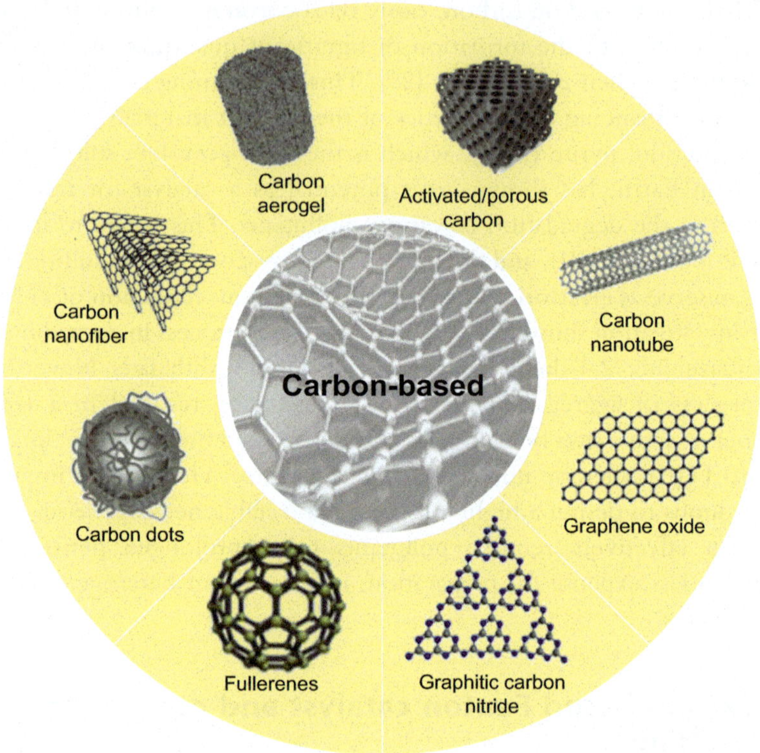

Figure 7.4 Carbon-based as a support for the Fenton catalyst.

conductivity, extensive SSA, and favorable chemical stability. For example, Wang and his colleagues successfully created S-scheme bismuth chromate/ bismuth oxide/ferroferric oxide@porous carbon microspheres ($Cr_2Bi_3O_{11}$–Bi_2O_3/Fe_3O_4@PCs) (F–CBFP) by utilizing calcium alginate as the carbon precursor, which demonstrated a 97.8% degradation efficiency for methyl orange in 60 min under PF conditions [32]. Because the PC has many functional groups, Bi_2O_3 and $Cr_2Bi_3O_{11}$ nanoparticles are directly deposited onto the surface of the highly conductive 3D magnetic porous Fe_3O_4@PC microspheres. This process creates strong connections between the nanoparticles and the microsphere surface, reducing potential barriers and preventing the nanoparticles from clustering together or detaching during Fenton reactions. In comparison to AC, PC materials often demonstrate superior electrical conductivity and structural integrity, although they may possess diminished adsorption capabilities for polar pollutants owing to their restricted surface functional groups. In addition,

CA is one of the most intriguing varieties of the recently developed carbon materials. Previous research has demonstrated that CA is an exceptional choice for both cathode material and metal support due to its high electrical conductivity, extensive surface area, and resistance to corrosion. For instance, Ye and colleagues have demonstrated that the use of FeNi alloy carbon aerogel (FeNi-CA) as the cathode allows for the simultaneous production and activation of H_2O_2 without requiring external aeration [31]. Furthermore, the use of CA as a supporting material has prompted significant attention from researchers seeking to decrease metal leaching in PF processes [44,45]. Another carbon-based material, CNF, is also an appealing type of electrode material that exhibits favorable properties such as effective adsorption, electrical conductivity, and catalytic activity [46]. The high adsorption/absorption capacity of CNF for organic pollutants can be attributed to their hydrophobic characteristics, fiber network structure, and large surface area-to-volume ratio. In addition, the exceptional electrical conductivity and mechanical integrity of free-standing CNFs make them highly desirable as stable electrodes for generating H_2O_2 through electrochemical means [43,47].

CNTs are one-dimensional forms of carbon that can be categorized into two types: single-walled CNTs (SWNTs) and multiwalled CNTs (MWNTs). In both types, the nanotubes exhibit inherent alignment resembling "ropes" and are bound together by van der Waals forces, namely through π–π stacking. These formations appear to be tubular in shape and have a high ratio of length to diameter, often measuring a few micrometers in length and a few nanometers in diameter. SWNTs are produced by precisely curling individual graphene sheets with a diameter ranging from 1 to 2 nm. Likewise, MWNTs consist of several coaxial layers of graphene that are separated by a distance of 0.34 nm. These MWNTs have diameters ranging in the hundreds of nanometers. Furthermore, MWNTs typically possess a zero-bandgap energy and a smaller SSA compared to SWNTs. CNTs are suitable for water treatment due to their thermal and electrical properties. Depending on their surface structures, CNTs can be semiconducting or metallic and have good electronic conductivity. By combining Fe ions with CNTs having negatively charged surface groups like carboxyl, hydroxyl, and ether, heterogeneous Fenton reactivity can be enhanced. The catalytic activity of CNTs is influenced by their reduced and oxidized catalyst states, as shown in multiple studies. It has been theoretically demonstrated that, apart from the direct introduction of external electrons into the catalysts, they can also facilitate the migration of electrons between the catalyst and H_2O_2, thereby influencing the redox cycle of Fe^{3+}/Fe^{2+}. Correspondingly, Seo et al. asserted that the reductive generation of Fe(II) by CNTs was the cause of the increased decomposition of H_2O_2 to ·OH [48], while Gao et al. discovered that without H_2O_2, the production of Fe(II) from Fe(III) remained constant as the MWCNT-COOH level increased, indicating that H_2O_2 is necessary for Fenton catalysis [49]. In addition, Li and coauthors stated that the presence of MWCNTs enhanced both the reactivity and stability of the Mn–FeS_2/MWCNT system through electron transfer involving the interaction of Fe–S–C [6]. While CNTs outperform other carbon-based materials in terms of electrical conductivity and electron transport, their cost and aggregation issues make them less favorable for large-scale industrial applications.

Graphene is the thinnest and most durable conductive nanomaterial currently available. Additionally, it exhibits a significant nonlinear antimagnetism, surpassing that of graphite. In the meantime, GO has a graphene framework with oxygen functional groups affixed to the surface. Zubir et al. synthesized Fenton-like catalysts using a combination of GO and Fe_3O_4 to degrade acid orange 7 (AO7). The GO facilitates electron transport, which helps to retain the active form of iron (Fe^{2+}) and promotes the splitting of H_2O_2 into ·OH, and these active species are responsible for the dye degradation [50]. Furthermore, the synthesis processes that were previously reported frequently relied on the utilization of hazardous organic compounds (such as hydrazine), elevated temperatures, or increased pressures. These factors have the potential to hinder the scalability and cost efficiency of the applications. Therefore, Liu and his colleagues have developed a simple and efficient technique to produce Fe_2O_3@graphene with enhanced longevity. This involves attaching α-Fe_2O_3 to a GO nanosheet, resulting in α-Fe_2O_3@GO [38].

In addition, g-C_3N_4 is a polymer-based substance composed primarily of carbon (C) and nitrogen (N). It is structurally connected by organized tris-triazine patterns, which were initially identified by Liebig in 1834 [51]. The structural constraints of CN prevent it from independently participating as the active center in the PF reaction. In the CN-based heterogeneous PF reaction, the active center of the reaction is typically created by the introduction of metal atoms. Consequently, because of the outstanding catalytic efficiency of composite catalysts, there has been a significant increase in research focusing on composite catalysts consisting of CN combined with other metal-containing compounds in recent years. Wang et al.

synthesized FeS_2@hollow tubular g-C_3N_4 heterojunction catalysts that had a greater number of electron transfer channels. This enhancement efficiently enhances the Fe^{3+}/Fe^{2+} cycle, leading to higher efficiency in the degradation of paint effluent [39]. Besides that, the creation of single-atom catalysts (SACs) is a highly effective method for enhancing the activation of H_2O_2. This is due to the presence of catalytic sites that are scattered at the atomic level, allowing for easy coordination with neighboring nonmetallic atoms in the CN framework. Nevertheless, the formation of nanoclusters or particles is inevitable during photoredox processes due to the reduction of metals involving electrons and the increased surface free energy. Hence, the task of creating stable SACs with a high level of atomic dispersion is a significant problem. A previous work has documented the creation of low-coordinated Fe–N_3 sites in g-C_3N_4 (FeCN) by a one-pot thermal poly-condensation process. These sites serve as extremely efficient PF catalysts [52]. GO and g-C_3N_4 are ideal for PF systems, where visible light activation is critical. GO offers better adsorption capabilities, whereas g-C_3N_4 is superior for electron transfer under light irradiation.

Fullerene, often referred to as soccerene, is a distinct form of carbon that exists alongside diamond and graphite. It possesses a distinctive three-dimensional structure in which π-electrons are spread out, allowing for efficient capture and transfer of electrons when exposed to light radiation. In fact, only C60 and C70 have been extensively investigated among the various forms of fullerenes, including C60, C70, C76, C82, and C84. The band gap energy (Eg) of fullerene ranges from 1.5 to 1.9 eV. Previously, Zou and colleagues utilized an uncomplicated and scalable impregnation process to create a composite of Fe_2O_3 doped onto fullerene C60, resulting in a C60–Fe_2O_3 composite [41]. The findings of the study indicated that there was minimal leaching of iron from the catalyst. CDs, which have diameters smaller than 10 nm, have become a highly attractive category of carbon nanomaterials. CDs possess a variety of surface functional groups, have excellent electron storage capacities, exhibit photoinduced electron transfer characteristics, and contain highly efficient and stable active centers. These qualities collectively give CDs significant potential for a wide range of applications in different scientific and technological fields. Shen and colleagues have created a highly effective PF system by applying a layer of Fe-doped carbon dots (Fe-CDs) onto the surface of highly crystalline C_3N_5 (HC–C_3N_5) nanosheets [42]. The findings of this work demonstrate that Fe-CDs serve as a source of solid-phase Fe species, which greatly enhance the activation of in situ produced H_2O_2. This activation leads to the production of highly oxidative radical O_2^- and ·OH, thereby facilitating the PF reaction.

As aforementioned, a variety of carbon-based compounds, including Fe and other metals, have been created to produce a catalyst with specific properties for use in Fenton technologies. Significantly, prior endeavors have successfully addressed certain deficiencies, including the issues of metal leaching and agglomeration, thus offering a beneficial point of reference for future investigations. However, it is important to consider other factors. For example, a better understanding of the mechanism of action of the Fenton/Fenton-like reagent oxidation reaction system is needed. It is necessary to conduct further research on the synergistic effects of external conditions on the catalyst surface, interfacial effects, and micro-kinetics. Carbon-based Fenton-like nanocomposite catalysts possess the attributes of rapid reaction rate, gentle reaction conditions, and the capacity to be reused. Consequently, it is expected to find broader applications in the domain of water treatment.

4. Carbon-based Fenton catalysts for wastewater treatment: mechanism and performances

The combination of carbon-based materials' ability to adsorb substances and their catalytic activity in the Fenton reaction is an appealing method for treating wastewater. To date, many researchers have taken the initiative to enhance Fenton/Fenton-like system performance by assisting with other systems, including PF, EF, MW-Fenton, and US-Fenton, as listed in Table 7.1. It is very crucial to understand the mechanism of each system in order to obtain the desired catalyst and improve the efficiency of the carbon-based Fenton catalyst.

4.1 Photo-Fenton system

The utilization of carbon in conjunction with a PF catalyst might augment the rate of catalytic degradation by enhancing the rate of electron transfer and accelerating the Fe(II)/Fe(III) redox cycle. There are several catalysts that have been fabricated for PF, as listed in Table 7.1. For instance, Fe_3O_4 nanoparticle-metal-free graphitic carbon nitride (FCN) achieved 90% removal efficiency of tetracycline in 60 min at a neutral pH of 7, with 2.45 mM of H_2O_2 and a catalyst dosage of 0.66 g L^{-1} [54]. Similarly, Fe–N_3 sites in carbon nitride (CN) exhibited 90.1% efficiency for the degradation of 4-chlorophenol under slightly acidic conditions (pH = 6),

Catalyst	Carbon	Pollutant	Fenton system	Performances	Ref
Fe_3O_4@starch-derived carbon	AC	Tetracycline	Photo	**98.2%** (35 min, $C_0 = 20$ mg L^{-1}, pH = 5, $C(H_2O_2) = 5$ mM, catalyst dosage = 0.8 g L^{-1})	[28]
Fe/S-biochar	Biochar	Phenol	Photo	**98%** (120 min, $C_0 = 100$ mg L^{-1}, pH = 5, $C(H_2O_2) =$ 10 mM, catalyst dosage = 4 g L^{-1})	[53]
Fe_3O_4 nanoparticle-metal-free graphitic carbon nitride (FCN)	g-C_3N_4	Tetracycline	Photo	**90%** (60 min, $C_0 = 10$ mg L^{-1}, pH = 7, $C(H_2O_2) = 2.45$ mM, catalyst dosage = 0.66 g L^{-1})	[54]
Fe–N_3 site in carbon nitride	CN	4-Chlorophenol	Photo	**90.1%** (60 min, $C_0 = 10$ mg L^{-1}, pH = 6, $C(H_2O_2) = 15$ mM, catalyst dosage = 0.6 g L^{-1})	[52]
S-scheme $Cr_2Bi_3O_{11}$–Bi_2O_3/Fe_3O_4@porous carbon microspheres	PC	Methyl orange tetracycline hydrochloride	Photo	**MO: 97.8%** (60 min) **TCH: 95.3%** (100 min) ($C_0 = 20$ mg L^{-1}, pH = 5, catalyst dosage = 0.015 g L^{-1})	[32]
Peanut shell-derived ferromagnetic activated carbon	AC	Methyl orange	Photo	**99.2%** (50 min, $C_0 = 50$ mg L^{-1}, pH = 2, $C(H_2O_2) = 0.5$ mL, catalyst dosage = 0.2 g L^{-1})	[34]

Continued

Table 7.1 Summary of different advanced Fenton systems for the removal of various pollutants.—cont'd

Catalyst	Carbon	Pollutant	Fenton system	Performances	Ref
Mn–FeS_2/MWCNT	MWCNT	Bisphenol A	Photo	**100%** (15 min, $C_0 = 50\ mg\ L^{-1}$, pH = 4, $C(H_2O_2) = 15\ mM$, catalyst dosage = $0.35\ g\ L^{-1}$)	[6]
CoFe-CNT	CNT	Bromaminic acid sodium salt (dye)	Electro	**100%** (40 min, $C_0 = 100\ mg\ L^{-1}$, pH = 3, current density = 10 mA/cm^2, catalyst dosage = $4\ g\ L^{-1}$)	[55]
FeNi-carbon aerogel	CA	Acetaminophen	Electro	**99.9%** (90 min, $C_0 = 10\ mg\ L^{-1}$, pH = 7, current density = 4.44 mA/cm^2)	[31]
Zero-valent iron nanoparticles loaded on activated carbon	AC	Antipyrine	Electro	**97%** (60 min, $C_0 = 40\ mg\ L^{-1}$, pH = 3, current density = 10 mA/cm^2, catalyst dosage = $1.4\ g\ L^{-1}$)	[33]
Fe_3O_4@ nitrogen-doped graphene oxide	GO	Dimethoate	Electro	**100%** (60 min, $C_0 = 50\ mg\ L^{-1}$, pH = 7, current density = 15 mA/cm^2, catalyst dosage = $0.5\ g\ L^{-1}$)	[26]

nZVI@N-rGOA	rGOA	Imidacloprid	Microwave	**100%** (2 min, $C_0 = 10$ mg L^{-1}, pH = 7, temperature = 90°C, catalyst dosage = 0.5 g L^{-1})	[56]
Fe@Carbon	AC	Norfloxacin	Microwave	**95.22%** (14 min, $C_0 = 50$ mg L^{-1}, pH = 4, MW power = 500 W, catalyst dosage = 0.1 g L^{-1})	[57]
Ferrihydrite/ porous AC	AC	Methyl orange	Microwave	**99.1%** (4 min, $C_0 = 20$ mg L^{-1}, pH = 4, MW power = 700 W, C (H_2O_2) = 40 mM, catalyst dosage = 10 g L^{-1})	[58]
Pb–$BiFeO_3$/rGO	rGO	Perfluorooctanoic acid	Microwave	**0.4** mg min^{-1} (5 min, $C_0 = 50$ mg L^{-1}, pH = 5, MW power = 300 W, C (H_2O_2) = 44 mM, catalyst dosage = 1.0 g L^{-1})	[59]

Continued

Table 7.1 Summary of different advanced Fenton systems for the removal of various pollutants.—cont'd

Catalyst	Carbon	Pollutant	Fenton system	Performances	Ref
Fe/N–C-2	CN	Tetracycline	Ultrasonic	**92.77%** (80 min, $C_0 = 100\ mg\ L^{-1}$, C (H_2O_2) = 60 mM, catalyst dosage = $0.2\ g\ L^{-1}$)	[60]
$MnFe_2O_4$/biochar	Biochar	Methylene blue	Ultrasonic	**95%** (20 min, $C_0 = 20\ mg\ L^{-1}$, pH = 5, C(H_2O_2) = 15 mM, catalyst dosage = $0.7\ g\ L^{-1}$)	[61]
Spinel ferrite cobalt on activated carbon	AC	Metronidazole	Ultrasonic-photo	**Ultrasound-photo: 95.7%** **Photo: 97%** (90 min, $C_0 = 30\ mg\ L^{-1}$, pH = 3, C(H_2O_2) = 4 mM, catalyst dosage = $0.3\ g\ L^{-1}$)	[62]

requiring a higher H_2O_2 concentration of 15 mM. These catalysts benefit from the graphitic carbon nitride structure, which enhances light absorption and promotes electron transfer, allowing for effective pollutant removal under near-neutral pH conditions. In comparison, the S-scheme $Cr_2Bi_3O_{11}$–Bi_2O_3/Fe_3O_4@PC microspheres achieved outstanding results for MO and TCH, with removal efficiencies of 97.8% in 60 min and 95.3% in 100 min, respectively [32]. Notably, this system operated at a low catalyst dosage of 0.015 g L^{-1} and pH 5, demonstrating the advantage of S-scheme heterojunctions in promoting charge separation and minimizing recombination. On the other hand, peanut shell-derived ferromagnetic AC achieved 99.2% efficiency for methyl orange within 50 min, though it required highly acidic conditions (pH = 2) and 0.5 mL of H_2O_2 with a catalyst dosage of 0.2 g L^{-1}. Despite its strong performance, the need for acidic pH may limit its industrial applicability.

Among all the reported systems, Mn–FeS_2/MWCNT showed the highest efficiency, achieving 100% removal of bisphenol A (BPA) in just 15 min at pH = 4, with a catalyst dosage of 0.35 g L^{-1} and 15 mM of H_2O_2. The synergy between Mn–FeS_2 and MWCNTs provided rapid generation of reactive oxygen species (ROS) and efficient electron transfer, making it highly effective for the degradation of recalcitrant pollutants in a short time (Fig. 7.5A) [6]. A previous study also comprehensively discussed the mechanism, where the sulfur compounds were essential in speeding up the regeneration of Fe(II), and the addition of Mn lowered the energy difference between the valence and conduction bands of Mn–FeS_2, thereby increasing the synthesis of $O_2^{\cdot -}$. This was achieved by the charge difference in the Fe–S–Mn bond, which facilitated the continuous activation of H_2O_2 and degradation of pollutants. Furthermore, the use of MWCNTs enhanced the catalyst's reactivity and stability by electron transfer of the Fe–S–C interaction. The mechanism has also been proposed as shown in Fig. 7.5B. The Mn–FeS_2/MWCNT system provides an excellent example of how catalysts facilitate electron transfer and enhance radical generation. The Fe site served as an "electron bridge pier" in the Mn–FeS_2/MWCNT catalytic process, facilitating electron transfer across the Fe–S–Mn and Fe–S–C interaction. Moreover, the addition of Mn to FeS_2 can enhance the production of $O_2^{\cdot -}$ by altering the electron distribution in the Fe–S–Mn bonds, thereby activating the H_2O_2 pathways. Simultaneously, the MWCNTs have the ability to transport electrons and inhibit the leaching of iron through the formation of Fe–S–C bonds. The electron flow, facilitated by the cocatalytic impact of Fe–S–Mn and

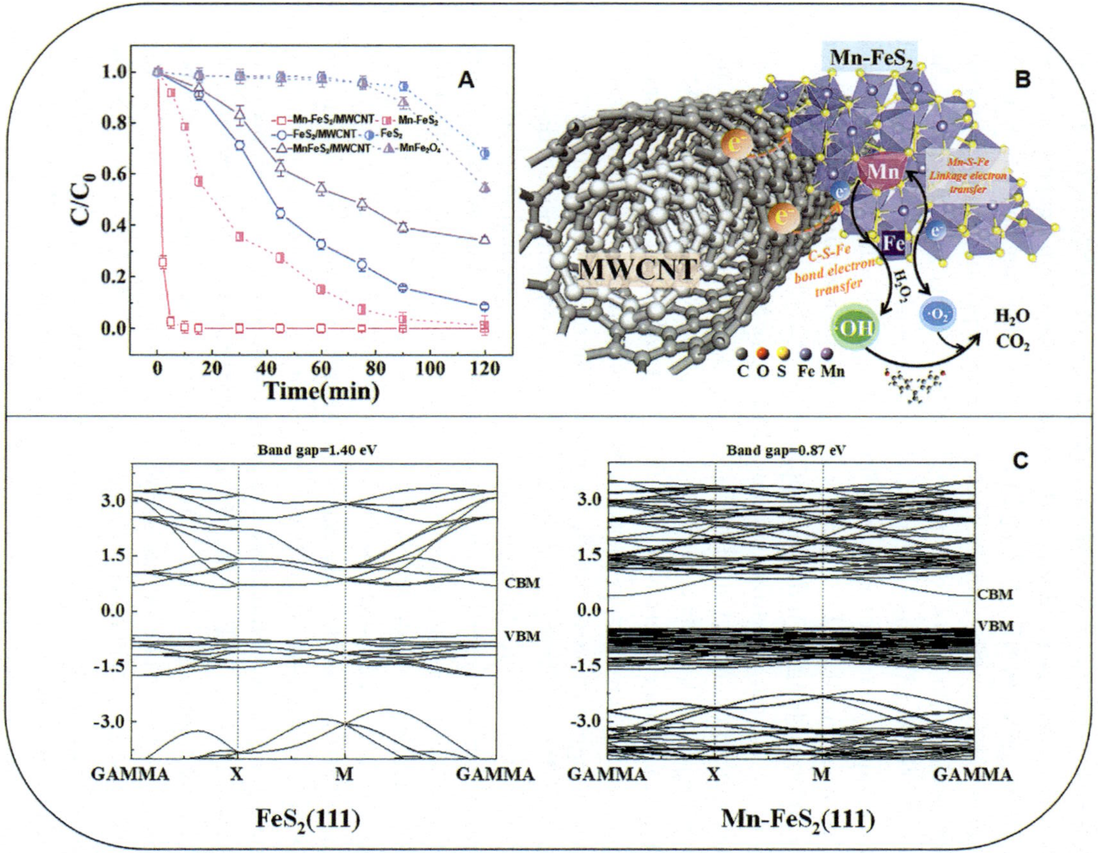

Figure 7.5 (A) Performance of BPA degradation, (B) proposed mechanism of Mn-FeS_2/MWCNT, and (C) band gap of FeS_2 (111) and Mn–FeS_2 (111) using density functional theory. *(Adapted with permission from Ref. [6].)*

Fe–S–C bonds, not only maintains the ongoing catalytic activity of Fe sites but also diversifies the types of free radicals, hence improving the destruction of pollutants. The Fe–S–C interaction minimizes Fe ion leaching, ensuring sustained H_2O_2 activation and catalytic efficiency over multiple cycles.

The Mn–FeS_2/MWCNT catalyst was studied to better comprehend its catalytic mechanism by theoretical calculations using density functional theory (DFT), specifically focusing on the contributions of Fe and Mn sites in the degrading processes. According to the information presented in Fig. 7.5C, the band gap of Mn–FeS_2 is 0.87 eV, which is narrower than the band gap of FeS_2, which is 1.40 eV. The findings indicate that the Mn–FeS_2 structure is more reactive, allowing for easier electron transfer between the electron donor and acceptor. This is because less energy is required to generate electrons. Therefore, Mn–FeS_2 theoretically exhibits higher performance and chemical reactivity as opposed to pristine FeS_2. The electronic location function (ELF) indicates that Mn–FeS_2 exhibits more electron delocalization compared to bare FeS_2. This enhanced electron delocalization could potentially promote electron transport between Mn–FeS_2 and H_2O_2. In addition, the Bader charge analysis was used to examine the transfer of charge between Mn–FeS_2/FeS_2 and H_2O_2. The charge transfer number from Mn–FeS_2 to H_2O_2 was 0.13 e, which was more than the bare FeS_2 (0.04 electrons). This indicates that the addition of Mn enhances the electron transport capacity of FeS_2. Overall, the DFT calculations offer important perspectives into the involvement of Mn and Fe sites in the activation of H_2O_2 and highlight the significant impact of Mn doping on improving the catalysts' performance.

In summary, the comparative analysis highlights that carbon-based catalysts, including g-C_3N_4, PC microspheres, AC, and MWCNT composites, significantly enhance the PF process. Catalysts such as S-scheme PC microspheres and Mn–FeS_2/MWCNT stand out for their rapid degradation efficiency, low catalyst dosage requirements, and robust mechanistic advantages. While some systems perform well under near-neutral pH, others require acidic conditions for optimal activity. These findings emphasize the importance of tailoring catalyst design and operational parameters to achieve efficient and cost-effective pollutant removal in wastewater treatment applications.

4.2 Electro-Fenton system

EF technology is an AOP derived from traditional Fenton technology as described in the previous part. The EF method is widely esteemed due to its ability to in situ produce H_2O_2 and successfully manage degradation expenses [12]. Furthermore, anodic oxidation and electrosorption, along with the oxidation of ·OH in the system, work together to speed up the removal process. Notably, most carbon-based materials used in EF listed in Table 7.1 exhibit superior performance within less than 60 min. Nevertheless, the restricted durability of electrocatalysts and the quantity of H_2O_2 produced in situ at the cathode are acknowledged as obstacles in their advancement and utilization.

The ZVI nanoparticle-loaded AC catalyst showed 97% removal efficiency for antipyrene after 60 min under acidic conditions (pH = 3), with a current density of 10 mA/cm^2 and a catalyst dosage of 1.4 g L^{-1} [33]. While the removal efficiency was slightly lower than that of other systems, the use of ZVI nanoparticles combined with AC enhances adsorption and electron transfer, contributing to pollutant degradation. The moderate catalyst dosage and operational parameters make this system a balanced choice in terms of efficiency and cost. On the other hand, Fe_2O_4@nitrogen-doped GO achieved 100% removal of dimethoate within 60 min at a neutral pH of 7, using a higher current density of 15 mA/cm^2 and a low catalyst dosage of 0.5 g L^{-1} [26]. This catalyst exhibited outstanding performance due to the synergistic effect between Fe_2O_4 nanoparticles and nitrogen-doped GO, which enhances catalytic activity, electron transfer, and stability. Notably, this system combines high efficiency with a lower catalyst dosage, making it highly practical and cost-effective for real-world applications.

In comparison, Li and colleagues successfully created a composite material consisting of CNT encapsulated with a combination of cobalt (Co) and iron (Fe). They achieved this by using a template called ZIF8, which is easy to synthesize, and then subjecting the material to high-temperature carbonization after introducing the Co and Fe [55]. A degradation efficiency of 100% was attained for a concentration of 100 mg L^{-1} of bromaminic acid sodium salt (BAA) within a time period of 40 min, resulting in a complete mineralization rate of 64% (Fig. 7.6A). The catalyst increased the production of H_2O_2 at the EF cathode and also enhanced its activation. The Co-CNT electrode demonstrated a remarkable buildup of H_2O_2, reaching 114.6 mg L^{-1} in just 60 min (Fig. 7.6B).

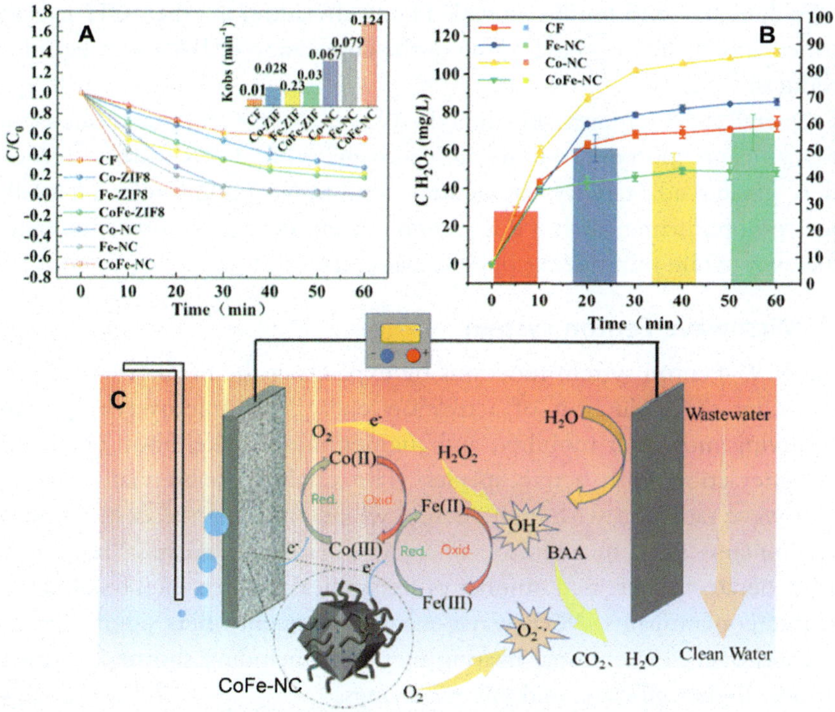

Figure 7.6 (A) Removal of BAA via EF methods, (B) H_2O_2 accumulation under various cathodic conditions, and (C) proposed mechanism. *(Adapted with permission from Ref. [55].)*

The key factors contributing to this phenomenon are primarily the inherent characteristics of the carbon felt and the carbonized CNT material obtained from the metal–organic framework structure postcarbonization.

Based on the proposed mechanism (Fig. 7.6C), oxygen molecules that are adsorbed initially move to the active sites on the surface of the electrode. At these sites, they undergo reduction in the same location, producing H_2O_2 by a pathway that involves the transfer of two electrons. Afterward, the CoFe-NC catalyst activates H_2O_2 or accepts electrons to generate ·OH, which aids in the breakdown of BAA. At the same time, the Fe^0 in the CoFe-NC catalyst transforms Fe^{3+} and Co^{3+} into Fe^{2+} and Co^{2+}, thus assuring catalytic reactivity. Fe^0 can undergo oxidation by H_2O_2 to produce Fe^{2+}, which subsequently combines with $H_2O_2$to generate ·OH. Co^{2+} is capable of reducing Fe^{3+} to Fe^{2+}, and the cathode can also provide electrons for the reduction of Co^{3+} and Fe^{3+}. The combined action of Co^{2+}/Co^{3+} and $Fe^0/Fe^{2+}/Fe^{3+}$ cycles allows

CoFe-NC to consistently activate H_2O_2. In addition, the ·OH species formed anodically may have a role in the conversion of BAA into harmless products.

Overall, Fe_2O_4@nitrogen-doped GO and CoFe-CNT stand out as the most efficient and rapid systems, respectively, while FeNi-Carbon aerogel offers greater adaptability for neutral conditions. Future research could focus on optimizing catalyst dosage and current density to further enhance efficiency while minimizing operational costs.

4.3 Microwave-Fenton system

The MW-Fenton system functions by expediting the breakdown of pollutants by the reduction of activation energy via microwave heating, amplifying molecular mobility, and augmenting the likelihood of molecular interaction with active species. The catalyst preferentially absorbs microwave radiation with high energy, which enhances the destruction of contaminants [63]. In addition, microwave radiation creates "hot spots" that generate an increased number of free radicals, which then break down organic contaminants. Microwave-assisted treatment offers significant advantages over conventional heating methods, including shorter treatment periods, higher efficacy, and energy savings.

Vieira and his colleagues successfully synthesized a novel microwave (MW)-sensitive catalyst by highly stable construction of $CuFeS_2$ (CFS) nanoparticles on AC [64]. The complete mineralization of all compounds occurred within a time frame of 20–30 min, following pseudo-first-order kinetics. The catalyst demonstrated excellent stability during up to 10 repeated cycles. According to the findings, three primary pathways are observed when activated carbon/$CuFeS_2$ (ACFS) is exposed to microwave irradiation (Fig. 7.7A). Initially, the H_2O_2 molecules undergo a quick conversion into ·OH species, a process that is expedited by the creation of localized areas of high reactivity known as hot spots. Secondly, the ·OH produced undergoes a series of consecutive reactions, resulting in the complete breakdown of all three compounds into their mineral components. Lastly, these conversions lead to a significant fluctuation in the flow of energy, which facilitates catalytic reductions by minimizing the rate of recombination. However, even with all of that considered, the ACFS catalyst demonstrated an impressive ability to be reused in up to 10 consecutive runs with minimal loss of CFS particles. Hence, it is indisputable that the catalyst being discussed exhibits exceptional heterogeneous

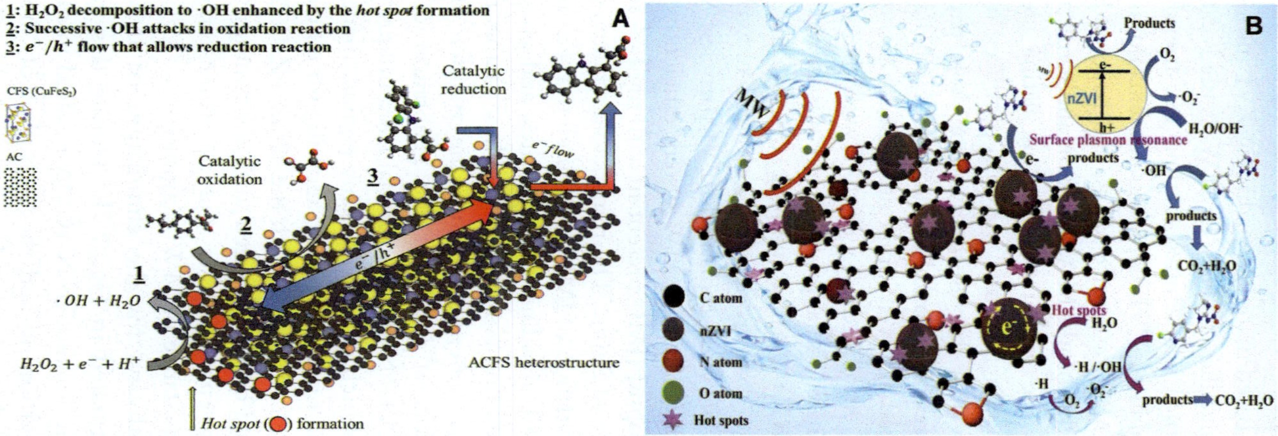

Figure 7.7 Proposed mechanism of (A) pharmaceutical compounds by activated carbon/$CuFeS_2$ (ACFS) and (B) imidacloprid removal by nZVI@N-rGOA when MW is irradiated. *(Adapted with permission from Refs. [56,64].)*

activity and maintains a highly stable surface shape, even when operating at elevated temperatures.

In addition, iron-based materials, including ZVI and iron minerals, are widely utilized as metallic catalysts in MW-assisted catalytic oxidation systems. This preference is attributed to their eco-friendly characteristics and abundant availability. ZVI demonstrates exceptional conductivity, remarkably low magnetic resonance losses, and large eddy current losses. Moreover, it efficiently absorbs MW radiation and forms a circular electric current, resulting in the formation of an oscillating electric field with charge carriers. Consequently, the material produces internal heat and forms a microplasma area that creates electron–hole pairs on its surface while catalytically breaking down contaminants. Zhang and colleagues successfully synthesized a novel composite called nZVI@N-rGOA. This composite was then employed as a catalyst with the assistance of MW technology to eliminate contaminants from wastewater [56]. The final product, nZVI@N-rGOA, had a complex structure like an open pomegranate with multiple compartments. It consisted of wrinkled and flaky reticular N-rGOA walls with implanted nZVI particles. When combined with MW, the nZVI@N-rGOA/MW system efficiently removed almost 100% of the imidacloprid (IMI) present in a 10 $mg\,L^{-1}$ solution within a remarkably short time of 2 min. The remarkable efficiency of microwave-assisted catalytic degradation of IMI is attributed to the combined effects of active centers of nZVI, pyridinic nitrogen sites, and open multicompartment structures resembling pomegranates. These factors facilitate the utilization of microwave energy, enhance the transfer of reactants, and promote the emission of heat and the generation of active species. Furthermore, the nZVI@N-rGOA material has a unique structure like an open pomegranate, which consists of many compartments. This structure offers a wide SSA and reaction cavities, allowing for efficient mass transfer of reactants and enhanced contact with the active nZVI. The catalyst may have heavily depended on the localized surface plasmon resonance (LSPR) caused by the eddy current losses and the three-dimensional porous structure, with the assistance of MW technology.

The reaction pathways for the elimination of IMI by nZVI@N-rGOA/MW are illustrated in Fig. 7.7B. During this investigation, the application of MW irradiation to N-rGOA resulted in the quick generation of localized overheating and the formation of "hot spots" due to inherent dielectric and conduction losses around the nZVI. This sudden increase in temperature led to the detachment of the iron oxides. The recently uncovered Fe^0 transferred electrons to IMI by LSPR, resulting in the production of Fe^{2+}. Subsequently, Fe^{2+} was either oxidized to Fe^{3+} or created other compounds. The strong dielectric properties of the produced iron oxides facilitated effective separation of electrons and h^+ ions when exposed to MW irradiation. Afterwards, in the presence of water, OH^- and O_2^- reacted with h^+ and electrons to produce ·OH and $O_2^{\cdot-}$, which are highly oxidizing. Additionally, the surrounding H_2O was transformed into ·OH and ·H by intrinsic dielectric and conduction losses as well as LSPR at the "hot spots" induced by the pyridinic N sites, FeO, and Fe_3O_4. The pollutant will be further degraded by the active species that are generated.

Besides that, the Fe@carbon catalyst removed 95.22% of norfloxacin in 14 min at acidic pH 4 with a lower catalyst dosage of $0.1\ g\ L^{-1}$ under 500 W microwave power, highlighting its efficiency with minimal material usage [57]. Ferrihydrite/porous AC exhibited exceptional performance, degrading 99.1% of methyl orange in just 4 min at pH 4 under high microwave power (700 W) [58]. However, this system required the highest catalyst dosage ($10\ g\ L^{-1}$), indicating a potential trade-off between speed and material efficiency. On the other hand, the Pb–$BiFeO_3$/rGO catalyst showed relatively low performance, achieving only $0.4\ mg\ L^{-1}$ degradation of perfluorooctanoic acid in 5 min under slightly acidic conditions (pH 5) with 300 W microwave power and $0.1\ g\ L^{-1}$ catalyst dosage [59].

The results highlight the influence of microwave power, catalyst dosage, and pollutant type on degradation performance. Systems with higher microwave power, such as ferrihydrite/porous AC, demonstrated faster degradation, whereas Pb–$BiFeO_3$/rGO, operating at lower power, showed limited efficiency. Additionally, simpler pollutants like IMI and methyl orange were degraded more efficiently compared to the more persistent perfluorooctanoic acid, underscoring the impact of pollutant complexity. The nZVI@N-rGOA catalyst emerges as the most efficient and practical choice, offering an ideal balance between rapid degradation, operational simplicity, and material usage, making it highly suitable for practical applications.

4.4 Ultrasonic-Fenton system

The US-Fenton method produces highly reactive hydroxyl radicals with great oxidation potential, which can effectively break down organic contaminants in wastewater. This process is characterized by its simplicity, low cost, and the ability to operate at high temperatures and pressures.

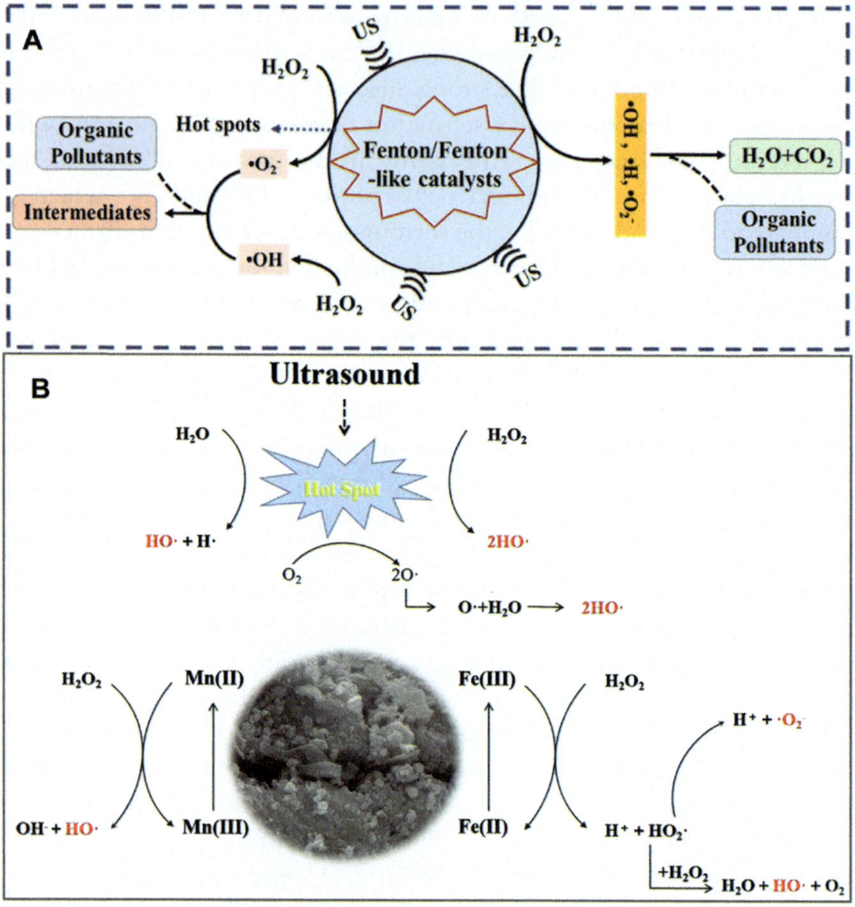

Figure 7.8 (A) US-Fenton/Fenton-like systems in water and (B) proposed mechanism of the $MnFe_2O_4$/BC-based ultrasound-assisted heterogeneous Fenton-like process. *(Adapted with permission from Refs. [60,65].)*

In the US-Fenton-like system (Fig. 7.8A), US-Fenton serves as both a reactant and a catalyst. It has the ability to chemically break down organic molecules when used as a reactant. As a catalyst, it induces sonolysis of the oxidant, such as H_2O_2, resulting in the formation of highly reactive oxidizing free radicals, such as ·OH and ·OOH. These radicals then launch a complicated series of oxidative degradation reactions on contaminants [65]. Furthermore, the US-Fenton promotes the spread of Fe^{2+} and ·OH produced during the reaction, thereby significantly boosting the interaction between the free radicals and contaminants [66].

As listed in Table 7.1, the Fe/N–C–2 system supported on CN, a degradation efficiency of 92.77% for tetracycline was achieved within 80 min under ultrasonic conditions [60]. CN enhances the catalytic performance by providing a large surface area and facilitating the adsorption of the pollutant and H_2O_2 molecules, thereby improving the generation of ROS. Furthermore, Cheng and his colleagues triumphantly synthesized a nanocomposite of $MnFe_2O_4$ ($MnFe_2O_4$/BC) using biochar as the carrier for the degradation of methylene blue (MB) by a US-Fenton-like heterogeneous reaction [61]. The findings demonstrated the effective deposition of $MnFe_2O_4$ with a spherical shape onto the surface of the biochar. Biochar injection suppressed the aggregation of $MnFe_2O_4$ and significantly enhanced the 41.4 $m^2 g^{-1}$ of SSA to 95.1 $m^2 g^{-1}$. When employing the $MnFe_2O_4$/BC composite as a catalyst, 95% of MB was decomposed during a span of 20 min. This resulted in a reaction rate constant of 0.09 min^{-1}, which is significantly higher than the rate seen with bare $MnFe_2O_4$ (0.00995 min^{-1}). The enhanced efficiency can be attributed to the mutual stimulation of $MnFe_2O_4$/BC and US-Fenton, which concurrently augment the rate of mass transfer at the solid–liquid interface and promote the production of active species. Biochar, derived from biomass, acts as an eco-friendly and cost-effective support material. It enhances the catalyst's dispersion and increases the active sites, thereby improving the interaction between the pollutant and the catalyst. Furthermore, its porous structure and stability make it suitable for facilitating mass transfer during the ultrasonic process.

In regard to the reaction process (Fig. 7.8B), the decomposition of water and H_2O_2 occurred concurrently during ultrasonication, resulting in the generation of ·OH. The catalytic process between $MnFe_2O_4$/BC and H_2O_2 oxidation oxidizes lattice oxygen, reducing the surface oxidation state [Fe(II) to Fe(III) or Mn(III) to Mn(II)] and producing ·OH. This work suggests a novel approach to the design and fabrication of catalysts for the degradation of organic contaminants by enhancing our knowledge of the mechanism of heterogeneous Fenton-like reactions facilitated by US-Fenton.

In summary, carbon-based materials like CN, biochar, and AC act as key contributors to the overall catalytic performance by providing structural support, enhancing the adsorption of pollutants, and facilitating the generation and transport of reactive species. While CN enhances electron transfer in the Fe/N–C–2 system, biochar contributes to the fast degradation kinetics in the $MnFe_2O_4$/biochar system. On the other hand, AC synergistically boosts the efficiency of the spinel ferrite cobalt catalyst under ultrasonic-photo conditions. These findings underscore the versatility and importance of carbon-based supports in AOPs, offering sustainable and effective solutions for wastewater treatment.

5. Challenges and perspectives

The sophisticated heterogeneous Fenton-like technologies have demonstrated their robustness and efficacy in eliminating diverse water contaminants. Carbon-based Fenton-like nanocomposite catalysts possess the attributes of rapid reaction rate, gentle reaction conditions, and the capacity to be reused, making them advantageous for future applications in industrial water treatment. However, to address the existing knowledge gaps and enhance their practical applicability, the following recommendations are proposed:

- To lessen the possibility of harm from acidic conditions that would damage the efficacy of the catalyst, composite catalysts should be loaded with acid-resistant functional materials to safeguard their active sites and structural integrity.
- Since real water matrices are highly complex and differ significantly from laboratory conditions, future studies should prioritize testing catalysts in actual wastewater systems containing diverse organic and inorganic components. This would improve the reliability of the results and their relevance to real-world applications. Challenges in fabricating carbon-based Fenton catalysts, including the attainment of uniform structural characteristics and the optimization of surface functionality for improved activity, must be addressed. Creating scalable and eco-friendly synthesis processes will be essential for industrial implementation.
- Emphasis should be placed on strategies to enhance catalyst recovery and reuse, including the integration of magnetic materials or surface changes to promote effective separation. Furthermore, the long-term performance under cyclical usage must be examined to guarantee cost-effectiveness and sustainability.
- Investigating the impact of innovative porous catalyst carriers on the stability and distribution of active components can yield significant insights for enhancing performance and longevity in practical applications.

These particular suggestions aim to offer explicit guidance for forthcoming research and tackle practical obstacles related to the synthesis, recovery, and reutilization of carbon-based Fenton catalysts, thereby promoting their wider implementation in industrial water treatment.

References

[1] L. Liu, R. Yu, S. Zhao, X. Cao, X. Zhang, S. Bai, Heterogeneous Fenton system driven by iron-loaded sludge biochar for sulfamethoxazole-containing wastewater treatment, J. Environ. Manag. 335 (2023) 117576, https://doi.org/10.1016/j.jenvman.2023.117576.

[2] S. Navalon, A. Dhakshinamoorthy, M. Alvaro, H. Garcia, Heterogeneous Fenton catalysts based on activated carbon and related materials, ChemSusChem 4 (2011) 1712–1730, https://doi.org/10.1002/cssc.201100216.

[3] H. Ghasemi, B. Aghabarari, M. Alizadeh, A. Khanlarkhani, N. Abu-Zahra, High efficiency decolorization of wastewater by Fenton catalyst: Magnetic iron-copper hybrid oxides, J. Water Process Eng. 37 (2020) 101540, https://doi.org/10.1016/j.jwpe.2020.101540.

[4] S. Mozaffari, W. Li, M. Dixit, S. Seifert, B. Lee, L. Kovarik, G. Mpourmpakis, A. M. Karim, The role of nanoparticle size and ligand coverage in size focusing of colloidal metal nanoparticles, Nanoscale Adv. 1 (2019) 4052–4066, https://doi.org/10.1039/c9na00348g.

[5] V. Guzsvány, N. Banić, Z. Papp, F. Gaál, B. Abramović, Comparison of different iron-based catalysts for photocatalytic removal of imidacloprid, React. Kinet. Mech. Catal. 99 (2010) 225–233, https://doi.org/10.1007/s11144-009-0106-1.

[6] W. Li, Q. Zhu, X. Yin, Z. Gao, K. Wei, S. Liu, X. Zhang, H. Chen, Y. Zhang, W. Han, Enhanced heterogeneous Fenton catalysis by carbon nanotube-loaded Mn doped FeS_2 catalysts for pollutant degradation: Co-enhancement effect of Fe-S-Mn and Fe-S-C linkages, Sep. Purif. Technol. 335 (2024) 126150, https://doi.org/10.1016/j.seppur.2023.126150.

[7] Z. Song, F.R. Xiu, Y. Qi, Degradation and partial oxidation of waste plastic express packaging bags in supercritical water: Resources transformation and pollutants removal, J. Hazard. Mater. 423 (2022) 127018, https://doi.org/10.1016/j.jhazmat.2021.127018.

[8] W. Shi, W. Sun, Y. Liu, K. Zhang, H. Sun, X. Lin, Y. Hong, F. Guo, A self-sufficient photo-Fenton system with coupling in-situ production H_2O^2 of ultrathin porous g-C_3N_4 nanosheets and amorphous FeOOH quantum dots, J. Hazard. Mater. 436 (2022) 129141, https://doi.org/10.1016/j.jhazmat.2022.129141.

[9] H. Guan, Y. Chen, Y. Xue, A novel Fe_3O_4@Pt synthesized via a self-assembly strategy as efficient Fenton-like@mimetic enzyme catalyst for degradation of tetracyclines in food processing wastewater, Colloids Surfaces A Physicochem. Eng. Asp. 695 (2024) 134260, https://doi.org/10.1016/j.colsurfa.2024.134260.

[10] O.M. Rodríguez-Narváez, L.S. Pérez, N.G. Yee, J.M. Peralta-Hernández, E. R. Bandala, Comparison between Fenton and Fenton-like reactions for l-proline degradation, Int. J. Environ. Sci. Technol. 16 (2019) 1515–1526, https://doi.org/10.1007/s13762-018-1764-1.

[11] M. Cao, P. Xu, K. Tian, F. Shi, Q. Zheng, D. Ma, G. Zhang, Recent advances in microwave-enhanced advanced oxidation processes (MAOPs) for environmental remediation: A review, Chem. Eng. J. 471 (2023) 144208, https://doi.org/10.1016/j.cej.2023.144208.

[12] F.C. Moreira, R.A.R. Boaventura, E. Brillas, V.J.P. Vilar, Electrochemical advanced oxidation processes: A review on their application to synthetic and real wastewaters, Appl. Catal. B Environ. 202 (2017) 217–261, https://doi.org/10.1016/j.apcatb.2016.08.037.
[13] A. Gopinath, L. Pisharody, A. Popat, P.V. Nidheesh, Supported catalysts for heterogeneous electro-Fenton processes: Recent trends and future directions, Curr. Opin. Solid State Mater. Sci. 26 (2022) 100981, https://doi.org/10.1016/j.cossms.2022.100981.
[14] V. Poza-Nogueiras, E. Rosales, M. Pazos, M.Á. Sanromán, Current advances and trends in electro-Fenton process using heterogeneous catalysts – A review, Chemosphere 201 (2018) 399–416, https://doi.org/10.1016/j.chemosphere.2018.03.002.
[15] Ö. Görmez, F. Görmez, B. Gözmen, Comparison of the heterogeneous GO-$FePO_4$/electro-Fenton against the homogeneous Fe(II) ion and Fe(III)-oxalate complex/electro-Fenton for the degradation of metronidazole, J. Water Process Eng. 43 (2021) 1–11, https://doi.org/10.1016/j.jwpe.2021.102265.
[16] D. Wang, Y. Li, S. Hu, J. Hu, H. Hou, B. Liu, H. Zheng, X. Luo, H. Li, A triple-cathode electron-Fenton System for efficient Fe^{2+} regeneration and in-situ H_2O_2 electro-activation, Sep. Purif. Technol. 299 (2022) 121704, https://doi.org/10.1016/j.seppur.2022.121704.
[17] S.O. Ganiyu, M. Zhou, C.A. Martínez-Huitle, Heterogeneous electro-Fenton and photoelectro-Fenton processes: A critical review of fundamental principles and application for water/wastewater treatment, Appl. Catal. B Environ. 235 (2018) 103–129, https://doi.org/10.1016/j.apcatb.2018.04.044.
[18] Q. Yang, H. Mou, X. Hu, S. Qu, Y.F. Tsang, Study on the mechanisms of DMP degradation by electro-Fenton method modified by single atomic-Co doped carbon-based catalyst, Sep. Purif. Technol. 331 (2024) 125580, https://doi.org/10.1016/j.seppur.2023.125580.
[19] Y. Liu, J. Pan, J. Zhang, A. Tang, Y. Liu, Study on mass transfer-reaction kinetics of NO removal from flue gas by using a UV/Fenton-like reaction, Ind. Eng. Chem. Res. 51 (2012) 12065–12072, https://doi.org/10.1021/ie300883f.
[20] H. Wang, S. Meng, Z. Zhang, C. Zhou, H. Zhang, Y. Liu, C. Shu He, P. Zhou, B. Lai, Mechanical agitation and ultrasound constructed in situ Fenton system: The role of atomic hydrogen for Fe(III) reduction, Sep. Purif. Technol. 340 (2024) 126859, https://doi.org/10.1016/j.seppur.2024.126859.
[21] B. Xi, Q. Shi, Removement of thiocyanate from industrial wastewater by microwave-Fenton oxidation method, J. Environ. Sci. (China) 25 (2013) S201–S204, https://doi.org/10.1016/S1001-0742(14)60656-2.
[22] L. Su, P. Wang, X. Ma, J. Wang, S. Zhan, Regulating local electron density of iron single sites by introducing nitrogen vacancies for efficient photo-Fenton process, Angew. Chem. Int. Ed. 60 (2021) 21261–21266, https://doi.org/10.1002/anie.202108937.
[23] X. Wei, H. Yi, C. Lai, X. Huo, D. Ma, C. Du, Synergistic effect of flower-like $MnFe_2O_4$/MoS_2 on photo-Fenton oxidation remediation of tetracycline polluted water, J. Colloid Interface Sci. 608 (2022) 942–953, https://doi.org/10.1016/j.jcis.2021.10.033.
[24] P. Mishra, S. Patnaik, K. Parida, An overview of recent progress on noble metal modified magnetic Fe_3O_4 for photocatalytic pollutant degradation and H_2 evolution, Catal. Sci. Technol. 9 (2019) 916–941, https://doi.org/10.1039/c8cy02462f.
[25] C. Lai, X. Shi, L. Li, M. Cheng, X. Liu, S. Liu, B. Li, H. Yi, L. Qin, M. Zhang, N. An, Enhancing iron redox cycling for promoting heterogeneous Fenton performance: A review, Sci. Total Environ. 775 (2021) 145850, https://doi.org/10.1016/j.scitotenv.2021.145850.

[26] D. Chai, Y. Chen, Z. Jiang, Z. Li, X. Zhang, X. Chen, Y. Wang, K. Cui, Insight on magnetic nitrogen-doped graphene oxide composite electrode constructing heterogeneous Electro-Fenton system for treating organophosphorus pesticide wastewater, Sep. Purif. Technol. 345 (2024) 127419, https://doi.org/10.1016/j.seppur.2024.127419.

[27] K. Benabbas, N. Zabat, I. Hocini, Facile synthesis of Fe_3O_4/CuO a core-shell heterostructure for the enhancement of photocatalytic activity under visible light irradiation, Environ. Sci. Pollut. Res. 28 (2021) 4329–4341, https://doi.org/10.1007/s11356-020-10749-5.

[28] C. Wu, T. Guo, Y. Chen, Q. Tian, Y. Zhang, Z. Huang, H. Hu, T. Gan, Facile synthesis of excellent Fe_3O_4@starch-derived carbon photo-Fenton catalyst for tetracycline degradation: Rapid Fe^{3+}/Fe^{2+} circulation under visible light condition, Sep. Purif. Technol. 329 (2024) 125174, https://doi.org/10.1016/j.seppur.2023.125174.

[29] Q. Wu, H. Yang, L. Kang, Z. Gao, F. Ren, Fe-based metal-organic frameworks as Fenton-like catalysts for highly efficient degradation of tetracycline hydrochloride over a wide pH range: Acceleration of Fe(II)/Fe(III) cycle under visible light irradiation, Appl. Catal. B Environ. 263 (2020) 118282, https://doi.org/10.1016/j.apcatb.2019.118282.

[30] H. Li, X. Shu, P. Tong, J. Zhang, P. An, Z. Lv, H. Tian, J. Zhang, H. Xia, Fe–Ni alloy nanoclusters anchored on carbon aerogels as high-efficiency oxygen electrocatalysts in rechargeable Zn–Air batteries, Small 17 (2021) 1–15, https://doi.org/10.1002/smll.202102002.

[31] Q. Ye, T.N. Hunter, H. Xu, D. Harbottle, G.M. Kale, M.R. Tillotson, Synergistic effect of Fe and Ni on carbon aerogel for enhanced oxygen reduction and H_2O_2 activation in electro-Fenton process, Sep. Purif. Technol. 353 (2024) 128436, https://doi.org/10.1016/j.seppur.2024.128436.

[32] Y. Wang, J. Yang, B. Wang, M. Chen, L. Ran, S. Liu, M. Zhou, L. Zhang, Y. Jiang, X. Dai, L. Lin, Y. Zhang, Fabrication of close-contact S-scheme $Cr_2Bi_3O_{11}$-Bi_2O_3/Fe_3O_4@porous carbon microspheres based on in-situ reaction: Enhanced photo-Fenton wastewater treatment, J. Colloid Interface Sci. 673 (2024) 690–699, https://doi.org/10.1016/j.jcis.2024.06.127.

[33] D. Tesnim, A.M. Díez, B. Amor Hédi, M.Á. Sanroman, M. Pazos, Sustainable removal of antipyrine from wastewater via an eco-friendly heterogeneous electro-Fenton-like process employing zero-valent iron nanoparticles loaded activated carbon, Chem. Eng. J. 493 (2024), https://doi.org/10.1016/j.cej.2024.152494.

[34] J.N. Ndongmo, J.L. Mabou, D.R. Tchuifon Tchuifon, S. Makota, C.G. Fotsop, M. Conde, P.G. Tchieta, Synthesis and characterization of peanut shell-derived ferromagnetic activated carbon: Application in the Fenton process for the degradation of methyl orange, Front. Environ. Chem. 5 (2024) 1–16, https://doi.org/10.3389/fenvc.2024.1375705.

[35] J.C. García, A.M. Pedroza, C.E. Daza, Magnetic Fenton and photo-Fenton-like catalysts supported on carbon nanotubes for wastewater treatment, Water Air Soil Pollut. 228 (2017), https://doi.org/10.1007/s11270-017-3420-7.

[36] F. Görmez, Ö. Görmez, B. Gözmen, D. Kalderis, Degradation of chloramphenicol and metronidazole by electro-Fenton process using graphene oxide-Fe_3O_4 as heterogeneous catalyst, J. Environ. Chem. Eng. 7 (2019) 102990, https://doi.org/10.1016/j.jece.2019.102990.

[37] S. Guo, G. Zhang, J.C. Yu, Enhanced photo-Fenton degradation of rhodamine B using graphene oxide-amorphous $FePO_4$ as effective and stable heterogeneous catalyst, J. Colloid Interface Sci. 448 (2015) 460–466, https://doi.org/10.1016/j.jcis.2015.02.005.

[38] Y. Liu, W. Jin, Y. Zhao, G. Zhang, W. Zhang, Enhanced catalytic degradation of methylene blue by A-Fe_2O_3/graphene oxide via heterogeneous photo-Fenton reactions, Appl. Catal. B Environ. 206 (2017) 642–652, https://doi.org/10.1016/j.apcatb.2017.01.075.
[39] C. Wang, B. Wei, H. Zhu, Y. He, G. Ran, Q. Song, Engineering FeS_2 nanoparticles on tubular g-C_3N_4 for photo-Fenton treatment of paint wastewater, Chin. Chem. Lett. 33 (2022) 3073–3077, https://doi.org/10.1016/j.cclet.2021.09.051.
[40] C. Li, Y. Lian, Q. Zhou, Synergistic effect between Ni and Ce dual active centers initiated by activated fullerene soot for electro−Fenton degradation of tetracycline, Catalysts 12 (2022), https://doi.org/10.3390/catal12050509.
[41] C.Y. Zou, Z. Da Meng, W.C. Ji, S.Q. Liu, Z. Shen, Y. Zhang, N.S. Jiang, Preparation of a fullerene[60]-iron oxide complex for the photo-Fenton degradation of organic contaminants under visible-light irradiation, Cuihua Xuebao/Chinese J. Catal. 39 (2018) 1051–1059, https://doi.org/10.1016/S1872-2067(18)63067-0.
[42] Y. Shen, Y. Wang, P. Shan, R. Xu, X. Sun, J. Hou, F. Guo, C. Li, W. Shi, Boosted photo-self-Fenton degradation activity by Fe-doped carbon dots as dual-function active sites for in-situ H_2O_2 generation and activation, Sep. Purif. Technol. 353 (2025) 128529, https://doi.org/10.1016/j.seppur.2024.128529.
[43] X. Li, C. Xiao, X. Ruan, Y. Hu, C. Zhang, J. Cheng, Y. Chen, Enrofloxacin degradation in a heterogeneous electro-Fenton system using a tri-metal-carbon nanofibers composite cathode, Chem. Eng. J. 427 (2021) 130927, https://doi.org/10.1016/j.cej.2021.130927.
[44] T. Yao, W. Jia, Y. Feng, J. Zhang, Y. Lian, J. Wu, X. Zhang, Preparation of reduced graphene oxide nanosheet/FexOy/nitrogen-doped carbon layer aerogel as photo-Fenton catalyst with enhanced degradation activity and reusability, J. Hazard. Mater. 362 (2019) 62–71, https://doi.org/10.1016/j.jhazmat.2018.08.084.
[45] F. Duarte, F.J. Maldonado-Hódar, A.F. Pérez-Cadenas, L.M. Madeira, Fenton-like degradation of azo-dye Orange II catalyzed by transition metals on carbon aerogels, Appl. Catal. B Environ. 85 (2009) 139–147, https://doi.org/10.1016/j.apcatb.2008.07.006.
[46] J. Wang, C. Liu, J. Li, R. Luo, X. Hu, X. Sun, J. Shen, W. Han, L. Wang, In-situ incorporation of iron-copper bimetallic particles in electrospun carbon nanofibers as an efficient Fenton catalyst, Appl. Catal. B Environ. 207 (2017) 316–325, https://doi.org/10.1016/j.apcatb.2017.02.032.
[47] A. Barhoum, T. Favre, S. Sayegh, F. Tanos, E. Coy, I. Iatsunskyi, A. Razzouk, M. Cretin, M. Bechelany, 3D self-supported nitrogen-doped carbon nanofiber electrodes incorporated co/coox nanoparticles: Application to dyes degradation by electro-Fenton-based process, Nanomaterials 11 (2021), https://doi.org/10.3390/nano11102686.
[48] J. Peng, J. Xue, J. Li, Z. Du, Z. Wang, S. Gao, Catalytic effect of low concentration carboxylated multi-walled carbon nanotubes on the oxidation of disinfectants with Cl-substituted structure by a Fenton-like system, Chem. Eng. J. 321 (2017) 325–334, https://doi.org/10.1016/j.cej.2017.03.133.
[49] S.H. Yoo, D. Jang, H.I. Joh, S. Lee, Iron oxide/porous carbon as a heterogeneous Fenton catalyst for fast decomposition of hydrogen peroxide and efficient removal of methylene blue, J. Mater. Chem. A 5 (2017) 748–755, https://doi.org/10.1039/c6ta07457j.
[50] N.A. Zubir, C. Yacou, J. Motuzas, X. Zhang, X.S. Zhao, J.C. Diniz Da Costa, The sacrificial role of graphene oxide in stabilising a Fenton-like catalyst GO-Fe_3O_4, Chem. Commun. 51 (2015) 9291–9293, https://doi.org/10.1039/c5cc02292d.

[51] K. Xie, J. Fang, L. Li, J. Deng, F. Chen, Progress of graphite carbon nitride with different dimensions in the photocatalytic degradation of dyes: A review, J. Alloys Compd. 901 (2022), https://doi.org/10.1016/j.jallcom.2021.163589.

[52] Y. Shi, Q. Zhu, Y. Li, Y. Tao, K. Li, K. Xu, H. Shang, M. Chen, H. Li, D. Zhang, G. Li, Low-coordinated Fe–N_3 sites in carbon nitride for efficient photo-Fenton wastewater treatment, Mater. Today Chem. 38 (2024), https://doi.org/10.1016/j.mtchem.2024.102125.

[53] A. Cheng, Y. He, X. Liu, C. He, Honeycomb-like biochar framework coupled with Fe_3O_4/FeS nanoparticles as efficient heterogeneous Fenton catalyst for phenol degradation, J. Environ. Sci. (China) 136 (2024) 390–399, https://doi.org/10.1016/j.jes.2022.08.037.

[54] K.H. Badagoppam Haroon, S.K. Bhunia, Fenton-like catalysts based on graphitic carbon nitride nanosheets decorated with Fe_3O_4 nanoparticles for removal of colorless tetracycline, ACS Appl. Nano Mater. (2023), https://doi.org/10.1021/acsanm.3c05542.

[55] Z.H. Li, H.L. Liu, D.F. Kong, C. Gong, X. Chen, Novel CoFe-NC bimetallic catalysts for heterogeneous electro-Fenton process to remove BAA in dye wastewater, J. Environ. Chem. Eng. 12 (2024) 112114, https://doi.org/10.1016/j.jece.2024.112114.

[56] P. Zhang, X. Meng, M. Fan, S. Wu, C. Wang, X. Shang, H. Jia, H. Sun, Customized design of nZVI supported on an N-doped reduced graphene oxide aerogel for microwave-assisted superefficient degradation of imidacloprid in wastewater, Appl. Catal. B Environ. 340 (2024) 123258, https://doi.org/10.1016/j.apcatb.2023.123258.

[57] Y. Wang, R. Wang, N. Lin, J. Xu, X. Liu, N. Liu, X. Zhang, Degradation of norfloxacin by MOF-derived lamellar carbon nanocomposites based on microwave-driven Fenton reaction: Improved Fe(III)/Fe(II) cycle, Chemosphere 293 (2022) 133614, https://doi.org/10.1016/j.chemosphere.2022.133614.

[58] D. Xu, Y. Zhang, F. Cheng, P. Dai, Efficient removal of dye from an aqueous phase using activated carbon supported ferrihydrite as heterogeneous Fenton-like catalyst under assistance of microwave irradiation, J. Taiwan Inst. Chem. Eng. 60 (2016) 376–382, https://doi.org/10.1016/j.jtice.2015.10.036.

[59] S. Li, G. Zhang, W. Zhang, H. Zheng, W. Zhu, N. Sun, Y. Zheng, P. Wang, Microwave enhanced Fenton-like process for degradation of perfluorooctanoic acid (PFOA) using Pb-$BiFeO_3$/rGO as heterogeneous catalyst, Chem. Eng. J. 326 (2017) 756–764, https://doi.org/10.1016/j.cej.2017.06.037.

[60] Y. Yang, X. Zhang, Q. Chen, S. Li, H. Chai, Y. Huang, Ultrasound-assisted removal of tetracycline by a Fe/N-C hybrids/H_2O_2 Fenton-like system, ACS Omega 3 (2018) 15870–15878, https://doi.org/10.1021/acsomega.8b02581.

[61] Z. Cheng, S. Luo, X. Li, S. Zhang, T. Thang Nguyen, M. Guo, X. Gao, Ultrasound-assisted heterogeneous Fenton-like process for methylene blue removal using magnetic $MnFe_2O_4$/biochar nanocomposite, Appl. Surf. Sci. 566 (2021) 150654, https://doi.org/10.1016/j.apsusc.2021.150654.

[62] B. Kakavandi, M. Ahmadi, J. Bedia, M. Hashamfirooz, A. Naderi, V. Oskoei, H. Yousefian, R. Rezaei Kalantary, R. Pelalak, R. Dewil, Metronidazole degradation mechanism by sono-photo-Fenton processes using a spinel ferrite cobalt on activated carbon catalyst, Chemosphere 358 (2024) 142102, https://doi.org/10.1016/j.chemosphere.2024.142102.

[63] N. Remya, J.G. Lin, Current status of microwave application in wastewater treatment-A review, Chem. Eng. J. 166 (2011) 797–813, https://doi.org/10.1016/j.cej.2010.11.100.

[64] Y. Vieira, G. Severo de Carvalho, J. Leichtweis, C.M. Mistura, E.L. Foletto, A. Nawaz, S. Manoharadas, R. Zanella, G.L. Dotto, $CuFeS_2$/activated carbon

heterostructure as a microwave-responsive catalyst for reductive and oxidative degradation of ibuprofen, ketoprofen, and diclofenac, Chem. Eng. J. 480 (2024), https://doi.org/10.1016/j.cej.2023.148060.
[65] A. Babuponnusami, K. Muthukumar, A review on Fenton and improvements to the Fenton process for wastewater treatment, J. Environ. Chem. Eng. 2 (2014) 557–572, https://doi.org/10.1016/j.jece.2013.10.011.
[66] S. Liu, Z. Long, H. Liu, Y. Wang, J. Zhang, G. Zhang, J. Liang, Recent advances in ultrasound-Fenton/Fenton-like technology for degradation of aqueous organic pollutants, Chemosphere 352 (2024) 141286, https://doi.org/10.1016/j.chemosphere.2024.141286.

CHAPTER 8

Carbon-based hybrid materials and composites for water and wastewater treatment

Norfarhana Abdul Samad
Centre of Chemical Technology, Department of Petrochemical Engineering, Politeknik Tun Syed Nasir Syed Ismail, Pagoh Education Hub, Johor, Malaysia

1. Introduction

Access to safe and clean water is mandatory for human survival and economic development. Despite the importance of water supplies, developing industrialization, agricultural runoff, and urbanization have caused serious pollution. Emerging contaminants such as heavy metals, pharmaceuticals, microplastics, and endocrine-disrupting compounds further complicate the challenge of water treatment [1–3]. Conventional treatment methods, such as coagulation, sedimentation, and filtration, frequently inadequately address these pollutants due to limited selectivity, high operational costs, and potential production of secondary pollutants [4,5]. Moreover, these conventional techniques are not always adaptable to the diverse and dynamic nature of water contaminants, necessitating the development of advanced and flexible treatment solutions.

The emergence of carbon-based hybrid materials and composites has revolutionized approaches to water and wastewater treatment [6]. These advanced materials leverage the exceptional properties of carbon allotropes such as graphene, carbon nanotubes (CNTs), cellulose and biochar, often in combination with metal oxides, polymers, or nanoparticles. Their large surface area, tunable functional groups, and outstanding mechanical and thermal stability make them ideal for eliminating a wide range of contaminants [7]. The integration of carbon-based hybrids with various functional substances improves their efficacy, facilitating applications in adsorption, catalytic degradation, and membrane technologies. These

Carbon-Based Materials and Composites for Water and Wastewater Treatment
ISBN 978-0-443-33681-2
https://doi.org/10.1016/B978-0-443-33681-2.00008-9

materials exhibit multifunctionality by simultaneously addressing different types of contaminants, including heavy metals, organic pollutants, and pathogenic microorganisms. This adaptability makes carbon-based hybrids invaluable for diverse water treatment applications, ranging from point-of-use systems in households to large-scale industrial wastewater treatment plants. The integration of carbon-based hybrids in membrane technologies has enhanced antifouling properties and separation efficacy, consequently extending their lifespan and reducing maintenance expenses [8]. A significant advantage of carbon-based hybrids is their ability to tackle complex and emerging water contaminants efficiently.

For example, graphene-based hybrids are known for their remarkable adsorption capacity and catalytic properties, enabling them to participate in advanced oxidation processes (AOPs) for the removal of persistent organic pollutants. Graphene oxide (GO) and reduced graphene oxide (rGO) hybrids can target heavy metals, pharmaceuticals, and biological toxins due to their abundance of oxygen-containing functional groups, which facilitate interactions with pollutants [9,10]. Similarly, CNT-based composites have shown excellent performance in electrocatalytic applications, such as electro-Fenton reactions and catalytic ozonation, which are crucial for degrading refractory organic pollutants [11]. Biochar-derived hybrids, on the other hand, offer a cost-effective and sustainable option for water treatment, particularly in resource-limited regions, due to their ease of production and high porosity [12,13]. These materials are not only efficient but also adaptable for various types of treatment, which aligns with global efforts toward sustainable and resilient water management systems. The adaptability of carbon-based hybrids allows them to target a wide spectrum of contaminants, ranging from heavy metals and organic pollutants to pathogenic microorganisms, making them highly versatile in diverse applications.

This chapter aims to provide a comprehensive understanding of carbon-based hybrid materials, offering insights into their potential to contribute to sustainable and efficient water treatment solutions, focusing on three prominent types: graphene-based hybrids, CNT hybrids, and biochar-derived hybrids. Each type is explored in terms of its synthesis, structural characteristics, and functional properties, emphasizing their unique contributions to water treatment technologies. Additionally, the role of these materials in AOPs, membrane filtration, and adsorption mechanisms is discussed in detail. Through this exploration, the chapter seeks to inspire further research and innovation in the field, ultimately advancing global efforts to ensure access to clean and safe water for all.

2. Types of carbon-based hybrids

2.1 Graphene-based hybrids

Graphene, a two-dimensional material with remarkable electrical, thermal, and mechanical properties, has revolutionized material science [14]. Its extraordinary surface area and chemical versatility make it an ideal candidate for water treatment applications. Functionalization with metal oxides, polymers, or nanoparticles amplifies its adsorption and catalytic capabilities. GO and rGO are particularly effective due to their abundance of oxygen-containing functional groups, facilitating robust interactions with a variety of contaminants such as heavy metals, dyes, organic pollutants, and biological toxins [9,15]. Applications of graphene-based hybrids extend to AOPs, electrocatalysis, and membrane technologies, offering versatility in addressing complex water purification challenges. Furthermore, graphene hybrids' capability to integrate into composite membranes improves their antifouling features and filtration effectiveness, making them useful in current wastewater treatment [16].

2.2 Carbon nanotube-based hybrids

CNTs, cylindrical nanostructures known for their exceptional mechanical strength, electrical conductivity, and chemical stability, are crucial in hybrid material development for water treatment. Multi-walled and single-walled CNTs, when combined with functional substances like metal nanoparticles, polymers, or organic molecules, demonstrate enhanced adsorption capacity and catalytic activity. These hybrids excel in removing heavy metals, organic contaminants, and pathogenic microorganisms from water. Additionally, CNT-based composites exhibit exceptional electrocatalytic activity, making them suitable for AOPs such as electro-Fenton reactions and catalytic ozonation [11]. The unique morphology of CNTs ensures efficient pollutant capture, while their hybridization further enhances their surface area, reactivity, and durability. Emerging research focuses on the scalability of CNT-based hybrids, exploring cost-effective synthesis routes and sustainable applications in industrial-scale water treatment systems.

2.3 Biochar-based hybrids

Biochar, a porous carbonaceous material derived from biomass pyrolysis, offers a sustainable and cost-effective alternative for water treatment applications [13]. Its intrinsic properties, including high porosity, large surface area, and chemical stability, make it a valuable platform for hybridization. Functionalization with metal ions, polymers, or other carbon-based materials significantly enhances its adsorption capacity and catalytic performance [17,18]. These hybrids are particularly effective in removing heavy metals, organic pollutants, and nutrient contaminants like nitrates and phosphates. Biochar-based hybrids also demonstrate potential in rural and resource-limited settings due to their low cost and simplicity of fabrication. Their application extends to constructing composite membranes with superior filtration efficiency and mechanical properties. Furthermore, biochar hybrids' ability with integrate renewable energy systems, such as solar-driven water purification, underscores their relevance in sustainable water management practices. Advanced research is exploring biochar hybridization with nanomaterials, aiming to develop multifunctional solutions for global water challenges. Table 8.1 shows carbon-based hybrid materials used in water and wastewater treatment.

3. Synthesis and fabrication techniques

3.1 Chemical functionalization

Chemical functionalization is a pivotal step in enhancing the reactivity and performance of carbon-based materials for water treatment applications. By introducing functional groups such as hydroxyl, carboxyl, and amine onto carbon surfaces, the hydrophilicity and binding affinity of these materials toward specific contaminants are significantly improved [29]. This process enables targeted interactions with pollutants like heavy metals, organic compounds, and biological toxins. Methods such as acid oxidation, plasma treatment, and grafting techniques are widely employed to achieve effective chemical functionalization [30]. For instance, acid treatment with sulfuric and nitric acids introduces oxygen-containing groups on graphene and CNTs, creating active sites for adsorption and catalysis. Functionalized carbon materials not only exhibit enhanced pollutant removal efficiencies but also provide a versatile platform for further hybridization with other functional agents, including metal oxides and polymers.

Table 8.1 Carbon-based hybrid materials used in water and wastewater treatment.

Carbon-based hybrid material	Properties	Performance	Applications	Ref.
Graphene oxide/metal oxide (e.g., GO-TiO_2)	High surface area, oxygen-rich functional groups, strong photocatalytic properties	Effective degradation of organic pollutants under UV light, adsorption of heavy metals	Photocatalysis, adsorption of dyes, pesticides, and pharmaceutical contaminants	[8,19]
Multi-walled carbon nanotubes (MWCNT) with Fe_3O_4	High mechanical strength, magnetic properties, functionalized surfaces	Efficient removal of arsenic, lead, and other heavy metals; high reusability	Heavy metal adsorption, magnetic separation, advanced oxidation processes	[20,21]
Biochar/silver nanoparticles (Ag-biochar)	Renewable, antimicrobial properties, high porosity	Strong antimicrobial activity, adsorption of nutrients and organic pollutants	Antibacterial water treatment, removal of nitrates and phosphates	[22,23]
Graphene oxide/chitosan (GO-Chitosan)	Biocompatible, enhanced mechanical strength, functionalized with amino groups	High adsorption efficiency for dyes, pharmaceuticals, and heavy metals	Adsorption and removal of pharmaceuticals, textile dyes, and heavy metals	[24,25]
CNT/polymer (e.g., CNT-polyaniline)	Conductive, high mechanical integrity, chemically versatile	Selective adsorption of organic pollutants; enhanced mechanical and thermal stability	Adsorption of polycyclic aromatic hydrocarbons (PAHs), oil–water separation	[26,27]
Biochar/ZnO nanoparticles	Low-cost, photocatalytic activity, high surface area	Effective degradation of organic pollutants under visible light, removal of dyes	Photocatalysis, adsorption in rural water treatment systems	[17,28]

3.2 Hybridization methods

3.2.1 Sol–gel technique

The sol–gel technique is a versatile method for synthesizing metal oxide-carbon hybrids, commonly used to integrate metal oxide nanoparticles into carbon-based frameworks. This process involves the preparation of a colloidal solution, typically containing metal precursors such as metal alkoxides or metal salts. The solution is then subjected to gelation, which involves the transition from a liquid sol to a solid gel, typically by adding a gelling agent or altering the pH or temperature conditions. During gelation, the metal precursors are evenly dispersed within the carbon matrix, leading to a homogeneous hybrid material. The resulting metal oxide-carbon composites exhibit enhanced properties, including improved photocatalytic activity, which is useful in applications like environmental remediation, and increased adsorptive capacity, beneficial for wastewater treatment or gas sensing. The sol-gel method allows for precise control over the composition, porosity, and structure of the hybrid material, making it particularly advantageous for applications that require specific surface characteristics or functionalization [31]. Additionally, the process is cost-effective and scalable, making it suitable for large-scale production of these advanced composites. Overall, the sol–gel technique offers significant advantages in the synthesis of metal oxide-carbon hybrids, enabling the development of materials with tailored properties for a range of environmental, energy, and industrial applications.

3.2.2 Electrochemical deposition

Electrochemical deposition is a highly controlled and versatile method used to load nanoparticles, including metal oxides, onto carbon-based materials such as graphene, CNTs, and other conductive carbon supports. This technique involves the application of an electric current to an electrolyte solution containing metal ions, which are reduced onto the carbon material's surface to form nanoparticles. The deposition process is governed by parameters such as voltage, current density, electrolyte composition, and temperature, allowing precise control over the size, morphology, and distribution of the nanoparticles [32]. In the context of carbon-based supports, electrochemical deposition is particularly advantageous because it ensures a uniform distribution of metal oxide nanoparticles (such as MnO_2, TiO_2, or NiO) on the surface of carbon materials like graphene or CNTs. These carbon materials serve as excellent conductive substrates, which not only enhance the electronic conductivity but also provide a high surface area for the deposition of nanoparticles. The metal oxide-carbon hybrids formed through electrochemical deposition exhibit superior catalytic properties due to the synergy between the carbon support and the metal oxide. This is especially useful in AOPs, where the goal is to degrade organic pollutants in wastewater or air.

For example, the deposition of TiO_2 or MnO_2 on graphene significantly enhances the material's photocatalytic and electrocatalytic properties, improving its ability to degrade organic pollutants under light or electrochemical conditions [33]. TiO_2, a well-known photocatalyst, can be activated by UV light to generate reactive oxygen species (ROS) that oxidize organic contaminants, while MnO_2 can facilitate oxidation reactions in electrochemical processes. Additionally, the electrochemical deposition method allows for precise control of the loading amount of metal oxides, ensuring optimal catalytic activity and stability in real-world applications. Overall, electrochemical deposition provides a highly efficient, scalable, and customizable approach to fabricating metal oxide-carbon hybrids with enhanced catalytic and adsorptive properties, making it a key technique in fields such as environmental remediation, energy storage, and sensors.

3.2.3 Thermal treatment

Thermal treatment, including methods such as pyrolysis, annealing, and calcination, is widely used to enhance the structural and surface properties of carbon-based hybrids [34,35]. These processes involve exposing the material to elevated temperatures under controlled atmospheres, such as inert gases, to induce physical and chemical transformations. Pyrolysis, for example, involves the thermal decomposition of organic precursors like biomass in the absence of oxygen, leading to the formation of carbonaceous materials with enhanced porosity, surface area, and functionalized surfaces [36]. This is particularly beneficial for applications like water treatment, where high adsorption capacity is critical. By treating biochar with activating agents like potassium hydroxide (KOH) or phosphoric acid (H_3PO_4) and heating it to high temperatures, the resulting activated carbon hybrids exhibit a well-developed pore structure, enabling efficient removal of heavy metals and organic pollutants [37].

Annealing, on the other hand, improves the crystallinity and thermal stability of carbon hybrids, often by eliminating surface defects and residual organics. This method also enhances the bonding between carbon materials and embedded nanoparticles, making the hybrids mechanically robust for use in harsh environments [38]. For instance, annealed composites containing metal oxides like TiO_2 or Fe_2O_3 show improved catalytic properties and durability, allowing them to perform effectively in AOPs. These thermally treated materials are particularly valuable in water treatment applications, as the increased pore volume and functional groups introduced during thermal processing enhance their adsorption and catalytic efficiency. Whether for removing heavy metals, degrading organic pollutants, or operating under extreme conditions, thermally treated carbon hybrids are tailored to meet the demands of challenging environmental applications. Table 8.2 shows the fabrication techniques of carbon-based hybrid materials used in water treatment.

3.3 Challenges in fabrication

The large-scale fabrication of carbon-based hybrids faces significant challenges despite their potential for various applications. One major obstacle is the high production cost, as many synthesis methods, such as chemical vapor deposition (CVD), sol-gel processes, or electrochemical deposition, require expensive precursors, advanced equipment, and stringent process control. For example, the preparation of graphene-based hybrids often involves costly steps like the reduction and functionalization of GO, while the addition of metal or metal oxide nanoparticles further raises production expenses. Another critical challenge is scalability, as many laboratory-scale synthesis techniques, such as sol-gel or pyrolysis, struggle to maintain uniformity when scaled up [47]. Larger reactor volumes often result in nonuniform pore structures, inconsistent nanoparticle dispersion, or thermal gradients that degrade material quality. Ensuring uniformity in material properties, such as pore size, surface area, and nanoparticle distribution, is another significant hurdle. Variations in these properties can lead to inconsistent performance, particularly in applications like water treatment or catalysis, where reproducibility is crucial. Furthermore, the energy-intensive nature of fabrication processes, such as high-temperature pyrolysis or annealing, raises concerns about sustainability and environmental impact. These methods often consume large amounts of energy and may generate harmful byproducts, complicating their scalability and cost-effectiveness.

To address these challenges, innovative approaches are needed. Strategies such as utilizing low-cost precursors, like agricultural or industrial waste, can help reduce material costs. Advances in scalable fabrication

Table 8.2 Fabrication techniques of carbon-based hybrid materials used in water treatment.

Synthesis/ fabrication technique	Description	Materials involved	Applications in water/wastewater treatment	Advantages	Limitations	Ref.
Sol–gel process	A chemical solution process to create nanomaterials and composites.	Carbon-based nanomaterials (e.g., CNTs, graphene) and metal oxides (e.g., TiO_2, Fe_2O_3)	Adsorption of heavy metals, organic contaminants, and dye removal	High surface area, controllable composition, cost-effective	Requires precise control of conditions, long processing time	[39,40]
Hydrothermal/ autoclave method	High-temperature and high-pressure water to synthesize composites.	Carbon nanotubes, graphene oxide, activated carbon, metal nanoparticles	Removal of organic pollutants, heavy metals, and bacteria	High yield, enhanced material properties, scalable	High energy consumption, long reaction time	[41]
Freeze-drying	Removal of water from materials at low temperature, forming porous structures.	Activated carbon, biochar, graphene oxide	Filtration of contaminants, adsorption of heavy metals	High porosity, excellent mechanical stability	Slow process, expensive equipment	[42]

Continued

Table 8.2 Fabrication techniques of carbon-based hybrid materials used in water treatment.—cont'd

Synthesis/ fabrication technique	Description	Materials involved	Applications in water/wastewater treatment	Advantages	Limitations	Ref.
Electrospinning	Uses an electric field to produce fibrous materials with high surface area.	Carbon-based nanofibers, polymer matrices (e.g., polyacrylonitrile, cellulose)	Water filtration, removal of pathogens and heavy metals	High surface area, customizable fiber morphology	Requires expensive equipment, low yield	[43]
Pyrolysis	Thermal decomposition in the absence of oxygen to form carbon-based materials.	Biochar, carbon nanotubes, activated carbon	Removal of heavy metals, toxic organics	High carbon yield, eco-friendly	Requires high temperatures, limited control over product properties	[36]
Chemical vapor deposition (CVD)	Deposition of carbon-based materials from gas-phase precursors.	Carbon nanotubes, graphene, graphene oxide	Adsorption, filtration, and degradation of pollutants	High-quality materials, controllable properties	High cost, energy-intensive process	[44]
Wet chemical synthesis	Chemical reactions in solution to synthesize hybrid materials.	Graphene oxide, metal nanoparticles, polymers	Catalysis, pollutant adsorption, dye removal	Low cost, scalable, simple process	Less control over material structure, slow process	[7]

In situ polymerization	Polymerization of monomers in the presence of carbon-based materials.	Graphene oxide, CNTs, conducting polymers	Membrane fabrication, removal of organic and inorganic pollutants	Improved material properties, good dispersion	Limited scalability, complex process	[45]
3D printing (additive manufacturing)	Layer-by-layer deposition to form carbon-based composites.	Graphene, carbon nanotubes, resins	Development of customized filtration systems, removal of pollutants	High customization, scalable	Slow process, expensive equipment	[46]

techniques, such as microwave-assisted synthesis, 3D printing, and roll-to-roll processes, show promise for improving uniformity and efficiency. Additionally, leveraging machine learning and automation in material design and process optimization can enhance reproducibility and reduce production variability. By tackling these issues, the large-scale fabrication of carbon-based hybrids can become more feasible, paving the way for their broader adoption in environmental, energy, and industrial applications.

4. Applications in water and wastewater treatment

Carbon-based hybrids represent a significant advancement in water and wastewater treatment technologies due to their exceptional physico-chemical properties [48,49]. Their large surface area, tunable porosity, high chemical stability, and customizable functional groups make them highly versatile materials [14,50]. These hybrids are widely used in adsorptive removal, catalytic degradation, and membrane filtration processes, offering innovative and sustainable solutions to address a broad range of water contamination challenges. The unique structural features of carbon-based hybrids enable them to interact effectively with various pollutants, from heavy metals and organic contaminants to emerging pollutants such as microplastics and pharmaceuticals. Their ability to integrate functionalities, such as photocatalysis, adsorption, and filtration, within a single material allows for multifunctional treatment processes, reducing the complexity and cost of water treatment systems [7]. Moreover, their adaptability for functionalization enables targeted pollutant removal, catering to specific water quality requirements.

In adsorptive removal applications, carbon hybrids excel due to their high affinity for contaminants through mechanisms such as π–π interactions, hydrogen bonding, and electrostatic forces [15,51]. For catalytic degradation, these materials enhance AOPs by acting as efficient supports for catalysts like metal oxides or noble metals, improving the breakdown of persistent organic pollutants. Additionally, their integration into membrane technologies has transformed filtration systems, offering enhanced permeability, fouling resistance, and selectivity for both large-scale industrial applications and decentralized rural water treatment solutions. Beyond conventional uses, carbon-based hybrids are increasingly employed in integrated treatment systems that combine adsorption, catalysis, and filtration into compact and efficient processes. For example, hybrid materials with magnetic properties enable easy separation and reuse after pollutant removal, while others are being tailored for emerging contaminants like per- and polyfluoroalkyl substances (PFAS). These advancements position carbon-based hybrids as critical enablers of next-generation water treatment technologies, addressing both existing and future water quality challenges.

Research and development in this field are continuously expanding the potential of carbon-based hybrids. Innovations such as incorporating renewable precursors (e.g., agricultural waste-derived biochar), optimizing synthesis techniques for scalability, and leveraging nanotechnology for enhanced functionality are driving their applicability across diverse water treatment scenarios. With their unparalleled versatility and efficacy, carbon-based hybrids are poised to become integral components of sustainable water management strategies globally.

4.1 Adsorptive removal

The high adsorption capacity of carbon-based hybrids is a cornerstone of their application in water and wastewater treatment. Their performance is attributed to a combination of large surface area, tunable porosity, and functionalized surfaces, which enable strong interactions with diverse pollutants [50,52]. These materials are particularly effective for removing inorganic contaminants, organic pollutants, and emerging contaminants, addressing both traditional and modern water quality challenges.

4.1.1 Heavy metals

Graphene oxide-metal (GO-metal) composites are among the most effective adsorbents for toxic heavy metals such as lead (Pb), cadmium (Cd), and mercury (Hg) [10,53]. The incorporation of metal nanoparticles or ions into the carbon framework enhances adsorption through multiple mechanisms such as ion exchange, electrostatic interactions and complexation. For ion exchange, the metal ions replace counterions present on the adsorbent surface, facilitating their capture. Electrostatic Interactions works by negatively charged functional groups on the GO surface attract positively charged metal ions meanwhile the complexation mechanism occurs when the metal ions form stable complexes with functional groups such as hydroxyl (–OH) and carboxyl (–COOH) present on the carbon surface [54]. GO-metal hybrids demonstrate rapid adsorption kinetics and high removal efficiencies, even at low concentrations, making them ideal for treating heavy metal-laden industrial effluents.

Additionally, their regeneration capability allows for multiple reuse cycles, enhancing cost-effectiveness.

4.1.2 Organic pollutants

CNT-based composites are highly effective in adsorbing organic pollutants, including dyes, pesticides, and pharmaceutical residues. Their unique graphitic structure facilitates π–π stacking interactions with aromatic pollutants, while functionalized surfaces improve binding with polar or charged contaminants. Main mechanisms include hydrophobic interactions shows the nonpolar organic molecules are adsorbed onto the hydrophobic surface of CNTs [55]. For hydrogen bonding mechanism, functional groups such as hydroxyl and amine enhance adsorption of polar contaminants, such as dye [56] meanwhile the Van der Waals Forces shows the weak but cumulative interactions contribute to the retention of smaller organic molecules. Functionalization with oxygen or nitrogen groups further boosts adsorption performance by introducing polar sites that interact strongly with specific contaminants, such as endocrine-disrupting chemicals. CNT hybrids also exhibit excellent mechanical and chemical stability, enabling their use in harsh treatment conditions.

4.1.3 Emerging contaminants

Emerging contaminants, including microplastics, perfluoroalkyl substances (PFAS), and pharmaceutical residues, pose significant risks to water quality. Carbon-based hybrids are proving to be promising materials for tackling these persistent pollutants [50,57]. Microplastics have high hydrophobicity and adsorption potential of carbon hybrids, particularly biochar composites, make them effective in capturing microplastic particles from wastewater streams. Carbon materials functionalized with fluorophilic groups selectively adsorb PFAS molecules, which are known for their environmental persistence and toxicity. Their tailored surface chemistry facilitates adsorption even in complex matrices. Carbon hybrids can adsorb trace levels of pharmaceuticals and personal care products (PPCPs), including antibiotics, hormones, and surfactants, preventing their release into water bodies.

4.1.4 Enhancing adsorption performance

Efforts to improve the adsorptive removal capacity of carbon-based hybrids focus on enhancing surface area, optimizing functionalization, and improving regeneration. Innovations include hierarchical structures, which creating hybrid materials with microporous and mesoporous architectures, increases adsorption sites and facilitates faster diffusion of pollutants. Furthermore, multi-functional hybrids are combining adsorption with catalytic degradation capabilities enables simultaneous pollutant removal and breakdown. Then, regenerable adsorbents has incorporating magnetic nanoparticles allows for easy separation and recovery, extending the lifespan of the adsorbent. The versatility and high performance of carbon-based hybrids in adsorptive removal underscore their potential for addressing diverse water treatment needs, from industrial wastewater remediation to tackling emerging contaminants in municipal water supplies. Future research into scaling up production and further functionalization will expand their applicability in global water treatment challenges.

4.2 Catalytic degradation

Carbon-based hybrids are extensively utilized in AOPs due to their exceptional ability to act as catalysts or catalyst supports. These processes generate ROS, such as hydroxyl radicals (•OH) and superoxide radicals (O_2^-•), which effectively degrade persistent organic pollutants into less harmful or biodegradable byproducts. This capability makes them highly versatile and efficient in treating complex wastewater streams.

4.2.1 Photocatalysis

rGO-TiO_2 hybrids are among the most advanced photocatalytic materials for wastewater treatment. The synergy between rGO and TiO_2 nanoparticles enhances photocatalytic efficiency through enhanced electron transfer and broadened light absorption. The enhanced electron transfer shows rGO sheets act as electron sinks, reducing electron-hole recombination and extending the lifetime of photogenerated charge carriers. The broadened light absorption works by the presence of rGO enables photocatalytic activity under visible light, complementing TiO_2's UV activity. These hybrids effectively degrade various pollutants, such as phenols, azo dyes, and pesticides, commonly found in industrial wastewater. Their efficiency, coupled with stability under repeated cycles, makes them a promising solution for large-scale applications. Efforts are ongoing to further functionalize rGO-TiO_2 hybrids to target emerging contaminants, such as pharmaceutical residues and endocrine disruptors.

4.2.2 Electrocatalysis

CNT-based electrodes have gained prominence in electrochemical degradation due to their high conductivity, large surface area, and tunable surface chemistry. Functionalized CNTs enhance the generation of ROS in electrochemical systems, which are crucial for breaking down nonbiodegradable pollutants. In pharmaceutical residues, CNT-based electrodes degrade complex pharmaceutical compounds into simpler, less toxic molecules meanwhile, for industrial dyes, these materials efficiently remove dyes such as methylene blue and rhodamine B, which are challenging to treat using conventional methods. Furthermore, CNT composites with embedded metal nanoparticles or doped with heteroatoms (e.g., nitrogen) have shown enhanced catalytic activity, offering selective degradation of specific pollutants. Their durability and low energy requirements make them a sustainable option for continuous-flow wastewater treatment systems.

4.2.3 Fenton and photo-Fenton reactions

Carbon-based hybrids incorporating iron-based nanoparticles are highly effective in catalyzing Fenton and photo-Fenton reactions. These reactions involve the in-situ generation of hydroxyl radicals from hydrogen peroxide (H_2O_2), enabling rapid oxidation of organic pollutants. The use of iron-doped biochar or graphene composites reduces the cost of the catalysts while maintaining high activity. These hybrids are particularly effective for heavily polluted wastewater containing phenolic compounds, hydrocarbons, and refractory organics. Under UV or visible light, photo-Fenton reactions further accelerate the degradation process, providing an additional pathway for pollutant breakdown.

4.2.4 Emerging innovations in catalytic degradation

Advancements in catalytic degradation include integrating multiple catalytic functionalities into a single hybrid material, such as combining photocatalysis and electrocatalysis or pairing Fenton reactions with membrane filtration. Magnetic carbon hybrids, for instance, allow for easy recovery and reuse, minimizing secondary pollution. Moreover, the development of hybrids from renewable precursors, such as agricultural residues, aligns with sustainability goals while offering cost-effective solutions. Carbon-based hybrids continue to evolve, demonstrating unparalleled versatility and efficiency in catalytic degradation processes. Their integration into industrial and municipal wastewater treatment systems has the potential to significantly enhance water quality while addressing the challenges posed by persistent and emerging contaminants.

4.3 Membrane filtration

Membrane filtration technologies are critical in modern water and wastewater treatment systems due to their ability to separate contaminants from water efficiently. Hybrid membranes incorporating carbon materials, such as graphene and biochar, are revolutionizing this field by significantly enhancing the performance of traditional membrane materials [58]. These carbon-based hybrids offer superior permeability, enhanced selectivity, and improved fouling resistance, addressing many of the challenges faced by conventional membranes in water treatment processes.

4.3.1 Graphene-based membranes

Graphene and its derivatives, such as GO and rGO, are among the most promising materials for next-generation membranes [59]. These membranes exhibit remarkable properties that make them ideal for applications such as desalination, wastewater treatment, and organic solvent filtration. The atomic thickness of graphene and its highly conductive structure provide exceptionally high permeability for water molecules, allowing for rapid filtration without sacrificing performance. This enables higher flux rates compared to traditional polymeric membranes. GO membranes are particularly effective in ion separation, where the interlayer spacing of GO sheets can be precisely tuned to selectively filter ions, such as salts, while allowing water molecules to pass through [60]. This property is especially useful in desalination processes where high salt rejection rates are required. Graphene-based membranes are chemically stable and resistant to degradation under harsh environmental conditions, such as high salinity, temperature, and extreme pH levels. This makes them suitable for a wide range of industrial wastewater treatment applications. Graphene-based membranes have been successfully applied in desalination processes to reduce the energy consumption of traditional reverse osmosis systems, as well as in organic solvent filtration, where they show high resistance to solvent degradation and high separation efficiency for complex mixtures of organic compounds. These characteristics make graphene-based membranes an excellent choice for high-performance water treatment technologies.

4.3.2 Biochar-polymer membranes

Biochar, a carbon-rich material produced from the pyrolysis of organic biomass, is increasingly used as a reinforcing component in hybrid polymer membranes [35]. Biochar-infused membranes offer several benefits, particularly in resource-limited and rural settings where low-cost solutions are essential. Biochar is inexpensive and abundant, especially when derived from agricultural or industrial waste, making biochar-polymer membranes highly affordable compared to conventional membranes. This is particularly advantageous for decentralized water treatment systems in rural or low-resource areas. The porous structure of biochar provides additional surface area for adsorbing contaminants such as heavy metals, organic compounds, and pathogens, improving the overall filtration performance. This property is particularly useful in treating polluted water sources, such as those containing agricultural runoff or industrial effluents. Furthermore, biochar-based membranes exhibit reduced fouling tendencies, as the hydrophilic nature of biochar helps mitigate the accumulation of organic matter and microorganisms on the membrane surface. This reduces cleaning frequency and extends membrane life, making biochar-polymer membranes suitable for long-term use in rural water treatment applications. Biochar-polymer hybrid membranes are particularly effective in decentralized water treatment systems, where high removal efficiency, cost-effectiveness, and sustainability are essential. These membranes can address multiple types of contamination, including pathogens, heavy metals, and organic pollutants, providing a comprehensive solution for safe drinking water in rural areas.

4.3.3 Emerging trends in hybrid membrane technology

In recent years, there have been several advances in the development of carbon-based hybrid membranes for membrane filtration processes include functionalization of carbon materials, composite membrane, magnetic membrane, bio-inspired membrane, and so on. Incorporating functional groups, such as amines, hydroxyls, or carboxyls, onto the carbon surfaces of graphene or biochar further enhances their ability to target specific pollutants, improving selectivity and efficiency [61–63]. For example, functionalized biochar has been shown to more effectively remove pollutants such as pesticides and pharmaceutical residues. Combining carbon-based materials like graphene or CNTs with polymeric matrices or other nanomaterials results in composite membranes that benefit from the combined properties of both materials, enhancing overall filtration performance. These composites often offer improved mechanical strength, chemical resistance, and thermal stability. Incorporating magnetic nanoparticles into carbon-based membranes allows for easy separation and recovery, reducing the risk of secondary contamination and improving sustainability. This technology is particularly beneficial for applications in wastewater treatment and the recycling of membrane materials. Leveraging natural filtration processes found in biological systems (e.g., natural ion channels in biological membranes) has led to the development of bio-inspired carbon-based hybrid membranes that mimic these processes for more selective and efficient contaminant removal.

4.3.4 Applications of carbon-based hybrid membranes

Carbon-based hybrid membranes are being increasingly applied in a variety of water and wastewater treatment processes such as desalination, organic solvent filtration, heavy metal removal ad rural water treatment. Graphene-based membranes have shown promise in seawater desalination, offering a more energy-efficient alternative to traditional reverse osmosis membranes by achieving higher water flux and lower fouling [64]. Organic solvent filtration membranes are used in industries where organic solvent recovery and separation are required, such as in pharmaceutical manufacturing or petrochemical industries. Biochar-based membranes are employed in the removal of heavy metals from contaminated water, especially in mining regions or industrial effluent treatment. In remote areas with limited access to advanced water treatment infrastructure, biochar-polymer membranes provide a low-cost solution for producing clean drinking water for rural water treatment.

The integration of carbon-based materials into membrane filtration technologies significantly improves the performance and sustainability of water treatment systems [65]. These hybrid membranes not only provide efficient filtration capabilities but also contribute to the development of affordable, scalable, and robust solutions for water treatment, especially in challenging or resource-limited settings. As research continues to optimize these materials and techniques, hybrid membranes are set to play a critical role in addressing global water contamination issues. Table 8.3 shows the application of carbon-based hybrid materials in water and wastewater treatment.

Table 8.3 Application of carbon-based hybrid materials in water and wastewater treatment.

Method of separation	Carbon-based hybrid materials	Mechanism	Performance	Ref.
Adsorption	Graphene oxide-metal (GO-metal) composites	Ion exchange, electrostatic interactions, and complexation with functional groups.	Effective for removing heavy metals (e.g., Pb, Cd, Hg) with high efficiency and rapid adsorption rates.	[64]
	Carbon nanotube (CNT)-based composites	π–π stacking interactions with aromatic pollutants and functional groups for selective adsorption.	Efficient in removing organic pollutants such as dyes, pesticides, and pharmaceuticals from wastewater.	[66]
	Biochar-based composites	High surface area, microporosity, and hydrophobicity facilitate the removal of contaminants like microplastics, PFAS, and heavy metals.	Effective in removing microplastics, PFAS, and heavy metals from wastewater and industrial effluents.	[65]

Membrane filtration	Graphene-based membranes	High permeability and selective ion separation due to tunable interlayer spacing in graphene-based membranes.	Suitable for desalination and selective filtration, offering high efficiency and energy savings.	[67]
	Biochar-polymer hybrid membranes	Adsorptive capacity for contaminants and reduced fouling due to hydrophilic interactions.	Effective in removing heavy metals and pathogens, with low fouling rates, making them suitable for decentralized systems.	[7]
	Magnetic carbon-based membranes	Magnetic properties allow for easy recovery and reuse of membranes.	High recovery rate and reusability; ideal for situations requiring sustainable material use.	[68]
Catalytic degradation	Reduced graphene oxide (rGO)-TiO_2 hybrids	Enhanced photocatalytic degradation due to rGO's ability to facilitate electron transfer and improve light absorption.	High efficiency in degrading organic pollutants like azo dyes and phenols under UV or visible light.	[69]
	Iron-doped biochar or graphene hybrids	Fenton and photo-Fenton reactions generating hydroxyl radicals to oxidize organic pollutants.	Effective in degrading phenolic compounds and hydrocarbons in industrial effluents, cost-effective and fast.	[35]

Continued

Table 8.3 Application of carbon-based hybrid materials in water and wastewater treatment.—cont'd

Method of separation	Carbon-based hybrid materials	Mechanism	Performance	Ref.
Electrochemical degradation	Carbon nanotube (CNT)-based electrodes	Electrochemical oxidation and reduction reactions generate reactive oxygen species (ROS) for pollutant degradation.	Excellent for degrading persistent organic pollutants, pharmaceuticals, and dyes through electrocatalysis.	[70]
Combined adsorption + catalytic degradation	Multi-functional hybrids (e.g., biochar-rGO)	Combination of adsorption, photocatalysis, and electrochemical reactions to remove a broad range of contaminants.	High efficiency in removing complex pollutants such as microplastics, PFAS, and pharmaceutical residues.	[71]
Membrane filtration + magnetic separation	Magnetic carbon-based membranes	Magnetic properties allow for easy recovery and reuse of membranes.	High recovery rate and reusability; ideal for situations requiring sustainable material use.	[72]
Membrane filtration	Bio-inspired membranes	Mimic natural filtration processes such as ion channels, allowing selective and efficient contaminant removal.	High selectivity, efficiency, and low energy consumption for diverse contaminants in complex mixtures.	[73]

5. Mechanisms driving efficiency

The efficiency of carbon-based hybrids in water and wastewater treatment systems is driven by a combination of adsorption mechanisms, catalytic pathways, and membrane performance enhancements. Understanding these mechanisms is key to optimizing the performance of these materials in various treatment processes.

5.1 Adsorption mechanisms

The high adsorption capacity of carbon-based hybrids is one of the primary factors that make them effective for water purification [74]. Adsorption occurs when contaminants are attracted and held onto the surface of a material. This process can be categorized into two main types: physical and chemical adsorption.

5.1.1 Physical adsorption

Physical adsorption, also known as physisorption, is driven by weak forces such as van der Waals interactions and electrostatic forces between the contaminant molecules and the carbon surface [75]. The large surface area and porous structure of materials such as GO, CNTs, and biochar enhance their ability to adsorb a wide variety of pollutants. These materials provide a large number of adsorption sites, including micropores, mesopores, and macropores, that allow for the capture of contaminants. The adsorption process is typically reversible, with adsorbed pollutants being desorbed under certain conditions, such as heat or changes in solvent polarity. This mechanism is effective for the removal of contaminants like dyes, organic molecules, and small inorganic ions.

5.1.2 Chemical adsorption

Chemical adsorption, or chemisorption, involves the formation of stronger chemical bonds between the contaminant molecules and the adsorbent surface [76,77]. This can include covalent bonding, ionic bonding, or complexation, leading to more selective and stable pollutant removal. For example, when metal ions like lead (Pb^{2+}) or cadmium (Cd^{2+}) are adsorbed onto carbon-based materials, they often form complexes or ionic bonds with functional groups (such as carboxyl or hydroxyl groups) on the surface of the material. This type of adsorption typically results in a more permanent removal of pollutants, as the contaminants are tightly bound to the surface. Chemical adsorption is particularly effective for heavy metal ions, complex organic pollutants, and persistent chemicals such as pesticides or pharmaceutical residues [78].

5.2 Catalytic pathways

In addition to adsorption, carbon-based hybrids can also serve as catalysts or catalyst supports, facilitating the degradation of pollutants through AOPs and electrochemical reactions. These catalytic pathways can be divided into two main processes: radical formation and electron transfer.

5.2.1 Radical formation

Photocatalytic processes, particularly those involving materials like rGO and titanium dioxide (TiO_2) composites, generate reactive radicals such as hydroxyl radicals (•OH) and superoxide radicals (O_2^-•) [17]. When exposed to light (UV or visible), these photocatalysts absorb energy, which excites electrons and causes the formation of electron-hole pairs. The electron-hole pairs then generate hydroxyl radicals (•OH) from water molecules or superoxide ions (O_2^-•) from oxygen, both of which are highly reactive. These reactive species degrade a wide range of organic pollutants, including phenols, azo dyes, pesticides, and pharmaceutical residues, by breaking down complex molecular structures into smaller, less harmful byproducts, such as water and carbon dioxide. This process is particularly valuable for treating persistent organic pollutants that are resistant to conventional biodegradation methods.

5.2.2 Electron transfer

Electrocatalytic processes are employed in electrochemical degradation systems, where CNTs and graphene-based electrodes play a key role. These materials facilitate electron transfer during redox reactions, which helps to break down complex organic pollutants. In these systems, contaminants undergo oxidation or reduction reactions at the electrode surface, generating ROS that contribute to the degradation of the pollutants. For example, CNT-based electrodes can promote the electrochemical oxidation of pharmaceutical residues, industrial dyes, and other nonbiodegradable pollutants by donating electrons to the contaminant molecules. This method is advantageous for the breakdown of pollutants that are difficult to remove using traditional methods, providing a more sustainable and energy-efficient approach to wastewater treatment.

5.3 Membrane performance enhancement

Hybrid membranes, which integrate carbon-based materials into polymeric matrices, are increasingly being used in membrane filtration processes for water treatment. The performance of these membranes is enhanced by several factors, including improved hydrophilicity, mechanical strength, and resistance to fouling.

5.3.1 Enhanced surface hydrophilicity

The surface properties of hybrid membranes can be tailored to enhance their hydrophilicity, or water-attracting ability. GO and biochar, for instance, are inherently hydrophilic, and their incorporation into polymeric membranes can significantly improve water permeability [8,79]. The increased affinity for water molecules reduces the surface tension, allowing water to pass through the membrane more easily while simultaneously rejecting contaminants. This improvement in hydrophilicity enhances the overall filtration efficiency, enabling faster water flux and reducing the energy requirements for water treatment processes. Hybrid membranes benefit from the reinforcement provided by carbon-based materials such as CNTs or graphene. These materials are known for their high mechanical strength, which enhances the durability and lifespan of the membranes. This increased mechanical strength allows the membranes to withstand higher pressures, which is particularly beneficial in high-flow applications or in processes where membrane integrity is crucial, such as in reverse osmosis or nanofiltration systems.

Membrane fouling, caused by the accumulation of organic and inorganic contaminants on the membrane surface, is one of the major challenges in membrane filtration [80]. Carbon-based hybrids, particularly those with functionalized surfaces or incorporated with biochar, exhibit enhanced fouling resistance. The modification of surface properties (such as creating superhydrophilic surfaces or introducing electrostatic repulsion) can reduce the tendency for contaminants to adsorb onto the membrane. Additionally, the porous structure of biochar allows for the better retention and removal of pollutants, preventing them from clogging the membrane pores. This improvement in fouling resistance extends the operational lifespan of the membranes and reduces the frequency and cost of cleaning procedures. Together, these mechanisms—adsorption, catalytic degradation, and membrane performance enhancement—drive the efficiency of carbon-based hybrids in water and wastewater treatment. By leveraging these processes, hybrid materials offer versatile, sustainable, and cost-effective solutions to tackle a wide range of water contamination challenges.

6. Challenges and future directions

Despite the promising potential of carbon-based hybrids in water and wastewater treatment, several challenges must be addressed to enhance their practical application. These challenges span across production, performance under harsh conditions, and scalability. At the same time, significant research opportunities exist to overcome these hurdles and expand the impact of carbon-based hybrids in environmental management. One of the primary challenges associated with carbon-based hybrid materials is their high production costs. The synthesis of materials such as GO, CNTs, and their composites with metals or other functional groups often requires complex and expensive processes, such as CVD or high-temperature treatments. These costs can limit the widespread application of these materials, especially in large-scale water treatment operations where cost-effectiveness is critical. Additionally, the stability of these hybrids under harsh environmental conditions—such as extreme pH, high temperatures, or in the presence of aggressive contaminants—remains a concern. Many carbon-based hybrids can undergo degradation or loss of functionality when exposed to such conditions, limiting their long-term performance in real-world wastewater treatment settings. Ensuring that these materials maintain their high adsorption and catalytic efficiencies over prolonged periods is essential for their industrial adoption.

Scalability is another significant challenge. While laboratory-scale studies often show promising results for carbon-based hybrids, the transition to large-scale applications is complicated. Issues related to uniformity, production volume, and material consistency need to be resolved to ensure that these materials can be produced on a commercial scale without compromising their properties. Standardization and upscaling processes for these materials must be developed to meet the demands of the global water treatment industry. Despite these limitations, research opportunities abound in several key areas to improve the performance and applicability of carbon-based hybrids. The development of low-cost, scalable fabrication methods is essential for the widespread adoption of carbon-based hybrids. Techniques such as roll-to-roll processing, large-area CVD, and sol–gel methods show promise for scaling up production. Furthermore, using renewable, waste-derived carbon sources (e.g., agricultural waste, biochar) can reduce material costs and environmental impact, making these technologies more accessible and sustainable.

Another promising direction is enhancing the multifunctionality of carbon-based hybrids to address a broader spectrum of contaminants. Researchers are focusing on developing hybrids that not only target conventional pollutants like heavy metals, organic chemicals, and dyes but also emerging contaminants such as microplastics, pharmaceuticals, and personal care products. Functionalizing carbon materials with a wider range of chemical groups, such as amines, thiols, or phosphates, can increase their affinity for different classes of pollutants, expanding their utility across various water treatment scenarios. Incorporating renewable energy sources, such as solar, wind, or bioenergy, into water treatment systems that utilize carbon-based hybrids can improve the overall sustainability and energy efficiency of the process. For example, photocatalytic degradation of organic pollutants using solar-powered carbon-based photocatalysts could reduce the dependence on external electricity sources, making the system more environmentally friendly. The development of hybrid systems that combine carbon-based materials with renewable energy technologies could significantly reduce the operational costs and environmental footprint of water treatment systems, making them more sustainable in the long term.

The environmental and economic implications of carbon-based hybrid materials are pivotal in determining their large-scale adoption. While these materials offer promising solutions for wastewater treatment, it is important to assess their ecological impact through life cycle assessments (LCAs). LCAs will help evaluate the overall environmental footprint of producing, using, and disposing of carbon-based hybrids. For instance, the energy consumption and waste generation during the synthesis of materials like graphene or CNTs need to be carefully considered. Additionally, the recyclability and disposal of these materials at the end of their life cycle must be examined to ensure that their use does not inadvertently create additional environmental problems, such as microplastic pollution or resource depletion. Economically, the implementation of carbon-based hybrids in water treatment systems may face initial high capital costs due to the expensive production processes. However, if the cost of production can be reduced through innovations in synthesis methods and the use of waste-derived materials, the long-term operational savings and environmental benefits could offset these initial costs. Policy frameworks that encourage the industrial adoption of carbon-based hybrids, such as providing subsidies or incentives for green technology adoption, could accelerate their integration into mainstream water treatment practices. Regulatory bodies could play a crucial role in setting standards for material performance, cost-effectiveness, and environmental sustainability, ensuring that these technologies are deployed in ways that maximize both their economic and ecological benefits.

In conclusion, while carbon-based hybrids show great potential in improving water and wastewater treatment, overcoming current limitations related to production costs, stability, and scalability will be crucial. Through continued innovation in synthesis methods, multifunctionality, and the integration of renewable energy sources, these materials can become a key component of future sustainable water treatment technologies. Furthermore, assessing their environmental and economic impacts will be essential for maximizing their global adoption and ensuring their long-term viability.

References

[1] Chong, Z.T., Soh, L.S., Yong, W.F., 2023. Valorization of agriculture wastes as biosorbents for adsorption of emerging pollutants: Modification, remediation and industry application. Results Eng. 17, 100960. https://doi.org/10.1016/j.rineng.2023.100960.

[2] Herrera-Morales, J., Turley, T.A., Betancourt-Ponce, M., Nicolau, E., 2019. Nanocellulose-block copolymer films for the removal of emerging organic contaminants from aqueous solutions. Materials 12, 230. https://doi.org/10.3390/ma12020230.

[3] Rout, P.R., Zhang, T.C., Bhunia, P., Surampalli, R.Y., 2021. Treatment technologies for emerging contaminants in wastewater treatment plants: A review. Sci. Total Environ. 753, 141990. https://doi.org/10.1016/j.scitotenv.2020.141990.

[4] dos Santos, D.F., Moreira, W.M., de Araújo, T.P., Bergamasco, R., Ostroski, I.C., de Barros, M.A.S.D., 2022. Non-conventional processes applied for the removal of pharmaceutics compounds in waters: A review. Process Saf. Environ. Prot. 167, 527–542. https://doi.org/10.1016/j.psep.2022.09.043.

[5] Fernandes, J., Ramísio, P.J., Puga, H., 2024. A comprehensive review on various phases of wastewater technologies: Trends and future perspectives. Eng 5, 2633–2661. https://doi.org/10.3390/eng5040138.

[6] Nasrollahzadeh, M., Sajjadi, M., Iravani, S., Varma, R.S., 2021. Carbon-based sustainable nanomaterials for water treatment: State-of-art and future perspectives. Chemosphere 263, 128005. https://doi.org/10.1016/j.chemosphere.2020.128005.

[7] Soffian, M.S., Abdul Halim, F.Z., Aziz, F., Rahman, M.A., Mohamed Amin, M.A., Awang Chee, D.N., 2022. Carbon-based material derived from biomass waste for wastewater treatment. Environ. Adv. 9, 100259. https://doi.org/10.1016/j.envadv.2022.100259.

[8] Kusworo, T.D., Dalanta, F., Aryanti, N., Othman, N.H., 2021. Intensifying separation and antifouling performance of PSf membrane incorporated by GO and ZnO

nanoparticles for petroleum refinery wastewater treatment. J. Water Process Eng. 41, 102030. https://doi.org/10.1016/j.jwpe.2021.102030.

[9] Anegbe, B., Ifijen, I.H., Maliki, M., Uwidia, I.E., Aigbodion, A.I., 2024. Graphene oxide synthesis and applications in emerging contaminant removal: A comprehensive review. Environ. Sci. Eur. 36, 15. https://doi.org/10.1186/s12302-023-00814-4.

[10] Deshwal, N., Singh, M.B., Bahadur, I., Kaushik, N., Kaushik, N.K., Singh, P., Kumari, K., 2023. A review on recent advancements on removal of harmful metal/metal ions using graphene oxide: Experimental and theoretical approaches. Sci. Total Environ. 858, 159672. https://doi.org/10.1016/j.scitotenv.2022.159672.

[11] Qin, X., Zhao, K., Quan, X., Cao, P., Chen, S., Yu, H., 2021. Highly efficient metal-free electro-Fenton degradation of organic contaminants on a bifunctional catalyst. J. Hazard Mater. 416, 125859. https://doi.org/10.1016/j.jhazmat.2021.125859.

[12] Hama Aziz, K.H., Fatah, N.M., Muhammad, K.T., 2024. Advancements in application of modified biochar as a green and low-cost adsorbent for wastewater remediation from organic dyes. R. Soc. Open Sci. 11, 232033. https://doi.org/10.1098/rsos.232033.

[13] Samuel Olugbenga, O., Goodness Adeleye, P., Blessing Oladipupo, S., Timothy Adeleye, A., Igenepo John, K., 2024. Biomass-derived biochar in wastewater treatment- a circular economy approach. Waste Manag. Bull. 1 (4), 1–14. https://doi.org/10.1016/j.wmb.2023.07.007.

[14] Mbayachi, V.B., Ndayiragije, E., Sammani, T., Taj, S., Mbuta, E.R., Khan, A. ullah, 2021. Graphene synthesis, characterization and its applications: A review. Results Chem. 3, 100163. https://doi.org/10.1016/j.rechem.2021.100163.

[15] Jahan, N., Roy, H., Reaz, A.H., Arshi, S., Rahman, E., Firoz, S.H., Islam, M.S., 2022. A comparative study on sorption behavior of graphene oxide and reduced graphene oxide towards methylene blue. Case Stud. Chem. Environ. Eng. 6, 100239. https://doi.org/10.1016/j.cscee.2022.100239.

[16] Fu, H., Gray, K.A., 2024. Science of the total environment graphene-encapsulated nanocomposites: Synthesis , environmental applications, and future prospects. Sci. Total Environ. 955, 176753.

[17] Gonçalves, N.P.F., Lourenço, M.A.O., Baleuri, S.R., Bianco, S., Jagdale, P., Calza, P., 2022. Biochar waste-based ZnO materials as highly efficient photocatalysts for water treatment. J. Environ. Chem. Eng. 10, 107256. https://doi.org/10.1016/j.jece.2022.107256.

[18] Weidner, E., Karbassiyazdi, E., Altaee, A., Jesionowski, T., Ciesielczyk, F., 2022. Hybrid metal oxide/biochar materials for wastewater treatment technology: A review. ACS Omega 7, 27062–27078. https://doi.org/10.1021/acsomega.2c02909.

[19] Nguyen, H.T.V., Ngo, T.H.A., Do, K.D., Nguyen, M.N., Dang, N.T.T., Nguyen, T.T.H., Vien, V., Vu, T.A., 2019. Preparation and characterization of a hydrophilic polysulfone membrane using graphene oxide. J. Chem. 2019, 15–20. https://doi.org/10.1155/2019/3164373.

[20] Ibrahim, Z.Z., Hamidi, H., Afzal, W., Huseyin, M., 2024. Investigating the effects of nano-Fe_3O_4 and MWCNTs on the filtration and rheological properties of water-based muds at elevated temperature and pressure. Colloids Surf. A Physicochem. Eng. Asp. 701, 134912. https://doi.org/10.1016/j.colsurfa.2024.134912.

[21] Liu, H., Wang, C., Qin, Y., Huang, Y., Xiao, C., 2021. Oriented structure design and evaluation of Fe3O4/o-MWCNTs/PVC composite membrane assisted by magnetic field. J. Taiwan Inst. Chem. Eng. 120, 278–290. https://doi.org/10.1016/j.jtice.2021.02.031.

[22] Che, W., Xiao, Z., Wang, Z., Li, J., Wang, H., Wang, Y., Xie, Y., 2019. Wood-based mesoporous filter decorated with silver nanoparticles for water purification.

ACS Sustain. Chem. Eng. 7, 5134–5141. https://doi.org/10.1021/acssuschemeng.8b06001.
[23] Tomczyk, A., Szewczuk-Karpisz, K., 2022. Effect of biochar modification by vitamin C, hydrogen peroxide or silver nanoparticles on its physicochemistry and tetracycline removal. Materials 15 (15), 5379. https://doi.org/10.3390/ma15155379.
[24] Nayl, A.E.A.A., Abd-Elhamid, A.I., Arafa, W.A.A., Ahmed, I.M., El-Shanshory, A. A., Abu-Saied, M.A., Soliman, H.M.A., Abdelgawad, M.A., Ali, H.M., Bräse, S., 2022. Chitosan-functionalized-graphene oxide (GO@CS) beads as an effective adsorbent to remove cationic dye from wastewater. Polymers 14, 4236. https://doi.org/10.3390/polym14194236.
[25] Sabzevari, M., Cree, D.E., Wilson, L.D., 2018. Graphene oxide-chitosan composite material for treatment of a model dye effluent. ACS Omega 3, 13045–13054. https://doi.org/10.1021/acsomega.8b01871.
[26] Daramola, M.O., Hlanyane, P., Sadare, O.O., Oluwasina, O.O., Iyuke, S.E., 2017. Performance of carbon nanotube/polysulfone (CNT/Psf) composite membranes during oil–water mixture separation: Effect of CNT dispersion method. Membranes 7, 14. https://doi.org/10.3390/membranes7010014.
[27] Rameetse, M.S., Aberefa, O., Daramola, M.O., 2020. Effect of loading and functionalization of carbon nanotube on the performance of blended polysulfone/polyethersulfone membrane during treatment of wastewater containing phenol and benzene. Membranes 10, 54. https://doi.org/10.3390/membranes10030054.
[28] Kaur, M., Kaur, H., 2022. Synergistic effect of biochar impregnated with ZnO nanoflowers for effective removal of organic pollutants from wastewater. Appl. Surf. Sci. Adv. 12, 100339. https://doi.org/10.1016/j.apsadv.2022.100339.
[29] Dong, X., Chu, Y., Tong, Z., Sun, M., Meng, D., Yi, X., Gao, T., Wang, M., Duan, J., 2024. Mechanisms of adsorption and functionalization of biochar for pesticides: A review. Ecotoxicol. Environ. Saf. 272, 116019. https://doi.org/10.1016/j.ecoenv.2024.116019.
[30] Speranza, G., 2019. The role of functionalization in the applications of carbon materials: An overview. J. Carbon Res. 5, 84. https://doi.org/10.3390/c5040084.
[31] Bokov, D., Turki Jalil, A., Chupradit, S., Suksatan, W., Javed Ansari, M., Shewael, I. H., Valiev, G.H., Kianfar, E., 2021. Nanomaterial by sol-gel method: Synthesis and application. Adv. Mater. Sci. Eng. 1. https://doi.org/10.1155/2021/5102014.
[32] Escorcia-Díaz, D., García-Mora, S., Rendón-Castrillón, L., Ramírez-Carmona, M., Ocampo-López, C., 2023. Advancements in nanoparticle deposition techniques for diverse substrates: A review. Nanomaterials 13, 2586. https://doi.org/10.3390/nano13182586.
[33] Chakravorty, A., Roy, S., 2024. A review of photocatalysis, basic principles, processes, and materials. Sustain. Chem. Environ. 8, 100155. https://doi.org/10.1016/j.scenv.2024.100155.
[34] Yang, Y., Zhong, Z., Li, J., Du, H., Li, Z., 2022. Efficient with low-cost removal and adsorption mechanisms of norfloxacin, ciprofloxacin and ofloxacin on modified thermal kaolin: Experimental and theoretical studies. J. Hazard Mater. 430, 128500. https://doi.org/10.1016/j.jhazmat.2022.128500.
[35] Zhong, C., Zhang, S., Yang, S., Yuan, B., Xu, Q., Xie, Z., Du, C., 2023. Mo and N co-doped iron biochar materials activating peroxymonosulfate for enhanced degradation of bisphenol A: Mechanism discussion and practical application. Chem. Eng. J. 466, 143298. https://doi.org/10.1016/j.cej.2023.143298.
[36] Viotti, P., Marzeddu, S., Antonucci, A., Décima, M.A., Lovascio, P., Tatti, F., Boni, M.R., 2024. Biochar as alternative material for heavy metal adsorption from groundwaters: Lab-scale (column) experiment review. Materials 17, 809. https://doi.org/10.3390/ma17040809.

[37] Qin, Y., Chai, B., Wang, C., Yan, J., Fan, G., Song, G., 2022. Removal of tetracycline onto KOH-activated biochar derived from rape straw: Affecting factors, mechanisms and reusability inspection. Colloids Surf. A Physicochem. Eng. Asp. 640, 128466. https://doi.org/10.1016/j.colsurfa.2022.128466.
[38] Zhu, Z., Wang, W., Qi, D., Luo, Y., Liu, Y., Xu, Y., Cui, F., Wang, C., Chen, X., 2018. Calcinable polymer membrane with revivability for efficient oily-water remediation. Adv. Mater. 30 (30), 1–8. https://doi.org/10.1002/adma.201801870.
[39] Abbas, N., Shao, G.N., Haider, M.S., Imran, S.M., Park, S.S., Kim, H.T., 2016. Sol–gel synthesis of TiO2-Fe2O3 systems: Effects of Fe2O3 content and their photocatalytic properties. J. Ind. Eng. Chem. 39, 112–120. https://doi.org/10.1016/j.jiec.2016.05.015.
[40] El Mouchtari, E.M., El Mersly, L., Belkodia, K., Piram, A., Lebarillier, S., Briche, S., Rafqah, S., Wong-Wah-Chung, P., 2023. Sol-gel synthesis of new TiO2 ball/activated carbon photocatalyst and its application for degradation of three hormones: 17α-ethinylestradiol, estrone, and β-estradiol. Toxics 11, 299. https://doi.org/10.3390/toxics11040299.
[41] Davies, G., McGregor, J., 2021. Hydrothermal synthesis of biomass-derived magnetic carbon composites for adsorption and catalysis. ACS Omega 6, 33000–33009. https://doi.org/10.1021/acsomega.1c05116.
[42] Zeng, H., Xu, K., Wang, F., Sun, S., Li, D., Zhang, J., 2021. Preparation of adsorbent based on water treatment residuals and chitosan by homogeneous method with freeze-drying and its As(V) removal performance. Int. J. Biol. Macromol. 184, 313–324. https://doi.org/10.1016/j.ijbiomac.2021.06.032.
[43] Ahmadi Bonakdar, M., Rodrigue, D., 2024. Electrospinning: Processes, structures, and materials. Macromolecules 4, 58–103. https://doi.org/10.3390/macromol4010004.
[44] Sheoran, K., Kaur, H., Siwal, S.S., Saini, A.K., Vo, D.V.N., Thakur, V.K., 2022. Recent advances of carbon-based nanomaterials (CBNMs) for wastewater treatment: Synthesis and application. Chemosphere 299, 134364. https://doi.org/10.1016/j.chemosphere.2022.134364.
[45] Wu, H., Zhang, Q., Xu, B., Liu, X., Lin, Y., Jiang, N., Yao, H., Tang, Y., Wang, L., Yu, L., Yang, Z., Wang, X., 2024. In-situ interfacial polymerization of polyamide TFN membranes by adding a diamino-silane coupling agent: Toward enhanced desalination performance. Desalination 580, 117508. https://doi.org/10.1016/j.desal.2024.117508.
[46] Scaffaro, R., Mistretta, M.C., Balsamo, M., 2024. Innovative 3D-printed devices for water pollutant removal: Comprehensive review on printing parameters, composition, properties and performances of the latest 3D-systems. Polym. Test. 140, 108627. https://doi.org/10.1016/j.polymertesting.2024.108627.
[47] Yusaf, T., Mahamude, A.S.F., Farhana, K., Harun, W.S.W., Kadirgama, K., Ramasamy, D., Kamarulzaman, M.K., Subramonian, S., Hall, S., Dhahad, H.A., 2022. A comprehensive review on graphene nanoparticles: Preparation, properties, and applications. Sustainability 14, 12336. https://doi.org/10.3390/su141912336.
[48] Fatema, K.N., Sagadevan, S., Cho, J.Y., Jang, W.K., Oh, W.-C., 2022. Graphene-based nanocomposite using new modeling molecular dynamic simulations for proposed neutralizing mechanism and real-time sensing of COVID-19. Nanotechnol. Rev. 11 (1), 1555–1569. https://doi.org/10.1515/ntrev-2022-0093.
[49] Shinde, P.A., Ukarde, T.M., Gogate, P.R., Pawar, H.S., 2021. An integrated approach of adsorption and membrane separation for treatment of sewage water and resource recovery. J. Water Process Eng. 40, 101795. https://doi.org/10.1016/j.jwpe.2020.101795.

[50] Nordin, A.H., Norfarhana, A.S., Noor, S.F.M., Paiman, S.H., Nordin, M.L., Husna, S.M.N., Ilyas, R.A., Ngadi, N., Bakar, A.A., Ahmad, Z., Azami, M.S., Nawawi, W.I., Nabgan, W., 2023. Recent advances in using adsorbent derived from agricultural waste for antibiotics and non-steroidal anti-inflammatory wastewater treatment: A review. Separations 10, 300. https://doi.org/10.3390/separations10050300.
[51] Ogunleye, D.T., Akpotu, S.O., Moodley, B., 2023. Crystalline nanocellulose anchored on reduced graphene oxide for the removal of pharmaceuticals from aqueous systems: Adsorbent characterization and adsorption performance. ChemistrySelect 8 (4), e202202533. https://doi.org/10.1002/slct.202202533.
[52] Izan, N.R., Zainol, M.M., Nordin, A.H., Asmadi, M., Wong, S.L., Azhar, M.A.I., Alias, N.H., 2022. Removal of methylene blue via adsorption using magnetic char derived from food waste. Malays. J. Chem. 24 (2), 283–292.
[53] Saeedi-Jurkuyeh, A., Jafari, A.J., Kalantary, R.R., Esrafili, A., 2020. A novel synthetic thin-film nanocomposite forward osmosis membrane modified by graphene oxide and polyethylene glycol for heavy metals removal from aqueous solutions. React. Funct. Polym. 146, 104397. https://doi.org/10.1016/j.reactfunctpolym.2019.104397.
[54] Wang, R.-F., Deng, L.-G., Li, K., Fan, X.-J., Li, W., Lu, H.-Q., 2020. Fabrication and characterization of sugarcane bagasse–calcium carbonate composite for the efficient removal of crystal violet dye from wastewater. Ceram. Int. 46 (17), 27484–27492. https://doi.org/10.1016/j.ceramint.2020.07.237.
[55] Bai, L., Bossa, N., Qu, F., Winglee, J., Li, G., Sun, K., Liang, H., Wiesner, M.R., 2017. Comparison of hydrophilicity and mechanical properties of nanocomposite membranes with cellulose nanocrystals and carbon nanotubes. Environ. Sci. Technol. 51 (1), 253. https://doi.org/10.1021/acs.est.6b04280.
[56] Kausar, A., Zohra, S.T., Ijaz, S., Iqbal, M., Iqbal, J., Bibi, I., Nouren, S., El Messaoudi, N., Nazir, A., 2023. Cellulose-based materials and their adsorptive removal efficiency for dyes: A review. Int. J. Biol. Macromol. 224, 1337–1355. https://doi.org/10.1016/j.ijbiomac.2022.10.220.
[57] Nordin, A.H., Ahmad, Z., Husna, S.M.N., Ilyas, R.A., Azemi, A.K., Ismail, N., Nordin, M.L., Ngadi, N., Siti, N.H., Nabgan, W., Norfarhana, A.S., Azami, M.S.M., 2023. The state of the art of natural polymer functionalized fe3o4 magnetic nanoparticle composites for drug delivery applications: A review. Gels 9, 121. https://doi.org/10.3390/gels9020121.
[58] Norfarhana, A.S., Ilyas, R.A., Ngadi, N., 2022. A review of nanocellulose adsorptive membrane as multifunctional wastewater treatment. Carbohydr. Polym. 291, 119563. https://doi.org/10.1016/j.carbpol.2022.119563.
[59] Bassyouni, M., Abdel-Aziz, M.H., Zoromba, M.S., Abdel-Hamid, S.M.S., Drioli, E., 2019. A review of polymeric nanocomposite membranes for water purification. J. Ind. Eng. Chem. 73, 19–46. https://doi.org/10.1016/j.jiec.2019.01.045.
[60] Ursino, C., Castro-Muñoz, R., Drioli, E., Gzara, L., Albeirutty, M.H., Figoli, A., 2018. Progress of nanocomposite membranes for water treatment. Membranes 8 (2), 1–40. https://doi.org/10.3390/membranes8020018.
[61] Bai, M., Qiang, L., Meng, M., Li, B., Wang, S., Wu, Y., Chen, L., Dai, J., Liu, Y., Pan, J., 2021. Upper surface imprinted membrane prepared by magnetic guidance phase inversion method for highly efficient and selective separation of Artemisinin. Chem. Eng. J. 405, 126899. https://doi.org/10.1016/j.cej.2020.126899.
[62] Yin, X., He, Y., Wang, Y., Yu, H., Chen, J., Gao, Y., 2020. Bio-inspired antifouling cellulose nanofiber multifunctional filtration membrane for highly efficient emulsion separation and application in water purification. Kor. J. Chem. Eng. 37 (10), 1751–1760. https://doi.org/10.1007/s11814-020-0568-4.

[63] Zhu, W., Han, M., Kim, D., Zhang, Y., Kwon, G., You, J., Jia, C., Kim, J., 2022. Facile preparation of nanocellulose/Zn-MOF-based catalytic filter for water purification by oxidation process. Environ. Res. 205, 112417. https://doi.org/10.1016/j.envres.2021.112417.
[64] Melchor-Durán, A.P., Moreno-Virgen, M.R., Bonilla-Petriciolet, A., Reynel-Ávila, H.E., Lucio Ortiz, E., González-Vázquez, O.F., 2024. Heavy metal removal from water using graphene oxide in magnetic-assisted adsorption systems: Characterization, adsorption properties, and modelling. Separations 11, 294. https://doi.org/10.3390/separations11100294.
[65] Bandara, T., Xu, J., Potter, I.D., Franks, A., Chathurika, J.B.A.J., Tang, C., 2020. Mechanisms for the removal of Cd(II) and Cu(II) from aqueous solution and mine water by biochars derived from agricultural wastes. Chemosphere 254, 126745. https://doi.org/10.1016/j.chemosphere.2020.126745.
[66] Athari, M., Fattahi, M., Khosravi-Nikou, M., Hajhariri, A., 2022. Adsorption of different anionic and cationic dyes by hybrid nanocomposites of carbon nanotube and graphene materials over UiO-66. Sci. Rep. 12, 20415. https://doi.org/10.1038/s41598-022-24891-2.
[67] Liu, P., Zhu, C., Mathew, A.P., 2019. Mechanically robust high flux graphene oxide - Nanocellulose membranes for dye removal from water. J. Hazard Mater. 371, 484–493. https://doi.org/10.1016/j.jhazmat.2019.03.009.
[68] Li, J., Liu, H., Qin, Y., Sun, Y., Wang, C., Zhao, J., Huang, Q., Xiao, C., 2024. Preparation and performance of magnetic carbon nanotubes modified PVC substrate composite nanofiltration membranes. J. Environ. Chem. Eng. 12, 112273. https://doi.org/10.1016/j.jece.2024.112273.
[69] Kocijan, M., Ćurkoví, L., Radoševí, T., Podlogar, M., 2021. Enhanced photocatalytic activity of hybrid rGO@TiO2/CN nanocomposite for organic pollutant degradation under solar light irradiation. Catalysts 11, 1023.
[70] Yang, H., Zhou, M., Yang, W., Ren, G., Ma, L., 2018. Rolling-made gas diffusion electrode with carbon nanotube for electro-Fenton degradation of acetylsalicylic acid. Chemosphere 206, 439–446. https://doi.org/10.1016/j.chemosphere.2018.05.027.
[71] Ma, C., Aryee, A.A., Zhu, K., Wang, R., Han, R., 2024. Adsorption and catalytic degradation of tetracycline hydrochloride by HCNTs/MnFe2O4. J. Environ. Chem. Eng. 12, 113156. https://doi.org/10.1016/j.jece.2024.113156.
[72] Humudat, Y.R., Abdul-Majeed, M.A., Khraibet, A.C., 2024. Membrane filtration enhanced by magnetic field for reducing endotoxin from dialysis water. Desalination Water Treat. 318, 100315. https://doi.org/10.1016/j.dwt.2024.100315.
[73] Goel, G., Hélix-Nielsen, C., Upadhyaya, H.M., Goel, S., 2021. A bibliometric study on biomimetic and bioinspired membranes for water filtration. NPJ Clean Water 4, 41. https://doi.org/10.1038/s41545-021-00131-4.
[74] Phan, H.T., Nguyen, K.D., Nguyen, H.H.M., Dao, N.T., Le, P.T.K., Le, H.V., 2023. Nata de coco as an abundant bacterial cellulose resource to prepare aerogels for the removal of organic dyes in water. Bioresour. Technol. Rep. 24, 101613. https://doi.org/10.1016/j.biteb.2023.101613.
[75] Choudhary, B., Paul, D., 2018. Isotherms, kinetics and thermodynamics of hexavalent chromium removal using biochar. J. Environ. Chem. Eng. 6 (2), 2335–2343. https://doi.org/10.1016/j.jece.2018.03.028.
[76] Madduri, S.B., Kommalapati, R.R., 2024. Harnessing novel reduced graphene oxide-based aerogel for efficient organic contaminant and heavy metal removal in aqueous environments. Nanomaterials 14, 1708. https://doi.org/10.3390/nano14211708.
[77] Tsade Kara, H., Anshebo, S.T., Sabir, F.K., Adam Workineh, G., 2021. Removal of methylene blue dye from wastewater using periodiated modified nanocellulose. Int. J. Chem. Eng. 2021, 9965452. https://doi.org/10.1155/2021/9965452.

[78] Zhang, D., He, Q., Hu, X., Zhang, K., Chen, C., Xue, Y., 2021. Enhanced adsorption for the removal of tetracycline hydrochloride (TC) using ball-milled biochar derived from crayfish shell. Colloids Surf. A Physicochem. Eng. Asp. 615 (2020), 126254. https://doi.org/10.1016/j.colsurfa.2021.126254.
[79] Karkooti, A., Yazdi, A.Z., Chen, P., McGregor, M., Nazemifard, N., Sadrzadeh, M., 2018. Development of advanced nanocomposite membranes using graphene nanoribbons and nanosheets for water treatment. J. Membr. Sci. 560, 97–107. https://doi.org/10.1016/j.memsci.2018.04.034.
[80] Jaspal, D., Malviya, A., El Allaoui, B., Zari, N., Bouhfid, R., El Kacem Qaiss, A., Bhagwat, S., 2023. Emerging advances of composite membranes for seawater pretreatment: A review. Water Sci. Technol. 88 (2), 408–429. https://doi.org/10.2166/wst.2023.220.

CHAPTER 9

The application of carbon-based materials and composites in point-of-use water filters

Novia Amalia Sholeha[1], Anees Ameera Fauzi[2], Abdul Wafi[3] and Nurul Sahida Binti Hassan[4]

[1]College of Vocational Studies, Bogor Agricultural University (IPB University), Bogor, Indonesia; [2]Faculty of Chemical and Energy Engineeing, Universiti Teknologi Malaysia, Johor, Malaysia; [3]Research Center for Food Technology and Processing, National Research and Innovation Agency (BRIN), Playen, Gunungkidul, Yogyakarta, Indonesia; [4]Centre of Hydrogen Energy, Institute of Future Energy, Johor, Malaysia

1. Introduction

Water is universally acknowledged as a vital worldwide substance that contributes to the advancement of sustainable development and is critical to various aspects of human existence [1–3]. Although water encompasses about 70% on the surface of Earth, only a limited fraction of it is accessible for practical use [4,5]. The problem of water scarcity is rapidly becoming a global concern. For almost 70% of population in the world would encounter the difficulties associated with water shortage by 2025, despite the fact that 50% of the global population lives within a distance of 200 km near the coastline [6]. The sixth sustainable development goal (SDG) of the United Nations emphasizes the importance of providing accessible and secure water services. Water shortage remains a significant problem characterized by inadequate water supply, resulting in restricted availability of safe and clean water from 2015 to 2022, as depicted in Fig. 9.1 from the UN's World Water Development Report 2024. Climate change, population increase, technological advancements, increasing water consumption, and limited water supplies persistently contribute to water stress, causing detrimental effects on both water quality and quantity [7,8]. In addition, the scarcity issue is exacerbated by water contamination resulting from industrial activity, agricultural runoff, and insufficient waste

Carbon-Based Materials and Composites for Water and Wastewater Treatment
ISBN 978-0-443-33681-2
https://doi.org/10.1016/B978-0-443-33681-2.00014-4

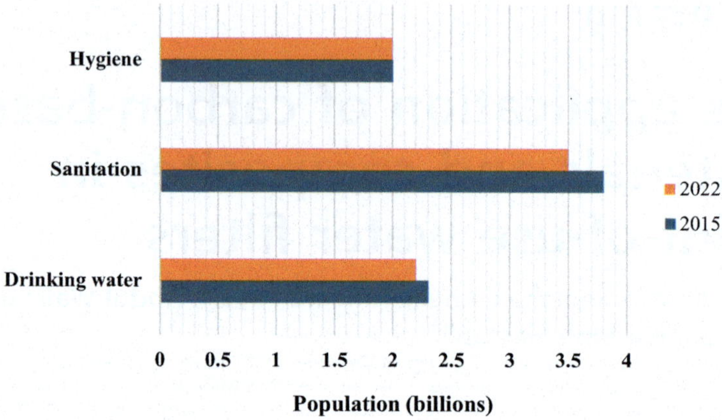

Figure 9.1 Global populations of cities and villages without access to properly regulated drinking water, regulated sanitation, and fundamental hygiene services, from 2015/17 to 2022. *(Source: UN World Water Development Report 2024 © 2023 United Nations. Reprinted with the permission of the United Nations).*

treatment. Various sources are available for the purpose of treating drinking water, including recycled/reclaimed water, surface water, and groundwater, and reclaimed. The susceptibility of these sources of water to various pollutants, such as silts, parasites, viruses, and bacteria is high [9,10]. The contamination characteristics of these pollutants exhibit substantial variations depending on the particular characteristics of the originating water [11].

Point-of-use (POU) filtration of water systems offer an environmental-friendly resolution for people lacking the availability of clean water, enabling people to effectively purify water within their own residences [12]. Various types of filtration systems can be directly installed onto home water sources. These systems include bio-sand filters, ceramic water filters, pitchers or bottles, faucet-mounted filters, under-the-sink filters, membrane filters, wood filters, bucket filters, and carbon-its composite filters, among others [6,13–15]. The systems possess a user-friendly interface, entail reduced initial expenses, exhibit simplified installation procedures, need less maintenance, and operate autonomously from the power grid. Furthermore, POU systems reduce the likelihood of re-contamination during the transportation and storage of water, as they treat water near the point of consumption [16]. Conversely, point-of-use (POU) systems have lower installed capacity, greater operating expenses, and less efficacy in eliminating some water pollutants like arsenic (As) and lead (Pb) with effectiveness decreasing to as low as 6.9% and 79.9% [17]. The point of use water filtrations are shown to significantly decrease the tap water's quality by acting as trapping or storage for sewage, scale, corrosion, algal blooms, or mucus deposits. These deposits promote the growth of microbes in household water distribution systems. Due to various performance difficulties, it is necessary to create durable POU filtration system of water that can be customized to accommodate a broad spectrum of source water conditions. An effective method to accomplish this is by implementing carbon and its composite in point-of-use (POU) water filter systems and by eliminating pollutants or emerging contaminants (ECs) that would otherwise remain in the water and wastewater after treatment [18–22]. The application of carbon materials in water filters has been published and illustrated in Fig. 9.2.

2. Process on drinking water treatment plant

Drinking water treatment plants (DWTPs) consist of a range of chemical, biological, and physical methods, predominantly due to source water's properties and the technology employed [23,24]. Drinking water treatment plants (DWTP) often utilize a blend of traditional and cutting-edge water treatment techniques to cleanse polluted water. These technologies include filtration, flocculation, water coagulation, sedimentation, desalination, disinfection, reverse osmosis (RO), nanofiltration (NF), and ultrafiltration (UF) depicted in Table 9.1 [27]. These techniques necessitate energy to

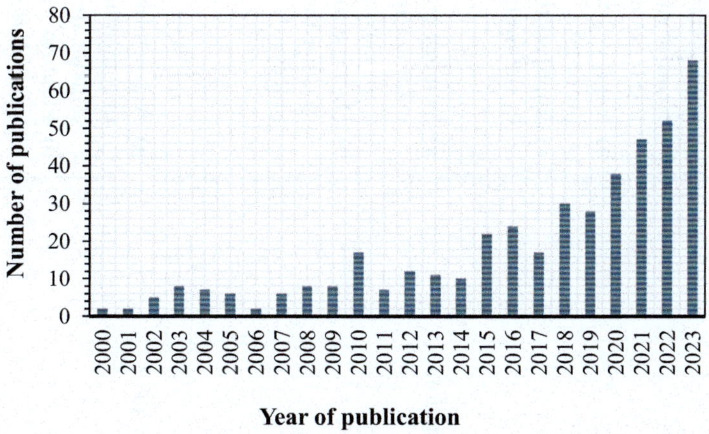

Figure 9.2 The publication of carbon for water filter from 2000–23. *(Sources: scopus.com with keywords of "carbon", "graphene oxide", "carbon nanotube", "water filter").*

Table 9.1 Detailed procedures in drinking water treatment plants (DWTPs) [23,25,26].

Process of treatments	Drawbacks	Advantages	Opportunities
Coagulation	**a)** The levels of natural organic compounds, pharmaceuticals, and PAHs are decreased. **b)** Improved the filtration procedure. **c)** Improves the longevity of the treatment system	**a)** Excessive need for energy. **b)** Agglomerates can disintegrate when subjected to vigorous mixing. **c)** The production of sludge results in further issues.	**a)** Utilizing advanced oxidation processes (AOPs) or membrane technologies in conjunction with coagulation. **b)** Exploring energy recuperation alternatives and renewable energy systems
Sedimentation	**a)** Minimal energy requirement resulting from gravitational forces. **b)** Decrease in the workload of other therapeutic processes.	**a)** The expenses of investment are high. **b)** A high flow velocity significantly decreases process efficiency. **c)** Generation of disinfection by-products (DBPs).	**a)** Enhance the efficiency of tank design for settling and improve the process of energy recovery. **b)** Incorporating membrane filtration (MF) or advanced oxidation processes (AOPs) and creating innovative sedimentation aids
Desalination	**a)** A dependable supply of uncontaminated water in difficult areas. **b)** Eliminate surplus salt and minerals from brackish or ocean.	**a)** Excessive need for energy. **b)** Generates brine, a very concentrated salt solution, as a secondary product. **c)** Demand substantial quantities of water for the purposes of cleaning and maintenance.	**a)** Utilizing cutting-edge RE and RO membrane technology that enhances both permeability and selectivity. **b)** Extensive study on forward osmosis and membrane distillation.
Filtration			

	a) Enhances chemical, microbiological, and physical water characteristics. **b)** Reduce turbidity by approximately 93%. **c)** System design that requires minimal maintenance. **d)** Can be used without specialized equipment	**a)** Exorbitant operation and maintenance expenses. **b)** Possibility of obstruction and decreased rate of fluid flow. **c)** Constraints in water pollution mitigation. **d)** Generates waste in the form of backwash filter media and water. **e)** Prone to mechanically malfunctions.	**a)** Employing machine learning algorithms to enhance the efficiency of filters, minimize energy consumption, and maximize energy recovery. **b)** Harnessing renewable energy sources to fulfill the demands of filtering operations.
Disinfection	**a)** Chlorination: Convenient to manage and cost-effective. **b)** Lingering toxicity of the wastewater.	**a)** Generating hazardous disinfection byproducts. **b)** Extended duration of contact. **c)** Highly corrosive.	**a)** Advancements in the creation of disinfection methods that consume less energy. **b)** Combining advanced oxidation processes (AOPs) and membrane filtration (MF) or electrochemical processes using renewable energy (RE).
Membrane technology	**a)** They produce a reduced amount of sludge and offer a high level of system flexibility. **b)** It is more convenient to have operational control.	**a)** Excessive need for energy. **b)** Installations are smaller in size. **c)** UF: Accumulation of substances on the	NF **a)** Creating membrane materials with improved selectivity and permeability.

Continued

Table 9.1 Detailed procedures in drinking water treatment plants (DWTPs) [23,25,26].—cont'd

Process of treatments	Drawbacks	Advantages	Opportunities
	c) Ozonation has a brief duration of contact. **d)** Membrane processes can be operated in a continuous manner. **e)** Ozone treatment reduces the intensity of watercolors and the amount of light absorbed. **f)** NF: Low-pressure operation, less need for extra treatment stages, and increased water recovery.	membrane surface that impairs its performance. **d)** Exorbitant administrative expenses. **e)** RO: Difficulties in disposing of brine waste. **f)** NF: Loss of water, fouling, deterioration in membrane efficiency, and selective removal of pollutants.	**b)** Advancement of intelligent monitoring and control systems for NF plants UF Conduct research on the impacts of backwashing, chemical cleaning, and feed water pre-treatment. b) Incorporating renewable energy sources. RO **a)** Integration with ultrafiltration (UF) or nanofiltration (NF). **b)** Investigating renewable energy sources for powering the reverse osmosis process.

purify water and render it potable by removing pollutants [28]. The global energy consumption of roughly 2.5% is dedicated to the purification of polluted water and resources, while energy usage contributes to 60% of the overall greenhouse gas emissions [25,26].

Coagulation is the act of introducing substances like coagulants or polymers into water to disrupt the stability of particles and promote the creation of bigger clusters called flocs [29]. The coagulation method typically utilizes mechanical agitation to facilitate mixing, which leads to the expenditure of energy to treat contaminants. The effectiveness of mixing is influenced by variables such as the dimensions and configuration of the mixing apparatus, the velocity at which mixing occurs, and the tank's structure of coagulation [25,26]. Higher energy consumption is typically necessary when there is an increase in blending speed and bigger container sizes, leading to an increase in carbon emissions. Furthermore, the primary source of energy consumption in the coagulation process is the use of chemicals, typically ranging from 0.4 to 0.7 kWh/m^3 [29].

Flocculation is a subsequent phase after coagulation that enhances the colliding and agglomeration of particles that have been unstable, resulting in the formation of larger flocs by introducing a coagulant such as alum or ferric chloride. Uniformly mixing the coagulant in water ensures that it is evenly distributed and can effectively interact with particles, so initiating the process of flocculation. To effectively treat water, the flocculation process necessitates the utilization of power at rate of rate of 0.012 kWh/m^3 to promote the blending and stirring. The water treatment plant in Tampa Bay, Florida had an average output over 250,000 cubic meters per day, for example [25,26,30]. Moreover, the quantity of energy utilized during the stage of sedimentation directly influences the speed when the particles descend and coalesce to create aggregates. The energy consumption in the sedimentation process is affected by various factors, particularly the configuration and functioning of sediment tanks. Consequently, sedimentation tanks in wastewater treatment facilities exhibit distinct energy requirements and differing degrees of efficacy [27]. Choosing the suitable filtration system, considering the energy input ranging from 0.005 to 0.014 kWh/m^3 [25,26], is an important factor to consider. Membrane filters, like filtration systems, may necessitate a substantial amount of energy for efficient operation due to its operational requirements and intricate design. The upkeep of the filtration system is also an additional component that impacts energy usage [31].

Various techniques for disinfection include ozonation, chlorination, and UV therapy. The chlorine exposure procedure entails the utilization of chloride in water to achieve effective disinfection by injectors or pumps [23,24]. These processes necessitate energy to provide enough flow velocities and forces for the procedure to occur. UV sterilization methods depend on energy usage to power the UV lights that emit the necessary radiation. Ozone surpasses chlorination in terms of disinfection effectiveness since it does not generate persistent byproducts, unlike chlorine [32].

Desalination is an effective treatment of water technique which necessitates a significant energy input for its execution that includes various techniques such as multi-effect distillation (MED), seawater reverse osmosis (SWRO), brackish water reverse osmosis (BWRO), reverse osmosis (RO), multistage flash distillation (MSF), thermal vapor compression (TVC), and mechanical vapor compression (MVC) [23,24]. Water at a plant for desalination is commonly subjected to electrical currents, high pressure, or elevated temperatures to attain the appropriate degree of purification. The amount of energy used is influenced by the quantity of water processed and the salinity degree of the originating water [33].

Reverse osmosis (RO) is a commonly utilized method for desalination that utilizes elevated pressure to compel water to pass through a partially permeable membrane [34]. Reverse osmosis (RO) employs pumps with high pressure to propel water across a semi-permeable membrane, whereas thermal techniques utilize heat to separate substances. A thorough investigation demonstrated that the SWRO approach may reduce energy consumption from a range of 3.5–4.0 kWh/m^3 to a precise figure of 2.1 kWh/m^3, indicating that traditional techniques for treatment are less energy-intensive than desalination methods [25,26,35].

Ultrafiltration (UF) is a membrane technique that operates at low pressure and is frequently implemented as an alternative to sedimentation and medium filtration [36]. It typically necessitates pre-treatment procedures such as coagulation and flocculation. The energy required for the blending and transportation of chemicals during pre-treatment might add to the overall energy demands of the treatment process. Utilizing membranes that have narrow pore diameters, experiencing membrane fouling, and operating on a big scale might result in increased energy requirements [36,37].

The point-of-use water filter effectively removes many substances including dyes, heavy metals, pharmaceutical waste, organic pollutants, and emerging contaminants (ECs) as listed in Table 9.2. Various contaminants could infiltrate the environment and can cause detrimental effects on

ecosystems and human health, even when present in extremely low amounts. The range concentrations of contaminants in water were microgrammes per liter (μg/L) or nanograms per liter (ng/L) [40,41]. The removal of contaminants from water using carbon primarily occurs through three mechanisms: i) Adsorption on its outer surface of based on carbon materials, ii) removing the particle pollution through physical filtration, and iii) facilitating the natural breakdown of the degradable component by forming an active biofilm on the filter's carbon surface [42]. The efficiency of pollutant elimination is contingent upon the adsorption capability of the carbon, together with parameters such as the quality of the source water, the existence of competing species, and numerous operational variables like flow rate, contact length, and filter bed depth [38,39].

3. Carbon-based materials and composites in water filtration system

Carbon-based materials i.e., activated carbon provide advantageous properties i.e., broad light absorption range, extensive specific surface area, high conductivity, adjustable architectures, and stable chemical properties [6,43]. Additionally, they are cost-effective, easily accessible, and environmentally sustainable. Graphene possesses a diverse spectrum of π-π conjugate structure and a notable theoretical particular surface area of $2600\ m^2\ g^{-1}$, both of which are beneficial for the absorption of organic pollutants and the improvement of catalytic activity [44]. Carbon-based materials possess unique and exceptional characteristics that can greatly improve the photoelectrochemical performance of semiconductors and achieve extraordinary catalytic efficiency and quantum efficiency [45]. Carbon-based compounds function as photosensitizers, expanding the spectrum of light absorption and enhancing light absorption. (iv) It serves as a stabilizer to prevent the sticking, isolation, and rusting of semiconductors; (v) It operates as a semiconductor to generate electron-hole pairs or create semiconductor-carbon (S–C) hetero-junctions to improve the separation of electron-hole pairs [20,46].

Advanced techniques, including chemical vapor deposition, hydrothermal, solvothermal, and microwave-assisted synthesis, can be utilized to manufacture carbon-based materials. These procedures require either assembling small components or disassembling bigger structures [47]. The use of microwave radiation to synthesize carbon-based compounds is a very effective, fast, ecologically friendly, and uncomplicated technology, in contrast to conventional approaches. The microwave facilitates and accelerates the process, often resulting in higher relative yields. The synthesis of carbon-based nanomaterials using microwave assistance necessitates low temperatures for development and provides significant advantages such as large surface areas, high adsorption capabilities, and purity, as well as energy efficiency [48,49]. Therefore, it offers a hopeful method for tackling the difficulties faced in industrial and municipal water treatment. Table 9.3 delineates the applications of carbon as water filtration systems.

Carbon nanotubes (CNTs) possess a sp^2 hybridized structure composed of carbon hexagons arranged in a concentric manner with a hollow cylindrical shape and exhibit 1:132,000,000 for diameter/length ratio [60]. A diverse range of techniques, such as arc-discharge method, chemical vapor deposition, laser ablation, and several others, were employed to generate the CNTs [61]. Carbon nanotubes (CNTs) have been used as photoelectrode materials in photoelectrochemical (PEC) water splitting because to its large surface area, capacity to maintain high constant current densities, excellent mechanical properties, high conductivity, and exceptional chemical and thermal durability. Furthermore, carbon nanotubes (CNTs) enhance the ability of semiconductive thin films to respond to light and have the capacity to retain significant specific surface areas. Their strong structural integrity and lack of reactivity make them very resistant to oxidation [14,62,63].

Graphene is an ideal choice for a coating or covering material due to its ability to provide substantial stability and protection to the photoanode material, such as the process of splitting water using sunlight [64]. Graphene is a solitary layer of carbon atoms that are sp^2-hybridized. This layer has a high level of electronic mobility due to the feeble interaction between electrons and phonons and electron delocalization. Graphene sheets enable them to efficiently accept and release electrons from excited semiconductors, hinder charge recombination, enhance interfacial charge transfer, provide a multitude of active adsorption sites, and provide additional photocatalytic reaction centers [65].

$g\text{-}C_3N_4$ (Graphitic Carbon Nitride)materials are exceptional heterogeneous catalysts and photocatalysts for a diverse array of processes due to their cost-effectiveness, exceptional stability, favorable energy-band position, and adjustable band gap. $g\text{-}C_3N_4$ has recently attracted significant attention as a potential material for the next generation of photoelectrodes. This is primarily attributable to its unique characteristics, which include an appropriate band structure for water splitting, a specific spectrum of visible light absorption, non-toxicity, and exceptional stability. In the longer

Table 9.3 Carbon and its composite for water filter and wastewater treatment.

Materials	Activity	Ref.
Heavy metals removal and photocatalyst		
Carbon/iron composite FexOy-BC (RM) from red mud	Efficiency of Cd (II) removal is 92.59%.	[5]
Carbon/sand from polystyrene foam waste	Water desalination process in maximum water collection rate was 1.51 $kg\,m^{-2}\,h^{-1}$-.	[1]
Carbon/zeolite from polystyrene foam waste	Water desalination process in maximum water collection rate was 1.24 $kg\,m^{-2}\,h^{-1}$.	
Xylan derived carbon dots (XMR-CDs)	The photothermal conversion process demonstrates an outstanding photothermal conversion rate of 3.3895 $kg\,m^{-2}h^{-1}$ and fast detection of Hg^{2+}.	[50]
Poly-N-phenylglycine @multi-walled carbon nanotubes	The adsorption capacity of Au (III) reached 1262.6 mg g^{-1} at a pH of 1.	[51]
Dyes removal and photocatalyst		
C-quantum dots/TiO_2	The degrading efficiency of RhB after the fifth cycle run was approximately 75%.	[52]
Activated carbon spheres from maize cobs	The Langmuir model of rose bengal dye removal had a q_{max} of 303.6 mg/g.	[3]
MOF/Carbon from wood	The removal efficiency is 99.98% of Congo red.	[53]
Reduced graphene oxide/carbon nitride	The degradation rates for methylene blue and rhodamine B were measured to be 99.20% and 91.07% respectively.	[54]

Continued

Table 9.3 Carbon and its composite for water filter and wastewater treatment.—cont'd

Materials	Activity	Ref.
Organic pollutant removal		
Activated carbon	The procedure effectively eliminated 87% of the total organic carbon (TOC) from actual high-concentration pharmaceutical effluent (starting TOC of 1625 mg/L).	[17]
Activated carbon from spent mushroom substrate	The percentage of contaminants removed from raw sewage water was around 84%, while the percentage of contaminants removed from treated sewage water was approximately 68%.	[55]
Biological oxygen-dosed activated carbon (BODAC)	Over the course of the 2-year observation period, the filters continuously eliminated certain organic micropollutants (OMPs) including hydrochlorothiazide, metoprolol, sotalol, and trimethoprim, with an efficacy of at least 70%.	[56]
Others (microplastics and pharmaceuticals waste)		
2D graphene and MXenes	The removal of microplastics (MPs) was 70%–99%	[57]
Activated carbon from biomass pollen typhae	The removal of ciprofloxacin was 93.1% and has ultra-high flux of 1200 L m^{-2} h^{-1} bar^{-1}.	[58]

Table 9.3 Carbon and its composite for water filter and wastewater treatment.—cont'd

Materials	Activity	Ref.
Cu nanoparticles/carbon from coconut coir	The Cu-ACC nanohybrid material demonstrated the ability to effectively eliminate 63% of fluoride and 69% of hardness in synthetic water.	[59]
Carbon-nanotube/ $Ni@TiO_2$:W	The diclofenac (DCF) anti-inflammatory medication was completely degraded through photocatalysis under ultraviolet-visible light.	[24]

wavelength ranges, the defects include a reduced electronic conductivity, an increased rate of recombination of photogenerated charges, and restricted absorption of visible light [65].

CQDs (Carbon quantum dots) have recently been employed in a variety of optoelectronic devices, including solar cells, photo redox catalysis, biosensors, batteries, and luminescent solar concentrators, due to their exceptional optical and electrical specifications. Recently, CQDs have been employed to increase the light absorption of metal oxides to generate hydrogen [66]. The efficiency of converting solar thermal energy (STH) in the modified devices can be substantially improved by either improving charge separation and transport or enhancing sunlight absorption [67].

4. Future perspective

Developing countries possess significant amounts of agricultural residues and biomass wastes that can be utilized to produce carbon-based materials for point-of-use water purification. Improving the design of water filtration technologies for point-of-use applications necessitates careful consideration of both suitability and cost. A substantial portion of individuals residing in low-income countries, impacted by water contaminated with heavy metals, often lack the financial means to purchase costly products. Investigations are necessary to explore cost-effective and environmentally sustainable methods for the disposal of contaminated materials identified in the local vicinity. Despite proposed strategies aimed at mitigating the issue, ongoing obstacles persist in obstructing the recontamination of the environment. Additional investigation is advised to determine solutions for these issues. Due to the changing environmental conditions in drinking water sources, future studies must prioritize the development of local technologies capable of adapting to these variations, particularly concerning pH levels and heavy metal concentrations. Additional research is necessary to investigate local technological innovations capable of simultaneously removing multiple metal ions.

References

[1] Yang, M., Zhang, H., Fu, Y., Wang, Z., Zhang, Y., Ding, Z., 2024. Transforming waste polystyrene foam to carbon on zeolite for efficient solar water desalination. J. Environ. Chem. Eng. 12, 113486. https://doi.org/10.1016/j.jece.2024.113486.

[2] Tong, T., Zhang, M., Chen, W., Huo, X., Xu, F., Yan, H., et al., 2024. Recent advances in carbon-based material/semiconductor composite photoelectrocatalysts: Synthesis, improvement strategy, and organic pollutant removal. Coord. Chem. Rev. 500, 215498. https://doi.org/10.1016/j.ccr.2023.215498.

[3] Sivaranjanee, R., Kumar, P.S., Rangasamy, G., 2024. Hydrothermally produced activated carbon spheres from discarded maize cobs for efficient removal of rose bengal dye from water environment. Desalin. Water. Treat. 317, 100123. https://doi.org/10.1016/j.dwt.2024.100123.

[4] Ma, X., Dong, B., Xie, F., Yang, H., Wang, C., Bittencourt, C., et al., 2025. A novel cost-effective kapok fibers and regenerated cellulose-based carbon aerogel for continuous oil/water separation. Sep. Purif. Technol. 353, 128435. https://doi.org/10.1016/j.seppur.2024.128435.

[5] Qi, J., Zhu, H., Yang, T., Wang, X., Wang, Z., Lei, X., et al., 2024. Biomass-derived carbon/iron composite (FexOy-BC (RM)) with excellent Cd (II) adsorption from wastewater – Red mud resource utilization. Arab. J. Chem. 17, 105411. https://doi.org/10.1016/j.arabjc.2023.105411.

[6] Patton, H., Krometis, L., Ling, E., Cohen, A., Sarver, E., 2024. Faucet-mounted point-of-use drinking water filters to improve water quality in households served by private wells. Sci. Total Environ. 906, 167252. https://doi.org/10.1016/j.scitotenv.2023.167252.

[7] Ulfa, M., Hanif, R.A., Sholeha, N.A., 2025. Surface chemistry and adsorption behavior of methylene blue on functionalized carbon materials: A comprehensive study. Commun. Sci. Technol. 10 (2), 276–291. https://doi.org/10.21924/cst.10.2.2025.1711.

[8] Bhattacharjee, T., Konwar, A., Sarmah, J., Chowdhury, D., 2023. A sustainable approach for heavy metal remediation from water using carbon dot based composites: A review. J. Hazard. Mater. Adv. 10, 100295. https://doi.org/10.1016/j.hazadv.2023.100295.

[9] Gusain, R., Kumar, N., Sinha, S., 2020. Recent advances in carbon nanomaterial-based adsorbents for water purification. Coord. Chem. Rev. 405, 213111. https://doi.org/10.1016/j.ccr.2019.213111.

[10] Veeman, D., Shree, M.V., Sureshkumar, P., Jagadeesha, T., Natrayan, L., Ravichandran, M., et al., 2021. Sustainable development of carbon nanocomposites: Synthesis and classification for environmental remediation. J. Nanomater. 1, 9182. https://doi.org/10.1155/2021/5840645.
[11] Selvaraj, M., Hai, A., Banat, F., Haija, M.A., 2020. Application and prospects of carbon nanostructured materials in water treatment: A review. J. Water Process Eng. 33, 100996. https://doi.org/10.1016/j.jwpe.2019.100996.
[12] Menya, E., Jjagwe, J., Mpagi, H., 2023. Progress in deployment of biomass-based activated carbon in point-of-use filters for removal of emerging contaminants from water : A review. Chem. Eng. Res. Des. 192, 412–440. https://doi.org/10.1016/j.cherd.2023.02.045.
[13] Njewa, J.B., Shikuku, V.O., 2023. Recent advances and issues in the application of activated carbon for water treatment in Africa: A systematic review (2007 – 2022). Appl. Surf. Sci. Adv. 18, 100501. https://doi.org/10.1016/j.apsadv.2023.100501.
[14] Mishra, S., Sundaram, B., 2023. Efficacy and challenges of carbon nanotube in wastewater and water treatment. Environ. Nanotechnol. Monit. Manag. 19, 100764. https://doi.org/10.1016/j.enmm.2022.100764.
[15] Jaspal, D., Malviya, A., 2020. Composites for wastewater purification: A review. Chemosphere 246, 125788. https://doi.org/10.1016/j.chemosphere.2019.125788.
[16] Abulkhair, H., Nallakukkala, S., Saad, M., Ahmed, I., Almatrafi, E., Bamaga, O., et al., 2024. Desalination of produced water via carbon dioxide hydrate using filter-based hydrate desalination reactor. Sep. Purif. Technol. 332, 125849. https://doi.org/10.1016/j.seppur.2023.125849.
[17] Qin, W., Dong, Y., Jiang, H., Hao, W., Imbrogno, J., Swenson, M., et al., 2024. A new approach of simultaneous adsorption and regeneration of activated carbon to address the bottlenecks of pharmaceutical wastewater treatment. Water Res. 252, 121180. https://doi.org/10.1016/j.watres.2024.121180.
[18] Mahadik, S., Surendran, S., Kim, J.Y., Lee, D., Park, J., Kim, T.-H., et al., 2023. Recent progress in the development of carbon-based materials in lead carbon batteries. Carbon Neutraliz. 2, 510–524. https://doi.org/10.1002/cnl2.78.
[19] Kazi, O.A., Chen, W., Eatman, J.G., Gao, F., Liu, Y., Wang, Y., et al., 2023. Material design strategies for recovery of critical resources from water. Adv. Mater. 35, 2300913. https://doi.org/10.1002/adma.202300913.
[20] Kumar, R., Sahoo, S., Joanni, E., Singh, R.K., Kian, W., Kar, K.K., et al., 2021. Recent progress on carbon-based composite materials for microwave electromagnetic interference shielding. Carbon N Y 177, 304–331. https://doi.org/10.1016/j.carbon.2021.02.091.
[21] Djakovi, T., Ivan, I., Panglisch, S., 2024. In silico modelling of selected organic substances adsorption from water onto activated carbon. Chem. Eng. Sci. 287, 119765. https://doi.org/10.1016/j.ces.2024.119765.
[22] Lan, J., Wang, B., Bo, C., Gong, B., Ou, J., 2023. Progress on fabrication and application of activated carbon sphere in recent decade. J. Ind. Eng. Chem. 120, 47–72. https://doi.org/10.1016/j.jiec.2022.12.045.
[23] Yateh, M., Li, F., Tang, Y., Li, C., Xu, B., 2024. Energy consumption and carbon emissions management in drinking water treatment plants: A systematic review. J. Clean. Prod. 437, 140688. https://doi.org/10.1016/j.jclepro.2024.140688.
[24] Valadez-renteria, E., Perez-gonzalez, R., Gomez-solis, C., Diaz-torres, L.A., Encinas, A., Oliva, J., et al., 2023. A novel and stretchable carbon-nanotube/Ni@TiO2:W photocatalytic composite for the complete removal of diclofenac drug from the drinking water. J. Environ. Sci. 126, 575–589. https://doi.org/10.1016/j.jes.2022.05.028.

[25] Zhao, H., Wang, Y., Yang, J., Zhang, J., Guo, Y., Li, S., 2023. Treatment of heavy metal ions-polluted water into drinkable water: Capture and recovery of Pb2+ by cellulose@CaCO3 bio-composite filter via spatial confinement effect. Chem. Eng. J. 454, 140297. https://doi.org/10.1016/j.cej.2022.140297.
[26] Senanu, L.D., Kranjac-berisavljevic, G., Cobbina, S.J., 2023. The use of local materials to remove heavy metals for household-scale drinking water treatment: A review. Environ. Technol. Innov. 29, 103005. https://doi.org/10.1016/j.eti.2023.103005.
[27] Cui, L., Wang, P., Che, H., Gao, X., Chen, J., Liu, B., et al., 2023. Co nanoparticles modified N-doped carbon nanosheets array as a novel bifunctional photothermal membrane for simultaneous solar-driven interfacial water evaporation and persulfate mediating water purification. Appl. Catal. B Environ. 330, 122556. https://doi.org/10.1016/j.apcatb.2023.122556.
[28] Ni, J., Wen, Y., Pan, D., Bai, J., Zhou, B., Zhao, S., et al., 2023. Light-driven simultaneous water purification and green energy production by photocatalytic fuel cell : A comprehensive review on current status , challenges , and perspectives. Chem. Eng. J. 473, 145162. https://doi.org/10.1016/j.cej.2023.145162.
[29] Fan, K., Chen, Q., Zhao, J., Liu, Y., 2023. Preparation of MnO_2-carbon materials and their applications in photocatalytic water treatment. Nanomaterials 13, 541. https://doi.org/10.3390/nano13030541.
[30] Monira, S., Roychand, R., Ibney, F., Bhuiyan, M., 2023. Nano and microplastics occurrence in wastewater treatment plants: A comprehensive understanding of microplastics fragmentation and their removal. Chemosphere 334, 139011. https://doi.org/10.1016/j.chemosphere.2023.139011.
[31] Negrete, A., Ramseier, S., Zimmermann, S., Le, P., Stoll, S., 2023. Contamination and removal efficiency of microplastics and synthetic fibres in a conventional drinking water treatment plant in Geneva, Switzerland. Sci. Total Environ. 880, 163270. https://doi.org/10.1016/j.scitotenv.2023.163270.
[32] Balaram, V., Copia, L., Kumar, U.S., Miller, J., Chidambaram, S., 2023. Pollution of water resources and application of ICP-MS techniques for monitoring and management — A comprehensive review. Geosyst. Geoenviron. 2, 100210. https://doi.org/10.1016/j.geogeo.2023.100210.
[33] Mensah, K., Shokry, H., Elkady, M., Hawash, H.B., Samy, M., 2024. Enhanced photocatalytic degradation of dyes using a novel waste toner-based TiO_2/Fe_2O_3@-nanographite nanohybrid: A sustainable approach. Water Sci. Eng. 1–10. https://doi.org/10.1016/j.wse.2024.01.005.
[34] Koseoglu-imer, D.Y., Oral, H.V., Calheiros, C.S.C., Krzeminski, P., Güçlü, S., Pereira, S.A., et al., 2023. Current challenges and future perspectives for the full circular economy of water in European countries. J. Environ. Manage. 345, 118627. https://doi.org/10.1016/j.jenvman.2023.118627.
[35] Peralta-Hernandez, J.M., Brillas, E., 2023. A critical review over the removal of paracetamol (acetaminophen) from synthetic waters and real wastewaters by direct, hybrid catalytic, and sequential ozonation processes. Chemosphere 313, 137411. https://doi.org/10.1016/j.chemosphere.2022.137411.
[36] Almanassra, I.W., Jaber, L., Namboorimadathil, S., Chatla, A., Kochkodan, V., Al-ansari, T., et al., 2023. Oxidized carbide-derived carbon as a novel filler for improved antifouling characteristics and permeate flux of hybrid polyethersulfone ultrafiltration membranes. Chemosphere 313, 137425. https://doi.org/10.1016/j.chemosphere.2022.137425.
[37] Manimegalai, S., Vickram, S., Raj, S., Rohini, K., Thanigaivel, S., Manikandan, S., et al., 2023. Carbon-based nanomaterial intervention and efficient removal of various

contaminants from effluents – A review. Chemosphere 312, 137319. https://doi.org/10.1016/j.chemosphere.2022.137319.
[38] Zhong, H., Guan, Y., Qiu, Z., Grady, B.P., Su, J., Huang, W., 2023. Application of carbon coated bentonite composite as an ultra-high temperature filtration reducer in water-based drilling fluid. J. Mol. Liq. 375, 121360. https://doi.org/10.1016/j.molliq.2023.121360.
[39] Kolya, H., Kang, C.-W., 2023. Next-generation water treatment: Exploring the potential of biopolymer-based nanocomposites in adsorption and membrane filtration. Polymers 15, 3421. https://doi.org/10.3390/polym15163421.
[40] Jonkers, W.A., Cornelissen, E.R., Vos, WM De, 2023. Hollow fiber nanofiltration: From lab-scale research to full-scale applications. J. Memb. Sci. 669, 121234. https://doi.org/10.1016/j.memsci.2022.121234.
[41] Sadia, M., Nollen, I., Helmus, R., Laak, T.L., Been, F., Praetorius, A., et al., 2023. Occurrence, fate, and related health risks of PFAS in raw and produced drinking water. Environ. Sci. Technol. 57 (8), 3062–3074. https://doi.org/10.1021/acs.est.2c06015.
[42] Jeong, Y., Gong, G., Lee, H., Seong, J., Won, S., Lee, C., 2023. Transformation of microplastics by oxidative water and wastewater treatment processes: A critical review. J. Hazard. Mater. 443, 130313. https://doi.org/10.1016/j.jhazmat.2022.130313.
[43] Quesada, H.B., Araújo, TP De, Vareschini, D.T., de Barros, M.A.S.D., Gomes, R.G., Bergamasco, R., 2020. Chitosan, alginate and other macromolecules as activated carbon immobilizing agents: A review on composite adsorbents for the removal of water contaminants. Int. J. Biol. Macromol. 164, 2535–2549. https://doi.org/10.1016/j.ijbiomac.2020.08.118.
[44] Fu, Q., Zhang, P., Zhuang, L., Zhou, L., Zhang, J., Wang, J., et al., 2022. Micro/nano multiscale reinforcing strategies toward extreme high-temperature applications: Take carbon/carbon composites and their coatings as the examples. J. Mater. Sci. Technol. 96, 31–68. https://doi.org/10.1016/j.jmst.2021.03.076.
[45] Almas, M., Asif, M., Nadeem, F., Nawaz, H., 2020. A review of advances in engineered composite materials popular for wastewater treatment. J. Environ. Chem. Eng. 8, 104073. https://doi.org/10.1016/j.jece.2020.104073.
[46] Ahmed, M.J., Hameed, B.H., Hummadi, E.H., 2020. Review on recent progress in chitosan/chitin-carbonaceous material composites for the adsorption of water pollutants. Carbohydr. Polym. 247, 116690. https://doi.org/10.1016/j.carbpol.2020.116690.
[47] Adeola, A.O., Duarte, M.P., Naccache, R., 2023. Microwave-assisted synthesis of carbon-based nanomaterials from biobased resources for water treatment applications: Emerging trends and prospects. Front. Carbon 2, 1220021. https://doi.org/10.3389/frcrb.2023.1220021.
[48] Li, Y., Shi, Y., Wang, H., Liu, T., Zheng, X., Gao, S., et al., 2023. Recent advances in carbon-based materials for solar-driven interfacial photothermal conversion water evaporation: Assemblies, structures, applications, and prospective. Carbon Energy 5, e331. https://doi.org/10.1002/cey2.331.
[49] Hassan, H.R., 2023. A review on different arsenic removal techniques used for decontamination of drinking water. Environ. Pollut. Bioavailab. 35, 2165964. https://doi.org/10.1080/26395940.2023.2165964.
[50] Zhang, Y., Xiao, H., Xiong, R., Huang, C., 2023. Xylan-based ratiometric fluorescence carbon dots composite with delignified wood for highly efficient water purification and photothermal conversion. Sep. Purif. Technol. 324, 124513. https://doi.org/10.1016/j.seppur.2023.124513.
[51] Wu, T., Lin, Z., Zhang, Y., Kanazawa, N., Komiyama, T., Zhu, C., et al., 2023. Poly-N-phenylglycine@multi-walled carbon nanotubes composite membrane for

improvement of Au (III) adsorption. Sep. Purif. Technol. 304, 122404. https://doi.org/10.1016/j.seppur.2022.122404.
[52] Falara, P.P., Ibrahim, I., Zourou, A., Sygellou, L., Emanuel, D., Em, G., et al., 2023. Bi-functional photocatalytic heterostructures combining titania thin films with carbon quantum dots (C-QDs/TiO2) for effective elimination of water pollutants. Environ. Sci. Pollut. Res. 30, 124976–124991. https://doi.org/10.1007/s11356-023-27285-7.
[53] Ma, X., Zhao, S., Tian, Z., Duan, G., Pan, H., Yue, Y., 2022. MOFs meet wood: Reusable magnetic hydrophilic composites toward efficient water treatment with super-high dye adsorption capacity at high dye concentration. Chem. Eng. J. 446, 136851. https://doi.org/10.1016/j.cej.2022.136851.
[54] Zhang, H., Li, L., Geng, L., Tan, X., Hu, Y., Mu, P., et al., 2023. Reduced graphene oxide/carbon nitride composite sponge for interfacial solar water evaporation and wastewater treatment. Chemosphere 311, 137163. https://doi.org/10.1016/j.chemosphere.2022.137163.
[55] Boulanger, N., Talyzin, A.V., Xiong, S., Hultberg, M., Grimm, A., 2024. High surface area activated carbon prepared from wood-based spent mushroom substrate for supercapacitors and water treatment. Colloids Surfaces A Physicochem Eng. Asp. 680, 132684. https://doi.org/10.1016/j.colsurfa.2023.132684.
[56] Bernadet, O., Larasati, A., Van Veelen, H.P.J., Jan, G., Euverink, W., Cristina, M., 2023. Biological Oxygen-dosed activated carbon (BODAC) filters – A bioprocess for ultrapure water production removing organics, nutrients and micropollutants. J. Hazard Mater. 458, 131882. https://doi.org/10.1016/j.jhazmat.2023.131882.
[57] Mehmood, T., Mustafa, B., Mackenzie, K., Ali, W., Irfan, R., Anum, W., et al., 2023. Recent developments in microplastic contaminated water treatment: Progress and prospects of carbon-based two-dimensional materials for membranes separation. Chemosphere 316, 137704. https://doi.org/10.1016/j.chemosphere.2022.137704.
[58] Zeng, F., Meng, Z., Xu, Z., Xu, J., Shi, W., Wang, H., et al., 2023. Biomass-derived porous activated carbon for ultra-high performance supercapacitor applications and high flux removal of pollutants from water. Ceram. Int. 49, 15377–15386. https://doi.org/10.1016/j.ceramint.2023.01.122.
[59] Madhusha, C., Jayasundara, T., Munaweera, I., Perera, C., Wijesinghe, G., 2023. Synthesis and structural characterization of copper nanoparticles doped activated carbon derived from coconut coir for drinking water purification. Mater. Today Chem. 27, 101312. https://doi.org/10.1016/j.mtchem.2022.101312.
[60] Khusnun, N.F., Arshad, A., Jalil, A.A., Firmansyah, L., Hassan, N.S., Nabgan, W., et al., 2023. An avant-garde of carbon-doped photoanode materials on photo-electrochemical water splitting performance : A review. J. Electroanal. Chem. 929, 117139. https://doi.org/10.1016/j.jelechem.2022.117139.
[61] Agasti, N., Gautam, V., Pandey, N., Genwa, M., Meena, P.L., Tandon, S., et al., 2022. Carbon nanotube based magnetic composites for decontamination of organic chemical pollutants in water: A review. Appl. Surf. Sci. Adv. 10, 100270. https://doi.org/10.1016/j.apsadv.2022.100270.
[62] Chen, X., Cheng, P., Tang, Z., Xu, X., Gao, H., Wang, G., 2021. Carbon-based composite phase change materials for thermal energy storage, transfer, and conversion. Adv. Sci. 8, 2001274. https://doi.org/10.1002/advs.202001274.
[63] Dubey, R., Dutta, D., Sarkar, A., Chattopadhyay, P., 2021. Functionalized carbon nanotubes: Synthesis, properties and applications in water purification, drug delivery, and material and biomedical sciences. Nanoscale Adv. 3, 5722–5744. https://doi.org/10.1039/D1NA00293G.

[64] Nasrollahzadeh, M., Sajjadi, M., Iravani, S., Varma, R.S., 2021. Carbon-based sustainable nanomaterials for water treatment: State-of-art and future perspectives. Chemosphere 263, 128005. https://doi.org/10.1016/j.chemosphere.2020.128005.
[65] Riegel, M., Gulde, B.H., Sacher, F., 2023. Sorptive removal of short-chain perfluoroalkyl substances (PFAS) during drinking water treatment using activated carbon and anion exchanger. Environ. Sci. Eur. 35, 12. https://doi.org/10.1186/s12302-023-00716-5.
[66] Richards, C.E., Tzachor, A., Alvin, S., Fenner, R., 2023. Rewards, risks and responsible deployment of artificial intellagence in water systems. Nat. Water 1, 422–432. https://doi.org/10.1038/s44221-023-00069-6.
[67] Wee, S.Y., Aris, A.Z., 2019. Revisiting the "forever chemicals", PFOA and PFOS exposure in drinking water. NPJ Clean Water 6, 57. https://doi.org/10.1038/s41545-023-00274-6.

CHAPTER 10

The application of carbon-based materials and composites for oily wastewater treatment

Sandeep Yadav[1], Pallavi Jain[1] and Prashant Singh[2]
[1]Department of Chemistry, SRM Institute of Science and Technology, Delhi NCR Campus, Modinagar, Uttar Pradesh, India; [2]Department of Chemistry, Atma Ram Sanatan Dharma College, University of Delhi, New Delhi, India

1. Introduction

Oily wastewater is a pretty complicated mix of water and different hydrocarbons, and it's a big problem for the environment. This type of wastewater comes from many different sources, like factories, oil refineries, petrochemical plants, car industries, and even runoff from urban areas. For example, the oil and gas industry alone creates over 250 million barrels of wastewater each year, and a lot of it's contaminated with hydrocarbons, especially during oil extraction and refining. The rules around oil and grease in wastewater vary from country to country, with the U.S. EPA setting a limit of 15 mg/L for oil and grease in wastewater discharges. In the car industry, wastewater is also full of oils, detergents, and solvents, which makes it hard to treat. With all these different contaminants, it's clear that we need stricter treatment solutions to protect our ecosystems and our health [1–5]. The intricate blend of oil, grease, and other organic and inorganic substances makes it difficult to treat using conventional methods. Left untreated, oily wastewater can wreak havoc on aquatic ecosystems, harm wildlife, and contaminate drinking water sources, posing severe risks to human health and the environment [6,7]. Given these adverse effects, regulatory bodies worldwide have set stringent standards for the discharge of oily wastewater. Meeting these regulations is crucial, not only to comply with legal requirements but also to protect our natural resources and ensure a sustainable future [8,9]. Traditional treatment methods like mechanical skimming, chemical coagulation, and biological

Carbon-Based Materials and Composites for Water and Wastewater Treatment
ISBN 978-0-443-33681-2
https://doi.org/10.1016/B978-0-443-33681-2.00007-7

treatments struggle with oily wastewater's complex contaminants. Mechanical skimming removes only 50%–60% of free oil, making it ineffective for emulsified oils. Chemical coagulation generates toxic sludge and increases costs by 30%, while biological treatments are inefficient for high oil concentrations, unable to treat wastewater with oil levels above 500–1000 mg/L. These methods also have slow processing times (up to 10–15 h), high operational costs, and can introduce secondary pollutants, further complicating wastewater treatment [10,11]. To address these challenges, the search for advanced materials and innovative technologies has intensified. Among the various emerging solutions, carbon-based materials have shown remarkable promise. Their unique properties, such as high adsorption capacity, large surface area, and chemical stability, make them ideal candidates for tackling the complexities of oily wastewater treatment [12]. Carbon-based materials, including activated carbon, carbon nanotubes, graphene, graphene oxide, and carbon quantum dots, have garnered significant attention for their potential applications in environmental remediation. These materials offer a versatile platform for developing highly efficient adsorption systems capable of capturing and removing oil contaminants from water [13]. Activated carbon, known for its porous structure and high surface area, has been widely used in various purification processes. Carbon nanotubes, with their exceptional mechanical strength and electrical properties, provide unique opportunities for enhancing adsorption processes [14]. Graphene and its derivatives, celebrated for their remarkable two-dimensional structure and functionalization possibilities, open new avenues for creating advanced composite materials [15]. Meanwhile, carbon quantum dots, with their tunable surface properties, add another layer of sophistication to the treatment arsenal.

In this chapter, we will delve into the application of these carbon-based materials and their composites in treating oily wastewater. We will explore their adsorption mechanisms, examine the factors influencing their efficiency, and discuss real-world case studies showcasing their practical applications. Additionally, we will address the challenges associated with their deployment and highlight future prospects for large-scale implementation.

2. Types of carbon-based materials for oily wastewater treatment

2.1 Activated carbon

Activated carbon is characterized by its highly porous structure, which provides an extensive surface area for adsorption. Typically, activated carbon can have a surface area ranging from 500 to 1500 m^2/g, allowing it to trap a wide range of contaminants [16]. The pore size distribution, including micropores (<2 nm), mesopores (2–50 nm), and macropores (>50 nm), plays a crucial role in determining its adsorption capacity and selectivity. This hierarchical pore structure ensures that activated carbon can capture various sizes of oil droplets and emulsified oil particles, making it a versatile material for oily wastewater treatment [17].

The primary adsorption mechanism of activated carbon involves physical adsorption, driven by van der Waals forces. The high surface area and porous structure facilitate the capture of oil molecules. Additionally, the presence of various functional groups on the surface of activated carbon, such as carboxyl, hydroxyl, and phenolic groups, enhances its chemical adsorption capabilities. These functional groups can interact with oil molecules through hydrogen bonding and electrostatic interactions, improving the overall adsorption efficiency [18,19].

Activated carbon has been extensively used in various industries for oily wastewater treatment. In a study conducted within the petrochemical industry, granular activated carbon (GAC) was found to remove up to 90% of oil and grease from wastewater, significantly reducing the concentration to meet regulatory standards. Additionally, powdered activated carbon (PAC) has been used in batch treatment processes to achieve high removal efficiencies. The reusability of activated carbon after regeneration through methods like thermal treatment or chemical washing further enhances its cost-effectiveness for long-term applications. Various other carbon-based materials are also used for the water treatment, as shown in Fig. 10.1.

2.2 Carbon nanotubes

Carbon nanotubes (CNTs) come in two primary forms: single-walled carbon nanotubes (SWCNTs) and multi-walled carbon nanotubes (MWCNTs). SWCNTs consist of a single graphene sheet rolled into a cylindrical shape, while MWCNTs are composed of multiple concentric graphene cylinders. These nanotubes can be synthesized through various methods, including chemical vapor deposition (CVD), arc discharge, and laser ablation. Among these, CVD is the most widely used technique due to its scalability and ability to produce high-quality CNTs [20]. To enhance their adsorption capabilities, CNTs are often functionalized with various chemical groups. Surface modifications, such as oxidation or amination, introduce functional groups like hydroxyl, carboxyl, or amine groups onto the surface of CNTs. These modifications increase the affinity

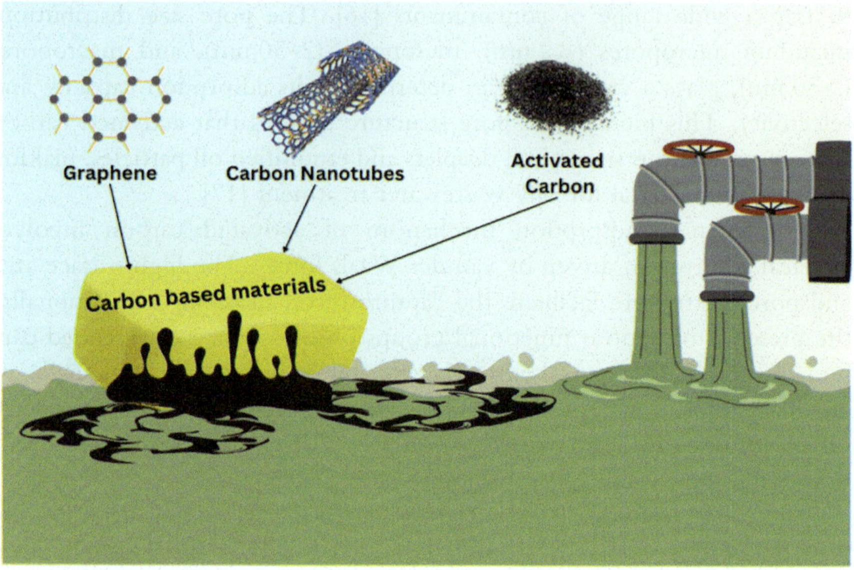

Figure 10.1 Carbon-based materials for oily wastewater treatment.

of CNTs for oil molecules, improving their adsorption efficiency. Fig. 10.2 mentions some classification of CNT functionalization mechanisms.

For instance, oxidized MWCNTs have shown enhanced oil removal due to the presence of oxygen-containing functional groups that facilitate interactions with oil contaminants through hydrogen bonding and electrostatic interactions [22,23]. Tian et al. developed a polyacrylonitrile membrane enhanced with CNTs-COOH and TiO_2, which achieved efficient separation of oil-in-water emulsions with a flux of 1180–3160 $L\,m^{-2}\,h^{-1}$ and oil rejection rates exceeding 99% [24]. In another study by Zhao and group, a PVDF membrane coated with a DA@CNTs/CS hydrogel achieved exceptional separation efficiency of 99.7% for anionic dyes and 99.4% for oil–water emulsions, with a high pure water flux of 1470 $L\,m^{-2}\,h^{-1}$ [25]. CNTs have demonstrated impressive performance in oily wastewater treatment. In a study, functionalized MWCNTs achieved an oil removal efficiency of up to 98%, even in the presence of other competing contaminants. The high surface area, combined with the tailored surface chemistry, allows CNTs to adsorb oil molecules over other organic compounds selectively. This selectivity makes CNTs particularly suitable for applications where the removal of specific contaminants is critical [26,27]. CNTs can be a cost-effective

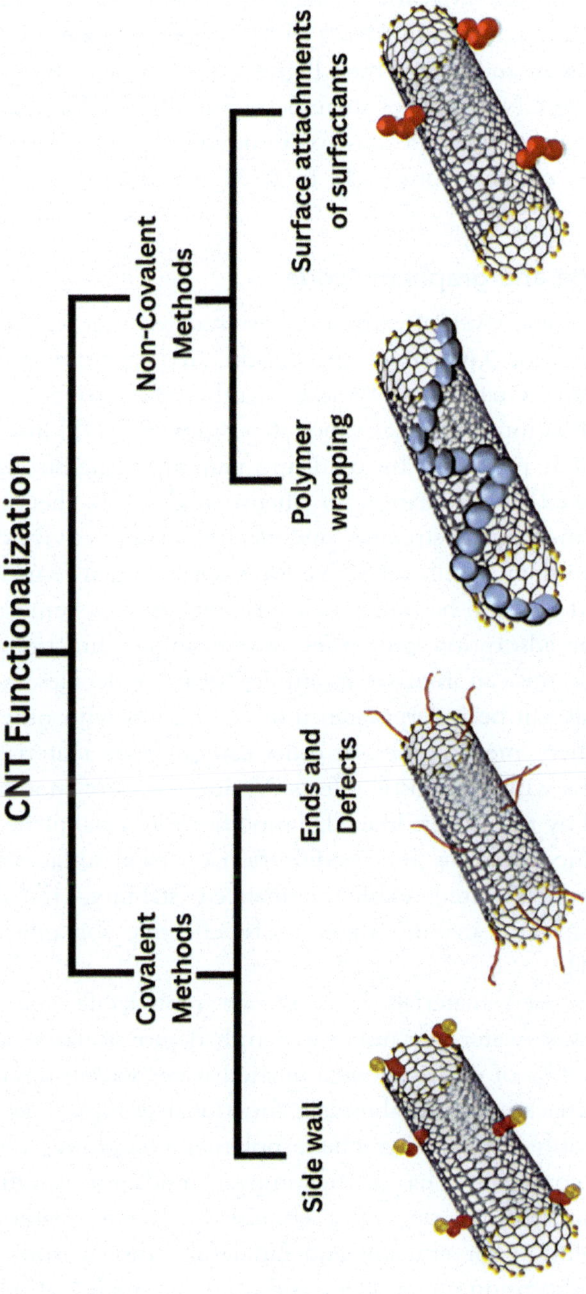

Figure 10.2 Various methods of CNT functionalization [21]. *(From Ref. [21].)*

solution for oil removal from wastewater due to their high adsorption capacity and the potential for oil recovery. Although the initial cost of CNTs is high, ranging from $1000 to $1500 per kilogram depending on the size, diameter, and purity, their high adsorption capacity—7.1–7.2 g of oil per gram of CNT—makes them economically feasible for wastewater treatment. For instance, the cost of removing 1 g of oil from water is relatively low, ranging from $0.14 to $0.21 per gram of oil using CNTs [28].

2.3 Graphene and graphene oxide

Graphene, a single layer of carbon atoms arranged in a hexagonal lattice, and graphene oxide (GO), its oxidized form, are revolutionary materials in water treatment. Graphene is typically synthesized through methods like chemical exfoliation, chemical vapor deposition (CVD), and mechanical cleavage. GO is produced by oxidizing graphite using strong oxidizing agents, followed by exfoliation to obtain single-layer sheets. The high surface area, mechanical strength, and electrical conductivity of graphene and GO make them highly attractive for environmental remediation [29]. Graphene and GO can be functionalized with various chemical groups to enhance their adsorption properties. For example, functionalizing GO with amine groups can improve its affinity for oil molecules. Additionally, these materials can be incorporated into composites with other materials, such as polymers, metal oxides, or other carbon-based materials, to create hybrid systems with synergistic effects. Various types of modifications are also explored by researchers like edge modification, basal plane alterations, asymmetric modifications and modification on basal surfaces (Fig. 10.3). These composites often exhibit enhanced stability, reusability, and adsorption capacity, making them more effective for oily wastewater treatment [31].

Graphene-based materials have shown remarkable performance in treating oily wastewater. For instance, a study demonstrated that GO could remove over 99% of oil from contaminated water within minutes, thanks to its high surface area and abundant functional groups. The use of graphene composites, such as graphene-polymer composites or graphene-metal oxide composites, has further improved adsorption efficiency and stability under various operating conditions. These materials offer a promising solution for efficient and rapid oil removal from wastewater [32,33]. Oil adsorption on graphene and graphene oxide (GO) is driven by

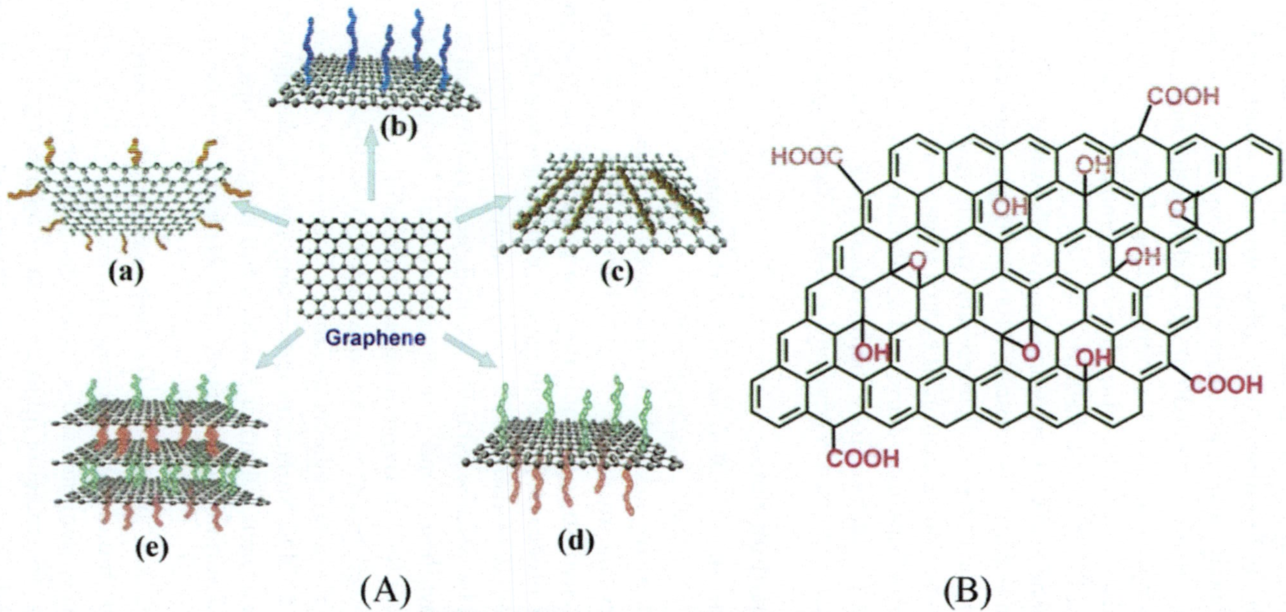

Figure 10.3 (A) Graphene functionalization can occur through (a) edge modifications, (b) basal plane alterations, (c) noncovalent molecular adsorption on the basal surface, (d) asymmetric changes to the basal plane, or (e) self-assembly of modified graphene layers. (B) Graphene oxide's chemical structure features a carbon network decorated with oxygenated groups like hydroxyl, epoxy, and carboxyl groups across its surface [30]. *(From Ref. [30].)*

hydrophobic interactions, π-π stacking, and surface functionalities. Graphene's hydrophobicity and sp^2-hybridized structure enable effective interaction with hydrocarbons, especially aromatic compounds, while GO's oxygen-containing groups enhance the adsorption of emulsified oils. The high surface area (up to 2630 m^2/g) and porosity increase their oil adsorption capacity, with GO adsorbing 60.9 g/100 g. These properties make graphene-based materials efficient for oil removal, though challenges like cost and scalability remain [34].

While graphene and GO are promising for oily wastewater treatment, their production is expensive, costing $100–$400 per gram. Additionally, their synthesis involves toxic chemicals and metals, leading to waste that risks environmental contamination. Residual solvents and heavy metals, if unmanaged, can pollute ecosystems, and accumulated nanomaterials from large-scale GO production may harm aquatic life. These challenges highlight the need for cost-effective and eco-friendly alternatives [35,36].

2.4 Carbon quantum dots

Carbon quantum dots (CQDs) are tiny carbon nanoparticles with sizes below 10 nm. They possess unique optical properties, such as photoluminescence, which make them attractive for various applications, including wastewater treatment. CQDs have a high surface area and tunable surface chemistry, allowing them to participate in photoinduced electron transfer processes. These properties can be leveraged to enhance the adsorption and degradation of oily contaminants under light irradiation [37]. CQDs can be synthesized using methods such as chemical oxidation, hydrothermal synthesis, and microwave-assisted synthesis. Chemical oxidation involves the oxidation of carbon sources like graphite or carbon nanotubes, followed by exfoliation to obtain CQDs. Hydrothermal synthesis uses high-pressure, high-temperature conditions to convert organic precursors into CQDs. Surface modifications, such as doping with heteroatoms (e.g., nitrogen, sulfur) or attaching functional groups, can further improve their adsorption capabilities and selectivity for oil molecules [38,39].

Although research on CQDs for oily wastewater treatment is still emerging, initial studies are promising. For example, CQDs synthesized from biomass have demonstrated effective removal of oil droplets from water, achieving efficiencies of up to 95% under optimal conditions. Their small size and high reactivity offer unique advantages for treating complex wastewater matrices. Additionally, CQDs can be incorporated into composite materials to enhance their stability and reusability, making them a promising candidate for large-scale applications [40,41].

3. Mechanisms of oil adsorption by carbon-based materials

Adsorption plays a pivotal role in removing oil pollutants, with carbon-based materials serving as effective adsorbents due to their unique structural and chemical properties. The process can occur via two primary mechanisms: physisorption and chemisorption. Physisorption relies on weak van der Waals forces, while chemisorption involves strong chemical bonds, each distinguished by their energy requirements, reversibility, and adsorption layers. Fig. 10.4 highlights these contrasting mechanisms, providing a visual framework for understanding how oil interacts with carbon-based materials.

3.1 Physical adsorption

Physical adsorption played a significant role in the removal of oil from wastewater by carbon-based materials. This process, also known as

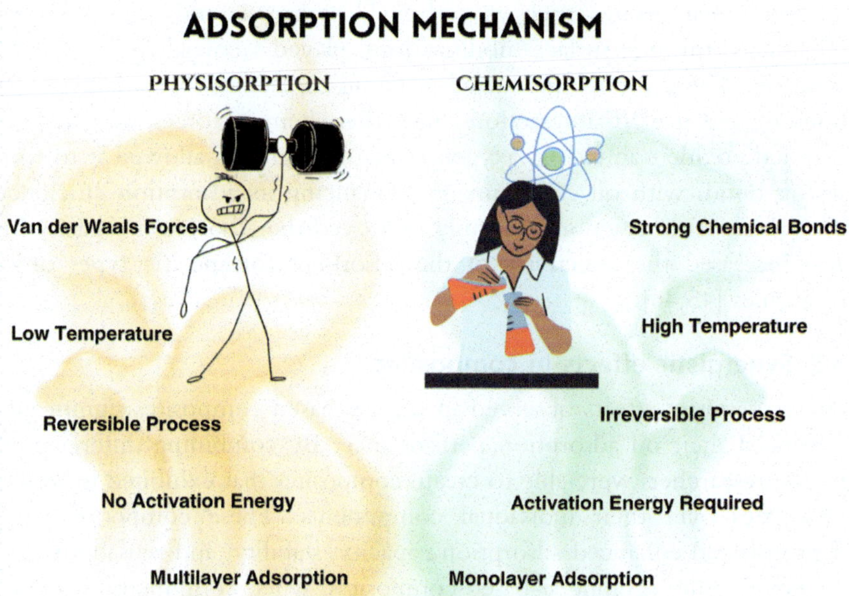

Figure 10.4 Major types of adsorption mechanisms.

physisorption, relied heavily on van der Waals forces to attract and hold oil molecules onto the surface of the adsorbent. Other characteristics, as mentioned in Fig. 10.4 are also observed in physisorption. The extensive surface area and porous structure of materials like activated carbon and carbon nanotubes were pivotal in enhancing this adsorption capability [42,43].

The intricate network of pores in activated carbon, ranging from micropores to macropores, provided numerous sites for oil molecules to adhere. These pores facilitated capillary action, drawing oil into the material's internal structure. Similarly, the high aspect ratio and large surface area of carbon nanotubes allowed for the efficient trapping of oil droplets. The van der Waals interactions between the oil molecules and the carbon surfaces were relatively weak but numerous, resulting in a significant cumulative adsorption capacity [44].

3.2 Chemical adsorption

Chemical adsorption, or chemisorption, involves the formation of stronger chemical bonds between the oil molecules and the surface of the carbon-based materials. This process was particularly important for materials like graphene oxide and functionalized carbon nanotubes, which possessed various reactive surface groups.

Functional groups such as carboxyl, hydroxyl, and amine groups, introduced through surface modifications, played a crucial role in chemisorption. These groups interacted with oil molecules through hydrogen bonding, electrostatic interactions, and other chemical forces. For example, graphene oxide's abundant oxygen-containing groups allowed it to form strong bonds with oil contaminants, enhancing its adsorption efficiency. The chemical interactions not only improved the adsorption capacity but also increased the selectivity of the adsorbent for specific types of oil molecules [45,46].

3.3 Synergistic effects in composites

The synergistic effects observed in carbon-based composites significantly enhanced their oil adsorption performance. By combining different materials, researchers were able to create composites that exhibited improved properties over their individual components. These composites often demonstrated enhanced adsorption capacity, stability, and reusability.

For instance, graphene-based composites, when combined with polymers or metal oxides, showed superior performance due to the synergistic interactions between the components. The high surface area of graphene provided numerous adsorption sites, while the polymer or metal oxide added functional groups that facilitated chemical adsorption. This combination resulted in a composite material that could effectively capture oil molecules through both physical and chemical adsorption mechanisms [47–49].

Similarly, composites of carbon nanotubes and activated carbon leveraged the strengths of both materials. The high mechanical strength and surface area of carbon nanotubes, combined with the extensive porosity of activated carbon, created a robust and highly efficient adsorbent. These composites were able to maintain their structural integrity and adsorption capacity even under demanding operational conditions, making them suitable for large-scale applications [50].

4. Factors influencing adsorption efficiency

The effectiveness of oil adsorption by carbon-based materials depends on several key factors: solution chemistry, material properties, and operational conditions. These factors work together to influence how well the adsorbent captures oil, shaping both the speed and extent of the process. Fig. 10.5 highlights these essential elements.

4.1 Material properties

4.1.1 Surface area and pore size distribution

The adsorption efficiency of carbon-based materials was greatly influenced by the properties of the material, characteristics of the solution and operational conditions (Fig. 10.5). In material properties, their surface area and pore size distribution play a significant role. Materials with larger surface areas provided more sites for oil molecules to adhere to, enhancing their adsorption capacity. Activated carbon, with its extensive network of micropores and mesopores, demonstrated this characteristic effectively. Similarly, carbon nanotubes and graphene derivatives, known for their high surface areas, exhibited superior adsorption properties due to the ample contact points available for oil molecules [51].

4.1.2 Functionalization and surface charge

The presence of functional groups on the surface of carbon-based materials significantly affected their adsorption efficiency. Functionalization, such as the introduction of carboxyl, hydroxyl, and amine groups, improved the chemical interaction between the adsorbent and oil molecules. These groups facilitated chemical adsorption through hydrogen bonding and

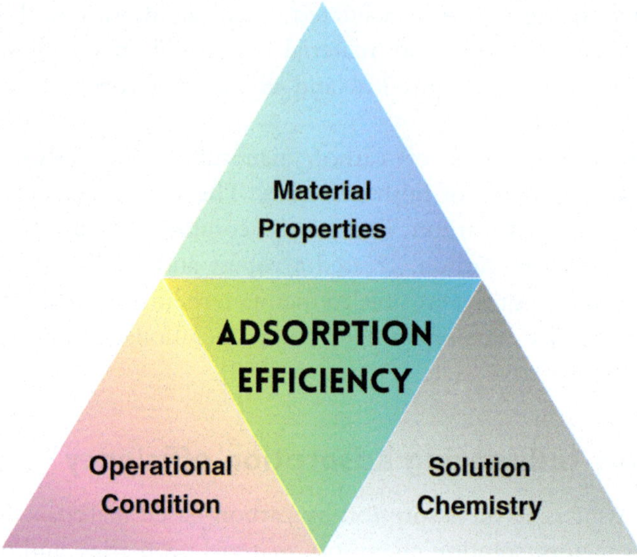

Figure 10.5 Factors affecting the efficiency of adsorption.

electrostatic interactions, enhancing the overall adsorption capacity. The surface charge of the material also played a crucial role, as it influenced the attraction or repulsion of oil molecules, which are often charged or polar [52].

4.2 Solution chemistry

4.2.1 pH and ionic strength

The pH of the solution impacted the adsorption process by altering the surface charge of the carbon-based materials and the ionization state of the oil molecules. At different pH levels, the functional groups on the adsorbent's surface could either gain or lose protons, affecting their interaction with oil molecules. For example, at low pH, the surface of graphene oxide might become positively charged, enhancing the adsorption of negatively charged oil droplets through electrostatic attraction [53,54].

Ionic strength, or the concentration of salts in the solution, also played a significant role. High ionic strength could shield the charges on the adsorbent and the oil molecules, reducing electrostatic interactions and thus decreasing adsorption efficiency. Conversely, in low ionic strength environments, electrostatic interactions were more pronounced, potentially enhancing adsorption [55,56].

4.2.2 Presence of competing ions and contaminants

The presence of competing ions and other contaminants in the solution could interfere with the adsorption process. These competing species might occupy adsorption sites on the carbon-based materials, reducing the availability of these sites for oil molecules. For instance, in wastewater containing heavy metals or organic pollutants, these contaminants could compete with oil molecules for the adsorption sites, decreasing the overall efficiency of the treatment process [57].

4.3 Operational conditions

4.3.1 Temperature and pressure

Temperature affected the adsorption capacity and kinetics of carbon-based materials. Generally, higher temperatures increase the mobility of oil molecules, enhancing their diffusion to the adsorbent surface. However, excessively high temperatures could also desorb already adsorbed molecules, reducing overall efficiency. Optimal temperature ranges were necessary to balance these effects and maximize adsorption [58].

Pressure, particularly in the context of gas-phase adsorption or in systems utilizing pressurized flow, influenced the adsorption process. Higher pressures could force more oil molecules onto the adsorbent surface, increasing adsorption capacity. This principle was especially relevant in applications involving carbon nanotubes and activated carbon, where pressurized systems were used to enhance contact between the oil and the adsorbent.

4.3.2 Flow rate and contact time

The flow rate of the wastewater through the adsorption system was critical. Higher flow rates reduced the contact time between the oil molecules and the adsorbent, potentially leading to lower adsorption efficiency. Conversely, slower flow rates allowed for more extended interaction, improving the chances of oil molecules adhering to the adsorbent surface. Optimizing flow rates was essential for balancing treatment efficiency and operational throughput.

4.3.3 Scale-up considerations for industrial applications

Scaling up the use of carbon-based materials from laboratory settings to industrial applications introduced additional factors that influenced adsorption efficiency. Factors such as the uniform distribution of the adsorbent, the design of the adsorption columns, and the handling of larger volumes of wastewater needed careful consideration. In industrial settings, ensuring consistent performance of the adsorbent material and maintaining its structural integrity under high flow rates and pressures were critical for effective treatment.

5. Comparison with traditional methods

5.1 Advantages and limitations of carbon-based materials

Carbon-based materials offered several advantages over traditional methods for oily wastewater treatment. Their high surface area and porosity provided a large number of adsorption sites, enhancing their capacity to remove oil and other contaminants. Functionalization of these materials introduced specific reactive groups, improving selectivity and chemical adsorption capabilities. Additionally, carbon-based materials such as CNTs and graphene showed excellent mechanical strength and stability, making them suitable for demanding environmental conditions [59].

However, there were also limitations. The production and functionalization of advanced carbon materials like CNTs and graphene could be costly and complex. Some materials, such as activated carbon, require regular regeneration to maintain their effectiveness, which could add to operational costs. Despite these challenges, the overall benefits often outweighed the limitations, particularly in applications requiring high performance and sustainability [60].

5.2 Cost-effectiveness and sustainability

Carbon-based materials demonstrated significant cost-effectiveness and sustainability benefits. The ability to regenerate activated carbon and reuse it multiple times reduced long-term costs. Graphene and CNTs, despite their higher initial costs, offered durability and longevity, reducing the need for frequent replacements. Additionally, the high efficiency of these materials in removing contaminants meant that lower quantities were required compared to traditional adsorbents, further enhancing their cost-effectiveness. Sustainability was also a key advantage, as many carbon-based materials could be produced from renewable sources and recycled after use [61].

6. Deployment in large-scale facilities

6.1 Design and engineering considerations

Deploying carbon-based materials in large-scale facilities required careful design and engineering considerations. Ensuring uniform distribution of the adsorbent, optimizing flow rates, and maintaining structural integrity under high-pressure conditions were critical factors. For instance, in designing a large-scale activated carbon filter system, engineers had to account for the pressure drop across the filter bed and ensure adequate contact time between the wastewater and the adsorbent. Similarly, integrating CNTs and graphene into existing treatment infrastructure required modifications to accommodate their unique properties and enhance their performance [62,63].

6.2 Case studies of industrial implementations

Several industrial implementations showcased the practical benefits of carbon-based materials in treating oily wastewater. A chemical manufacturing plant in China integrated CNT-based filters into its effluent treatment system, achieving a significant reduction in oil content from 200 mg/L to less than 10 mg/L. In Germany, a wastewater treatment plant used graphene-polymer composites to treat high oil loads from industrial discharges, successfully reducing oil concentrations from 500 mg/L to less than 1 mg/L. These case studies highlighted the scalability and effectiveness of carbon-based materials in real-world industrial applications, demonstrating their potential to enhance wastewater treatment processes on a large scale [64–66].

7. Challenges and future prospects

7.1 Technical challenges

One of the primary technical challenges associated with the use of carbon-based materials for oily wastewater treatment was the durability and lifecycle of these materials. Over time, the adsorption efficiency of materials such as activated carbon and carbon nanotubes (CNTs) tends to decrease due to fouling and saturation. This reduction in performance necessitated frequent regeneration or replacement of the adsorbent, which posed practical difficulties in continuous industrial applications. Moreover, the mechanical stability of materials like graphene oxide (GO) under harsh operational conditions was a concern, as prolonged exposure to high temperatures and pressures could compromise their structural integrity [65].

7.2 Regeneration and disposal issues

The regeneration and disposal of used carbon-based materials also presented significant challenges. While activated carbon could be thermally regenerated, this process required considerable energy input and could result in the partial degradation of the material. For advanced materials like

CNTs and graphene, regeneration methods were less developed and often involved complex chemical treatments. Disposal of spent adsorbents posed environmental risks, as the adsorbed oil and other contaminants could leach back into the environment if not properly managed. Developing efficient and sustainable regeneration techniques was crucial to address these issues and ensure the long-term viability of carbon-based materials in wastewater treatment [67].

7.3 Economic and environmental considerations

The cost of producing and deploying carbon-based materials was a significant economic consideration. While materials like activated carbon were relatively inexpensive, the production of advanced materials such as CNTs and graphene involved sophisticated and costly synthesis processes. Scaling up these production methods to meet industrial demands further increased costs. Additionally, the deployment of these materials required specialized equipment and infrastructure, adding to the overall expense. Despite their high initial costs, the long-term economic benefits of carbon-based materials, such as reduced operational costs and enhanced treatment efficiency, need to be evaluated to justify their widespread adoption [68,69].

The use of carbon-based materials in wastewater treatment can lead to operational savings of up to 30% compared to traditional methods. Additionally, the cost of carbon-based adsorbents is approximately $0.50 per kilogram, which is significantly lower than other advanced materials [70]. Carbon-based materials, particularly those derived from biomass, have a lower environmental footprint due to their renewable nature and potential for recycling. Advances in recycling spent adsorbents, such as regenerating activated carbon through thermal or chemical methods, can further reduce environmental impact. Activated carbon derived from agricultural waste like coconut shells and sawdust has shown high efficiency in removing pollutants from wastewater. Additionally, the use of green activation agents such as potassium carbonate in the synthesis of activated carbon is an innovative method that enhances sustainability [70,71].

The environmental impact of synthesizing and disposing of carbon-based materials also raised concerns. The production processes for CNTs and graphene often involve hazardous chemicals and generate waste, contributing to environmental pollution. Furthermore, the disposal of spent adsorbents, if not managed properly, could lead to secondary contamination. Sustainable synthesis methods, such as using biomass-derived carbon materials, and environmentally friendly disposal practices, were essential to mitigate these impacts. Researchers and industry practitioners needed to prioritize green chemistry principles and develop closed-loop systems to minimize the environmental footprint of carbon-based materials [72,73].

8. Future research directions

Future research was directed toward the development of emerging carbon-based materials and composites with enhanced properties. Innovations in the synthesis of carbon quantum dots (CQDs) and their integration into composite materials opened new avenues for efficient and selective oil adsorption. Additionally, research focused on functionalizing existing materials to improve their adsorption capacity, selectivity, and reusability. The exploration of hybrid materials, combining carbon-based adsorbents with polymers or metal oxides, aimed to leverage the synergistic effects for superior performance.

Integrating carbon-based materials with other wastewater treatment technologies was another promising research direction. Combining adsorption with advanced oxidation processes, membrane filtration, or biological treatment could enhance overall treatment efficiency and address a broader range of contaminants. For instance, coupling graphene oxide membranes with photocatalytic degradation allowed for simultaneous oil adsorption and degradation, improving the treatment outcomes. Developing modular treatment systems that incorporate multiple technologies could provide flexible and scalable solutions for diverse wastewater treatment challenges.

Addressing the policy and regulatory implications of deploying carbon-based materials in wastewater treatment was essential for their widespread adoption. Ensuring that these materials met safety and environmental standards required comprehensive evaluation and certification processes. Policymakers needed to establish clear guidelines and regulations for the synthesis, use, and disposal of carbon-based adsorbents to prevent environmental contamination and ensure public safety. Additionally, incentivizing research and development through funding and grants could accelerate the advancement of sustainable and cost-effective carbon-based treatment technologies.

9. Conclusion

In summary, carbon-based materials demonstrated significant potential for treating oily wastewater due to their high adsorption capacities, large surface areas, and chemical stability. Activated carbon, CNTs, GO, and CQDs showed unique advantages in various applications, achieving high removal efficiencies and meeting stringent environmental standards. Key factors for success included proper material functionalization, optimized operational conditions, and effective management of solution chemistry. Carbon-based materials offered a sustainable alternative to traditional methods, reducing environmental impact and operational costs. Their regeneration and reuse capabilities further enhanced cost-effectiveness. Policymakers and industry stakeholders were encouraged to support these materials through clear regulations and incentives for innovation. Future research should focus on emerging carbon-based materials and their integration with other treatment technologies. Innovations in synthesis, functionalization, and hybrid systems could yield more efficient solutions. Continued collaboration between researchers, industry practitioners, and policymakers was essential. By fostering innovation and collaboration, the potential of carbon-based materials could be fully realized, leading to cleaner water and a healthier ecosystem.

References

[1] Ahmed, J., Thakur, A., Goyal, A., 2021. Industrial Wastewater and its Toxic Effects. Royal Society of Chemistry (RSC), pp. 1–14. https://doi.org/10.1039/9781839165399-00001.

[2] Jia, F., Wei, N., Ma, D., Rahi, M.N., Jaeel, A.J., Abbas, A.J., 2021. Treatment of petroleum refinery effluents and wastewater in Iraq: A mini review. IOP Conf. Ser. Mater. Sci. Eng. 1058. https://doi.org/10.1088/1757-899X/1058/1/012072.

[3] Singh, S., Shikha, 2018. Treatment and Recycling of Wastewater from Oil Refinery/Petroleum Industry. Springer Science and Business Media LLC, pp. 303–332. https://doi.org/10.1007/978-981-13-1468-1_10.

[4] Shahbaz, M., Rashid, N., Saleem, J., Mackey, H., McKay, G., Al-Ansari, T., 2023. A review of waste management approaches to maximise sustainable value of waste from the oil and gas industry and potential for the State of Qatar. Fuel 332. https://doi.org/10.1016/j.fuel.2022.126220.

[5] Fakhru'l-Razi, A., Pendashteh, A., Abdullah, L.C., Biak, D.R.A., Madaeni, S.S., Abidin, Z.Z., 2009. Review of technologies for oil and gas produced water treatment. J. Hazard Mater. 170 (2–3), 530–551. https://doi.org/10.1016/j.jhazmat.2009.05.044.

[6] Jan, S., Mishra, A.K., Bhat, M.A., Bhat, M.A., Jan, A.T., 2023. Pollutants in aquatic system: A frontier perspective of emerging threat and strategies to solve the crisis for safe drinking water. Environ. Sci. Pollut. Res. Int. 30 (53), 113242–113279. https://doi.org/10.1007/s11356-023-30302-4.

[7] Ghosh, A., 2023. Challenges and mitigation techniques for clean rural water supply in Himachal Pradesh, India. In: Water Resources Management for Rural Development: Challenges and Mitigation. Elsevier, India, pp. 237–248. https://doi.org/10.1016/B978-0-443-18778-0.00022-2.

[8] Hamidi, B., Makarem, M.A., Rahimpour, M.R., 2023. Generation and characteristics of mineral oils and hydrocarbons wastewater from petroleum industries. In: Advanced Technologies in Wastewater Treatment: Oily Wastewaters. Elsevier, Iran, pp. 1–27. https://doi.org/10.1016/B978-0-323-99916-8.00003-1.

[9] Erfani, H., Madhu, N.R., Khodayari, S., Qureshi, M.A., Swetanshu, Singh, P., Jadoun, S., 2024. Separation and removal of oil from water/wastewater in the oil industry: A review. Environ. Technol. Rev. 13 (1), 325–343. https://doi.org/10.1080/21622515.2024.2343129.

[10] Bora, J., Hazra, R., Nag, S., Malik, S., 2023. Traditional treatment methods for industrial waste. In: Advanced and Innovative Approaches of Environmental Biotechnology in Industrial Wastewater Treatment. Springer Nature, India, pp. 419–442. https://doi.org/10.1007/978-981-99-2598-8_19.

[11] Sabiani, N.H.Md, Alkarimiah, R., Ayub, K.R., Makhtar, M.M.Z., Aziz, H.A., Hung, Y.-T., Wang, L.K., Wang, M.-H.S., 2023. Treatment of Palm Oil Mill Effluent. Springer Science and Business Media LLC, pp. 227–284. https://doi.org/10.1007/978-3-031-44768-6_7.

[12] Abid, M., Garcia, R., Martinez-Escandell, M., Fullana, A., Silvestre-Albero, J., 2024. Exceptional performance of Fe@carbon-rich nanoparticles prepared via hydrothermal carbonization of oil mill wastes for H2S removal. Chemosphere 358, 142140. https://doi.org/10.1016/j.chemosphere.2024.142140.

[13] Maafa, I.M., 2024. Recycled activated carbon from plastic waste for effective oil removal from produced water. Desalination Water Treat. 317, 100106. https://doi.org/10.1016/j.dwt.2024.100106.

[14] Turco, A., Foscarini, A., Piccirillo, C., Primiceri, E., Chiriacò, M.S., Ferrara, F., 2024. A promising solution for water remediation: PDMS-(nano)carbon hybrid materials for oil removal. Appl. Mater. Today 38, 102218. https://doi.org/10.1016/j.apmt.2024.102218.

[15] Yang, K., Zhang, F., Chen, Y., Zhang, H., Xiong, B., Chen, H., 2022. Recent progress on carbon-based composites in multidimensional applications. Compos. Appl. Sci. Manuf. 157, 106906. https://doi.org/10.1016/j.compositesa.2022.106906.

[16] Boulanger, N., Talyzin, A.V., Xiong, S., Hultberg, M., Grimm, A., 2024. High surface area activated carbon prepared from wood-based spent mushroom substrate for supercapacitors and water treatment. Colloids Surf. A Physicochem. Eng. Asp. 680, 132684. https://doi.org/10.1016/j.colsurfa.2023.132684.

[17] Zhang, J., Peng, K., Xu, Z.-K., Xiong, Y., Liu, J., Cai, C., Huang, X., 2023. A comprehensive review on the behavior and evolution of oil droplets during oil/water separation by membranes. Adv. Colloid Interface Sci. 319, 102971. https://doi.org/10.1016/j.cis.2023.102971.

[18] Rugarabamu, J.R., Zhao, D., Li, S., Diao, R., Song, K., 2023. Structure modeling of activated carbons used for simulating methane adsorption – A review. Petrol. Res. 8 (1), 103–117. https://doi.org/10.1016/j.ptlrs.2022.06.004.

[19] Wen, C., Liu, T., Wang, D., Wang, Y., Chen, H., Luo, G., Zhou, Z., Li, C., Xu, M., 2023. Biochar as the effective adsorbent to combustion gaseous pollutants: Preparation, activation, functionalization and the adsorption mechanisms. Prog. Energy Combust. Sci. 99, 101098. https://doi.org/10.1016/j.pecs.2023.101098.

[20] Ge, L., Zuo, M., Wang, Y., Wang, R., Rong, N., Qi, Z., Zhao, C., Zhang, Y., Xu, C., 2024. A review of comprehensive utilization of biomass to synthesize carbon

nanotubes: From chemical vapor deposition to microwave pyrolysis. J. Anal. Appl. Pyrolysis 177, 106320. https://doi.org/10.1016/j.jaap.2023.106320.

[21] Abdullah, T.A., Juzsakova, T., Hafad, S.A., Rasheed, R.T., Al-Jammal, N., Mallah, M.A., Salman, A.D., Le, P.C., Domokos, E., Aldulaimi, M., 2022. Functionalized multi-walled carbon nanotubes for oil spill cleanup from water. Clean Technol. Environ. Policy 24 (2), 519–541. https://doi.org/10.1007/s10098-021-02104-0.

[22] Chen, S., Zhou, J., Li, K., Chen, Z., Li, X., Su, X., Ao, X., Xie, H., Chen, L., Wu, X., Wu, W., 2023. Superhydrophobic and robust photothermal/electrothermal PVDF-a/CNT-s@PDMS membrane for crude oil removal and dye adsorption. Compos. Sci. Technol. 244, 110267. https://doi.org/10.1016/j.compscitech.2023.110267.

[23] Parmar, K.R., Dora, D.T.K., Pant, K.K., Roy, S., 2019. An ultra-light flexible aerogel-based on methane derived CNTs as a reinforcing agent in silica-CMC matrix for efficient oil adsorption. J. Hazard Mater. 375, 206–215. https://doi.org/10.1016/j.jhazmat.2019.04.017.

[24] Tian, S., He, Y., Zhang, L., Li, S., Bai, Y., Wang, Y., Wu, J., Yu, J., Guo, X., 2023. CNTs/TiO2- loaded carbonized nanofibrous membrane with two-type self-cleaning performance for high efficiency oily wastewater remediation. Colloids Surf. A Physicochem. Eng. Asp. 656. https://doi.org/10.1016/j.colsurfa.2022.130306.

[25] Zhao, J., Liu, H., Qi, Y., Wang, R., Lv, Z., Yu, Y., Sun, S., 2024. Construction of robust DA@CNTs/CS hydrogel coatings on hydrophobic PVDF membrane by deposition of self-assembly for efficient separation of oil-water emulsion and dye wastewater. Sep. Purif. Technol. 333. https://doi.org/10.1016/j.seppur.2023.125956.

[26] Tang, W., Zou, C., Liang, H., Da, C., Zhao, Z., 2022. The comparison of interface properties on crude oil-water and rheological behavior of four polymeric nanofluids (nano-SiO2, nano-CaO, GO and CNT) in carbonates for enhanced oil recovery. J. Petrol. Sci. Eng. 214, 110458. https://doi.org/10.1016/j.petrol.2022.110458.

[27] Liu, Y., Sun, Z.N., Ji, S.Z., Yang, Y.H., Xu, J.Q., Jing, G.L., 2023. Effect of nanocomposite pour point depressant EVAL/CNT on flow properties of waxy crude oil. Pet. Sci. 20 (6), 3807–3818. https://doi.org/10.1016/j.petsci.2023.05.016.

[28] Kayvani Fard, A., Mckay, G., Manawi, Y., Malaibari, Z., Hussien, M.A., 2016. Outstanding adsorption performance of high aspect ratio and super-hydrophobic carbon nanotubes for oil removal. Chemosphere 164, 142–155. https://doi.org/10.1016/j.chemosphere.2016.08.099.

[29] Yadav, S., Singh Raman, A.P., Meena, H., Goswami, A.G., Bhawna, Kumar, V., Jain, P., Kumar, G., Sagar, M., Rana, D.K., Bahadur, I., Singh, P., 2022. An update on graphene oxide: Applications and toxicity. ACS Omega 7 (40), 35387–35445. https://doi.org/10.1021/acsomega.2c03171.

[30] Maio, A., Pibiri, I., Morreale, M., La Mantia, F.P., Scaffaro, R., 2021. An overview of functionalized graphene nanomaterials for advanced applications. Nanomaterials 11 (7). https://doi.org/10.3390/nano11071717.

[31] Periasamy, A.P., Wu, W.P., Ravindranath, R., Roy, P., Lin, G.L., Chang, H.T., 2017. Polymer/reduced graphene oxide functionalized sponges as superabsorbents for oil removal and recovery. Mar. Pollut. Bull. 114 (2), 888–895. https://doi.org/10.1016/j.marpolbul.2016.11.005.

[32] Yap, P.L., Hassan, K., Auyoong, Y.L., Mansouri, N., Farivar, F., Tran, D.N.H., Losic, D., 2020. All-in-one bioinspired multifunctional graphene biopolymer foam for simultaneous removal of multiple water pollutants. Adv. Mater. Interfac. 7 (18). https://doi.org/10.1002/admi.202000664.

[33] Yang, S., Li, J., Cheng, Z., Li, F., Sha, S., Hou, C., Lu, H., Wu, J., Sheng, Z., Ma, J., 2022. Graphene-based melamine sponges with reverse wettability for oil/water

separation through absorption and filtration. J. Environ. Chem. Eng. 10 (3), 107543. https://doi.org/10.1016/j.jece.2022.107543.
[34] Liu, R., Gong, T., Zhang, K., Lee, C., 2017. Graphene oxide papers with high water adsorption capacity for air dehumidification. Sci. Rep. 7 (1). https://doi.org/10.1038/s41598-017-09777-y.
[35] Ding, X., Pu, Y., Tang, M., Zhang, T., 2022. Environmental and health effects of graphene-family nanomaterials: Potential release pathways, transformation, environmental fate and health risks. Nano Today 42. https://doi.org/10.1016/j.nantod.2022.101379.
[36] Baysal, A., Saygin, H., Ustabasi, G.S., 2020. Risks of graphene nanomaterial contamination in the soil: Evaluation of major ions. Environ. Monit. Assess. 192 (10). https://doi.org/10.1007/s10661-020-08561-2.
[37] Rani, U.A., Ng, L.Y., Ng, C.Y., Mahmoudi, E., 2020. A review of carbon quantum dots and their applications in wastewater treatment. Adv. Colloid Interface Sci. 278. https://doi.org/10.1016/j.cis.2020.102124.
[38] Baragau, I.A., Lu, Z., Power, N.P., Morgan, D.J., Bowen, J., Diaz, P., Kellici, S., 2021. Continuous hydrothermal flow synthesis of S-functionalised carbon quantum dots for enhanced oil recovery. Chem. Eng. J. 405. https://doi.org/10.1016/j.cej.2020.126631.
[39] Nugraha, M.W., Wirzal, M.D.H., Ali, F., Roza, L., Sambudi, N.S., 2021. Electrospun polylactic acid/tungsten oxide/amino-functionalized carbon quantum dots (PLA/WO3/N-CQDs) fibers for oil/water separation and photocatalytic decolorization. J. Environ. Chem. Eng. 9 (5). https://doi.org/10.1016/j.jece.2021.106033.
[40] Almufarij, R.S., Mohamed, M.E., 2023. Green synthesis of a carbon quantum dots-based superhydrophobic membrane for efficient oil/water separation. Materials 16 (15). https://doi.org/10.3390/ma16155456.
[41] Carballo, G.V., Yang, H.L., Hsu, Y.X., Leron, R.B., Tsai, H.A., Lee, K.R., 2024. Incorporation of zwitterionic carbon quantum dots in cellulose acetate tubular membrane for oil/water separation. Sep. Purif. Technol. 337. https://doi.org/10.1016/j.seppur.2024.126301.
[42] Manimegalai, S., Vickram, S., Deena, S.R., Rohini, K., Thanigaivel, S., Manikandan, S., Subbaiya, R., Karmegam, N., Kim, W., Govarthanan, M., 2023. Carbon-based nanomaterial intervention and efficient removal of various contaminants from effluents – A review. Chemosphere 312 (Pt 1), 137319. https://doi.org/10.1016/j.chemosphere.2022.137319.
[43] Zaker, A., Hammouda, S.ben, Sun, J., Wang, X., Li, X., Chen, Z., 2023. Carbon-based materials for CO2 capture: Their production, modification and performance. J. Environ. Chem. Eng. 11 (3), 109741. https://doi.org/10.1016/j.jece.2023.109741.
[44] Raninga, M., Mudgal, A., Patel, V.K., Patel, J., Kumar Sinha, M., 2023. Modification of activated carbon-based adsorbent for removal of industrial dyes and heavy metals: A review. Mater. Today Proc. 77, 286–294. https://doi.org/10.1016/j.matpr.2022.11.358.
[45] Allangawi, A., Alzaimoor, E.F.H., Shanaah, H.H., Mohammed, H.A., Saqer, H., El-Fattah, A.A., Kamel, A.H., 2023. Carbon capture materials in post-combustion: Adsorption and absorption-based processes. C-J. Carbon Res. 9 (1). https://doi.org/10.3390/c9010017.
[46] Fu, T., Zhang, B., Gao, X., Cui, S., Guan, C.-Y., Zhang, Y., Zhang, B., Peng, Y., 2023. Recent progresses, challenges, and opportunities of carbon-based materials applied in heavy metal polluted soil remediation. Sci. Total Environ. 856 (Pt 1), 158810. https://doi.org/10.1016/j.scitotenv.2022.158810.

[47] Vareda, J.P., 2023. On validity, physical meaning, mechanism insights and regression of adsorption kinetic models. J. Mol. Liq. 376, 121416. https://doi.org/10.1016/j.molliq.2023.121416.
[48] Zhang, W., Liu, Y., Tao, F., An, Y., Zhong, Y., Liu, Z.C., Hu, Z., Zhang, X., Wang, X., 2023. An overview of biomass-based oil/water separation materials. Sep. Purif. Technol. 316, 123767. https://doi.org/10.1016/j.seppur.2023.123767.
[49] Xiang, B., Liu, Q., Yan, W., Wei, Y., Mu, P., Li, J., 2023. Advances in special wettable materials for adsorption separation of high-viscosity crude oil/water mixtures. Chem. Commun. 59 (49), 7559–7578. https://doi.org/10.1039/d3cc00984j.
[50] Shang, Y., Gao, R., Wang, Y., Dong, X., Li, X., Tang, J., Wang, X., Liu, J., Xie, Y., Li, J., Gan, W., 2023. Scalable fabrication of efficient and recycling wood residue-derived sponge for crude oil adsorption. J. Hazard Mater. 441, 129979. https://doi.org/10.1016/j.jhazmat.2022.129979.
[51] Kumar, N., Pandey, A., Rosy, Chandra Sharma, Y., 2023. A review on sustainable mesoporous activated carbon as adsorbent for efficient removal of hazardous dyes from industrial wastewater. J. Water Process Eng. 54, 104054. https://doi.org/10.1016/j.jwpe.2023.104054.
[52] Ahmed, M.J., Hameed, B.H., Khan, M.A., 2023. Recent advances on activated carbon-based materials for nitrate adsorption: A review. J. Anal. Appl. Pyrolysis 169, 105856. https://doi.org/10.1016/j.jaap.2022.105856.
[53] Zhu, X., Wan, Z., Tsang, D.C.W., He, M., Hou, D., Su, Z., Shang, J., 2021. Machine learning for the selection of carbon-based materials for tetracycline and sulfamethoxazole adsorption. Chem. Eng. J. 406. https://doi.org/10.1016/j.cej.2020.126782.
[54] Xiang, Y., Xu, Z., Wei, Y., Zhou, Y., Yang, X., Yang, Y., Yang, J., Zhang, J., Luo, L., Zhou, Z., 2019. Carbon-based materials as adsorbent for antibiotics removal: Mechanisms and influencing factors. J. Environ. Manag. 237, 128–138. https://doi.org/10.1016/j.jenvman.2019.02.068.
[55] Hu, Y., Guo, T., Ye, X., Li, Q., Guo, M., Liu, H., Wu, Z., 2013. Dye adsorption by resins: Effect of ionic strength on hydrophobic and electrostatic interactions. Chem. Eng. J. 228, 392–397. https://doi.org/10.1016/j.cej.2013.04.116.
[56] El Badawy, A.M., Luxton, T.P., Silva, R.G., Scheckel, K.G., Suidan, M.T., Tolaymat, T.M., 2010. Impact of environmental conditions (pH, ionic strength, and electrolyte type) on the surface charge and aggregation of silver nanoparticles suspensions. Environ. Sci. Technol. 44 (4), 1260–1266. https://doi.org/10.1021/es902240k.
[57] Barquilha, C.E.R., Braga, M.C.B., 2021. Adsorption of organic and inorganic pollutants onto biochars: Challenges, operating conditions, and mechanisms. Bioresour. Technol. Rep. 15. https://doi.org/10.1016/j.biteb.2021.100728.
[58] Striolo, A., Gubbins, K.E., Gruszkiewicz, M.S., Cole, D.R., Simonson, J.M., Chialvo, A.A., Cummings, P.T., Burchell, T.D., More, K.L., 2005. Effect of temperature on the adsorption of water in porous carbons. Langmuir 21 (21), 9457–9467. https://doi.org/10.1021/la051120t.
[59] Sabzehmeidani, M.M., Mahnaee, S., Ghaedi, M., Heidari, H., Roy, V.A.L., 2021. Carbon based materials: A review of adsorbents for inorganic and organic compounds. Mater. Adv. 2 (2), 598–627. https://doi.org/10.1039/d0ma00087f.
[60] Egbedina, A.O., Bolade, O.P., Ewuzie, U., Lima, E.C., 2022. Emerging trends in the application of carbon-based materials: A review. J. Environ. Chem. Eng. 10 (2). https://doi.org/10.1016/j.jece.2022.107260.
[61] Gopinath, K.P., Vo, D.V.N., Gnana Prakash, D., Adithya Joseph, A., Viswanathan, S., Arun, J., 2021. Environmental applications of carbon-based materials: A review. Environ. Chem. Lett. 19 (1), 557–582. https://doi.org/10.1007/s10311-020-01084-9.

[62] Fagiolari, L., Bella, F., 2019. Carbon-based materials for stable, cheaper and large-scale processable perovskite solar cells. Energy Environ. Sci. 12 (12), 3437–3472. https://doi.org/10.1039/c9ee02115a.
[63] Mateos, R., Alonso, R.M., Escapa, A., Morán, A., 2017. Methodology for fast and facile characterisation of carbon-based electrodes focused on bioelectrochemical systems development and scale up. Materials 10 (1). https://doi.org/10.3390/ma10010079.
[64] Khan, N., Tabasi, Z.A., Liu, J., Zhang, B.H., Zhao, Y., Husain, T., 2022. Recent advances in functional materials for wastewater treatment: From materials to technological innovations. J. Mar. Sci. Eng. 10 (4). https://doi.org/10.3390/jmse10040534.
[65] Abuhasel, K., Mohamed, K., Alquraish, M., Munusamy, Y., Jeng, Y.T., Wastewater Treatment, O., 2021. Overview of conventional and modern methods, challenges, and future opportunities. Water 13 (7), 980. https://doi.org/10.3390/w13070980.
[66] Nazal, M., 2020. An overview of carbon-based materials for the removal of pharmaceutical active compounds. In: Carbon-Based Material for Environmental Protection and Remediation. IntechOpen. https://doi.org/10.5772/INTECHOPEN.91934.
[67] Baskar, A.V., Bolan, N., Hoang, S.A., Sooriyakumar, P., Kumar, M., Singh, L., Jasemizad, T., Padhye, L.P., Singh, G., Vinu, A., Sarkar, B., Kirkham, M.B., Rinklebe, J., Wang, S., Wang, H., Balasubramanian, R., Siddique, K.H.M., 2022. Recovery, regeneration and sustainable management of spent adsorbents from wastewater treatment streams: A review. Sci. Total Environ. 822. https://doi.org/10.1016/j.scitotenv.2022.153555.
[68] El Gamal, M., Mousa, H.A., El-Naas, M.H., Zacharia, R., Judd, S., 2018. Bio-regeneration of activated carbon: A comprehensive review. Sep. Purif. Technol. 197, 345–359. https://doi.org/10.1016/j.seppur.2018.01.015.
[69] Budzianowski, W.M., 2012. Value-added carbon management technologies for low CO2 intensive carbon-based energy vectors. Energy 41 (1), 280–297. https://doi.org/10.1016/j.energy.2012.03.008.
[70] Al-Anzi, B.S., Siang, O.C., 2017. Recent developments of carbon based nanomaterials and membranes for oily wastewater treatment. RSC Adv. 7 (34), 20981–20994. https://doi.org/10.1039/c7ra02501g.
[71] Guo, H., Cao, Y., Li, H., Xu, J., Zhu, H., 2024. Recent advances in biomass conversion to sustainable chemicals and polymers. Front. Chem. 12. https://doi.org/10.3389/FCHEM.2024.1475162/BIBTEX.
[72] Muradov, N.Z., Veziroğlu, T.N., 2008. "Green" path from fossil-based to hydrogen economy: An overview of carbon-neutral technologies. Int. J. Hydrogen Energy 33 (23), 6804–6839. https://doi.org/10.1016/j.ijhydene.2008.08.054.
[73] Sabir, S., 2015. Approach of cost-effective adsorbents for oil removal from oily water. Crit. Rev. Environ. Sci. Technol. 45 (17), 1916–1945. https://doi.org/10.1080/10643389.2014.1001143.

CHAPTER 11

The application of carbon-based materials and composites for drinking water treatment

Azrul Nurfaiz Mohd Faizal[1,5], Nicky Rahmana Putra[2], Augustine Agi[3], Abu Hassan Nordin[4], Muhammad Abbas Ahmad Zaini[5,6], Ahmad Syahmi Zaini[5], Ahmad Ilyas Rushdan[6], Norfarhana Abdul Samad[7] and Hilmi Abdul Rahman[8]

[1]Institute of Ecosystem Science Borneo (IEB), Universiti Putra Malaysia Sarawak, Bintulu, Sarawak, Malaysia; [2]Faculty of Engineering Technology and Science, Higher College of Technology, Ruwais, United Arab Emirates; [3]Faculty of Chemical and Process Engineering Technology, College of Engineering Technology, Universiti Malaysia Pahang, Gambang, Pahang, Malaysia; [4]Faculty of Applied Sciences, Universiti Teknologi MARA (UiTM), Arau Campus, Perlis, Malaysia; [5]Centre of Lipid Engineering and Applied Research (CLEAR), Ibnu Sina Institute for Scientific and Industrial Research, Universiti Teknologi Malaysia, Skudai, Johor, Malaysia; [6]Department of Chemical Engineering, Faculty of Chemical and Energy Engineering, Universiti Teknologi Malaysia, Skudai, Johor, Malaysia; [7]Centre of Chemical Technology, Department of Petrochemical Engineering, Politeknik Tun Syed Nasir Syed Ismail, Pagoh Education Hub, Johor, Malaysia; [8]Chemical Management Unit, University Laboratory Management Centre, Universiti Teknologi Malaysia Skudai, Johor, Malaysia

1. Introduction

Access to clean water and the sustainable management of water resources are essential for fostering economic development and enhancing public health. These goals are at the core of Sustainable Development Goal 6 (SDG 6), set by the United Nations Environment Program (UNEP). However, physical, chemical, biological, and radioactive contaminants in water systems pose significant threats to human health and the environment, undermining efforts to achieve these critical objectives. Industrial growth, technological advancements, and human activities have exacerbated water contamination, leading to global clean water scarcity that now affects approximately one-sixth of the world's population [1]. Despite these international efforts, according to estimates, around two billion individuals still need access to safe and affordable drinkable water in 2020; 1.7 billion lacked basic hygiene [2]. About 1.2 million people lose their lives due to

Carbon-Based Materials and Composites for Water and Wastewater Treatment
ISBN 978-0-443-33681-2
https://doi.org/10.1016/B978-0-443-33681-2.00006-5

the consumption of contaminated water every year; furthermore, 6% of deaths in low-income countries are on account of unsafe water resources [3]. Hence, public health risks and disease outbreaks such as diarrhea, dysentery, typhoid, hepatitis A, polio, and cholera among impoverished and vulnerable communities are linked to the unavailability of safe drinking water and poor sanitation, excluding geogenic contaminations. Most community water treatment facilities rely on surface water and groundwater as their primary sources [4]. With the population on the rise, the demand for drinking water is increasing, and so is the risk of contamination with a variety of chemical and biological pollutants. The two key factors affecting the quality of surface and groundwater are natural causes (hydrogeologic conditions, water-rock interaction) and various human activities (mineral resource extraction and wastewater discharge) [5]. Inorganic pollutants (such as fluoride, iron, nitrates, and heavy metals like cadmium, arsenic, and lead [6] are persistent and tend to accumulate in living organisms through the food chain; continued consumption of these could seriously threaten human health [7]. However, organic contaminants, including antibiotics, pesticides, dyes, phenols, and xenohormones, pose a severe risk to human health due to their high persistence, difficulty in removal, extreme toxicity, and easy trophic transfer [8]. These chemical contaminants can enter community water systems through various point and non-point sources, such as artificial recharge, agricultural runoff, or effluents from wastewater treatment plants [9]. Furthermore, numerous emerging organic contaminants (EOCs) such as atrazine, metolachlor, caffeine, and carbamazepine are frequently found in drinking water supply chains and are widely used globally in multiple sectors such as pharmaceutical and personal care products (PPCPs), pesticides, surfactants, food, and industrial additives[4,10]. These EOCs and other organics can form carcinogenic disinfection by-products (DBPs) in community drinking water treatment facilities [11], highlighting the urgent need for improved treatment methods to ensure water safety.

Among contaminants, heavy metals are potentially carcinogenic and are not biodegradable. Therefore, their presence in water beyond the allowed level is a preventable issue that can be extremely harmful to living beings [12]. Emerging heavy metal contaminants, including uranium (U), lead (Pb), and cadmium (Cd), are frequently detected at elevated concentrations in water systems across various regions of the world. Reported concentrations of U, Pb, and Cd in water systems can reach up to 4 mg L−1 (average: 8 μg L−1), 222 mg L−1 (average: 5.81 μg L−1), and 71 mg L−1 (less than 1 μg L−1), respectively [13]. However, these heavy metals in concentrations exceeding the maximum contamination limits (MCL) of 30 μg L^{-1} for U, 10 μg L^{-1} for Pb, and 5 μg L^{-1} for Cd pose significant health risks [14]. Prolonged exposure to these contaminants can adversely impact critical organs, including the kidneys, lungs, brain, gonads, immune system, and nervous system. Additionally, uranium presents the added risk of radiotoxicity due to its radioactive properties.

Biological contamination, particularly from bacteria, viruses, and other pathogens, also contributes to the global water crisis. *Escherichia coli* (*E. coli*), a fecal indicator bacterium commonly found in the excrement of warm-blooded animals and reptiles, plays a crucial role as a widely used marker for assessing water quality and the effectiveness of treatment methods like filtration and chlorination [13].

Carbon-based materials and composites are widely used in drinking water treatment due to their exceptional adsorption capacity, chemical versatility, and ability to address various contaminants. Their high surface area and porous structures enable efficient removal of organic pollutants, heavy metals, and microorganisms, while their functionalizable chemical properties enhance selectivity for specific contaminants. Carbon-based composites, such as metal-carbon hybrids, exhibit catalytic and photocatalytic properties, enabling advanced degradation of persistent pollutants. These materials are often renewable and reusable, reducing costs and environmental impacts, with options like biochar providing sustainable alternatives. Their scalability and proven effectiveness in commercial and municipal applications further solidify their role as indispensable tools for improving water quality.

Over the past years, researchers have dedicated significant efforts to designing carbon-based materials and composites for drinking water treatment. Materials like activated carbon, carbon nanotubes (CNTs), graphene oxides (GO), biochar, metal-oxide biochar, Mxene-polymer composites, and carbon dots can adsorb pollutants in drinking water. Therefore, it is important to provide an up-to-date account of these advances over the past years.

This review aims to comprehensively summarize the key characteristics and performance of carbon-based materials and composites, including biochar, activated carbon, graphene family and its composites, carbon nanotube, and other carbon material composites applied as a medium for drinking water treatment, and highlight potential applications in the sustainable water treatment. Furthermore, this review discussed the disadvantages and challenges of carbon-based materials and composites in water treatment. It also critically assessed the prospects for utilizing carbon-based materials and composites in water treatment, providing the audience with a comprehensive understanding of the challenges ahead.

2. Carbon-based materials in water treatment

2.1 Activated carbon

Activated carbon is a versatile material with flexible properties broadly used in removing air and water pollutants due to its high adsorption capacity, large surface area, and versatile applicability. It is a rich carbon material with unique porous textures and surface features that facilitate the capture of target pollutants [15]. Furthermore, activated carbon can be prepared using carbonaceous precursors through chemical and physical activation [16]. Chemical activation is performed by heating the carbon material pre-impregnated with activating agents such as H3PO4 [17], KOH [18], and ZnCl2 [19]. It is a one-step or two-step method involving the impregnation of precursor with a chemical agent followed by heating under an inert atmosphere at a temperature between 400 and 800 °C [20]. Chemical activation tends to develop micropores with systematic pore radius due to the orderly distribution of activating agents within the carbons to be activated. Conversely, chemical activation needs pervasive washing of the resulting activated carbon, which results in a high amount of waste production [21]. On the other hand, physical activation is carried out by treating the carbon material at a high temperature in the presence of oxidative gases such as CO_2 to burn off a fraction of carbon to create new micropores [22]. Physical activation is a two-step process whereby the raw material is first carbonized in the absence of oxygen, followed by activation under oxidant gases such as CO_2 and steam.

Activated carbon is highly effective in removing many impurities, including organic pollutants, chlorine, volatile organic compounds (VOCs), and disinfection by-products such as trihalomethanes. It enhances the taste and odor of drinking water by adsorbing organic compounds and other substances responsible for undesirable flavors. This adsorption process involves the organic compounds sticking to the surface of the carbon, effectively removing them from the water. Its strong affinity for hydrophobic compounds enables the efficient removal of pesticides, herbicides, and pharmaceutical residues. Additionally, activated carbon can adsorb heavy metals like lead and mercury when impregnated with specific functional groups or integrated with complementary treatment processes. Activated carbon is available in two primary forms: powdered activated carbon (PAC) and granular activated carbon (GAC). PAC is typically added directly to water during treatment to eliminate dissolved contaminants, while GAC is utilized in fixed-bed filters for continuous filtration. GAC filters are widely employed in household water purification systems and municipal treatment facilities.

One significant advantage of activated carbon is its regenerability. Spent activated carbon can be reactivated through thermal or chemical processes, enabling multiple reuse cycles. This reduces operational costs and environmental impact, making it a cost-effective and sustainable solution. Despite its effectiveness, activated carbon has certain limitations. Its performance can decline in the presence of high particulate loads and cannot altogether remove specific contaminants, such as nitrates and some dissolved minerals. Regular maintenance and regeneration are essential to sustain its efficiency over time. Table 11.1 summarizes the activation strategies of carbonaceous materials into activated carbon from previous studies.

2.2 Carbon nanotubes (CNTs)

Carbon Nanotubes or CNTs are a form of carbon allotrope with unique properties that make them useful in nanotechnology applications such as electronic systems, optics, pharmaceuticals, drug delivery [32], wastewater and water treatment, and so on. CNTs can be employed to effectively address water scarcity and pollution issues due to their promising physical, electrical, chemical, and morphological features. Adsorbents, photocatalysts, membranes, and filters [33], microbial fuel cells, disinfection, and desalination are some of the wastewater and water treatment applications for CNTs-based nanotechnologies [32]. The two varieties of CNTs are multi-walled CNTs (MWCNTs) and single-walled CNTs (SWCNTs) [32]. MWCNTs, as well as SWCNTs, show different physical characteristics due to structural differences despite their apparent similarity.

Single-walled CNTs are one-dimensional (1D), cylindrically formed carbon allotropes with a higher surface area (SA) and length-to-diameter ratio. Because of their sizable length-to-diameter ratio and small diameter, SWCNTs are categorized as 1D nanomaterials. SWCNTs get their name from the number of walls and hollow structures they have. They comprise graphene sheets wrapped around one-atom-thick walls to produce hollow

Table 11.1 Activation strategies of carbonaceous materials into activated carbon from previous studies.

Precursor	Activation strategies	BET SA (m^2/g)	Type of activation	Reference
Paddy stalk residue	N_2; flowrate = 100 mL/min; T = 500 °C; t = 2 h	489	Physical	[23]
Strychnos Potatorum shells	CO_2; flowrate = 80 mL/min; T = 700 °C; t = 2 h	872	Physical	[24]
Resorcinol-formaldehyde	CO_2; flowrate = 100 mL/min; T = 850 °C; t = 2 h	800	Physical	[25]
Walnut shell	CO_2; flowrate = 100 mL/min; T = 900 °C; t = 2 h	1163	Physical	[26]
Palm date stones	CO_2; flowrate = 400 mL/min; T = 300 °C; t = 1 h	34	Physical	[27]
Plastic waste	NaOH; T = 800 °C; t = 1 h	2060	Chemical	[28]
Poplar branch	$ZnCl_2$; T = 500 °C; t = 30 min	1648	Chemical	[29]
Cellulose	KOH; T = 800 °C; t = 1 h	3196	Chemical	[30]
Olive pomace	H_3PO_4; T = 500 °C; t = 1 h	1400	Chemical	[31]

tubes. The lattice orientation determines the electrical properties of CNTs. This lattice orientation results in entirely different properties. Depending on the orientation of the lattice, CNTs can have either metallic or semiconducting characteristics. Metallic characteristics in zig-zag and armchair CNTs refer to the ability to conduct electricity with little resistance. In contrast, semiconducting characteristics found in chiral CNTs refer to the ability to conduct electricity under certain conditions [34].

Multi-walled CNTs, a unique structure of carbon nanotubes, are composed of numerous SWCNTs concentrically arranged within one another. While MWCNTs are still considered a 1D form of carbon, they deviate from the typical characteristics of SWCNTs due to a higher likelihood of defects. However, their enhanced dispersibility and their cost-effective synthesis and purification methods serve as a reassuring counterbalance to these drawbacks. A comprehensive comparison of various characteristics of SWCNT and MWCNT is presented in Table 11.2 [35]. Table 11.3 provides a concise summary of CNTs synthesis from previous studies.

2.3 Graphene oxides (GO)

Membrane technology is of considerable importance in the separation process, playing a crucial role in environmental protection. It is a highly effective method for treating wastewater, effluents, and seawater desalination due to its cost-efficiency, high productivity, exceptional removal efficiency, straightforward operation, and minimal reliance on chemical additives. Membranes are typically classified based on their surface

Table 11.2 Comparison of various characteristics of SWCNT and MWCNT.

Parameter	SWCNTs	MWCNTs
Graphene sheet	Single	Multiple
Diameter (nm)	1.2–2	5–100
Density (g/cm^3)	1.33–1.40	2.1
Tensile strength (GPa)	150	150
Electrical conductivity (S/cm^1)	102–106	103–105
Thermal conductivity (W/m^1K^1)	6000	3000
Purity	Poor	High
Twisting	Easy	Difficult
Defect level	High	Low

Table 11.3 CNTs synthesis from previous studies.

Precursor	Synthesis method[a]	Support/ Catalyst	Operating parameters T (°C)	Time (min)	Carrier gas	Type of CNTs	Reference
Natural condensate gas	P + CVD	Ni/Al_2O_3	1000	112	N_2	SWCNT	[36]
Ethylene	Floating catalyst CVD	Bimetallic (Co-Fe, Co-Ni)	1050	0.17	N_2	SWCNT	[37]
Plastic waste	P + CVD	Ni	800	30	N_2	MWCNT	[38]
Rice straw	P	Al_2O_3/Fe–Ni	830	30	N_2	MWCNT	[39]
Knitted fabrics	EAD	Fe_3O_4	65	30	–	MWCNT	[40]
Benzene	P	$C_{10}H_{10}Fe$	850	240	N_2	MWCNT	[41]

[a]P = Pyrolysis; CVD = chemical vapor deposition; EAD = Electric arc discharge.

properties as either hydrophilic (polar surface) or hydrophobic (non-polar surface). Additionally, the filtration process and pore size are also used to categorize membranes into Nano-Filtration (NF), Micro-Filtration (MF), Ultra-Filtration (UF), and Reverse Osmosis (RO) membranes. Despite the cost-effectiveness of polymer membranes, their selectivity is often limited due to the lack of tunable and ordered pore structures. To enhance selectivity, membrane thickness is typically increased, which unfortunately leads to a trade-off between permeability and selectivity. In order to overcome these obstacles, scientists have started investigating two-dimensional nanomaterials for the creation of high-performance separation membranes, and the standout material is graphene oxides (GO), a derivative of graphene due to its controlled surface chemistry, exceptional mechanical properties, chemical stability, atomic-level thickness, and excellent film-forming ability.

GO is ideal for adsorption, surface reactions, and chemical functionalization and as a superb substrate for immobilizing nanomaterials, making it a potential material for high-performance separation membranes. However, traditional GO membranes face challenges such as swelling when hydrated and narrow interlayer spacing due to strong interactions between nanoflakes, which restricts water flux and complicates the precise control of interlayer spaces and pore sizes. Interestingly, this issue can be overcome by adding MOF (materials self-assembled from organic ligands and inorganic metal ions or clusters via coordination bonds) as a regulator to GO membrane which offers several advantages such as (1) uniform-sized MOF with controllable pore structures effectively intercalate into layered GO membrane, controlling interlayer spacing to mitigate the permeability-selectivity trade-off [42]; (2) The cavity structure inherent in MOF crystal structures provides sufficient space for guest species, offering additional molecular transport channels to enhance membrane permeability; (3) MOF increase the specific surface area of GO membrane, providing more active sites and imparting new membrane properties such as hydrophilicity [43], zeta potential, and roughness [44]. Thus, the application of MOF is expected to enhance the performance of GO membranes in water treatment, resolving the trade-off dilemma of traditional GO membranes and potentially breaking through the limitations of single-material membranes to achieve effective preparation and application of truly high-performance membranes. Table 11.4 provides a summary of the preparation method of MOF tuning to Graphene oxide membrane.

Table 11.4 Preparation method of MOF tuning to Graphene oxides membrane.

Materials used	Method	Type of method	Key process	Outcome	Advantages	References
GO, ZIF-8	In-situ growth with ZIF-8	In situ growth modification	Crystallization growth in GO interlayers	Formation of nanochannels; ultra-fast solvent permeability (10,000 L/m^2/h/bar for acetone)	High separation selectivity; improved permeability	[45]
ZIF-8@f-GOm	Selective growth for stability	In situ growth modification	Filling microporous defects with ZIF-8	Increased interlayer spacing (0.75–0.93 nm); stable water flux (60 L/m^2/h/bar)	Enhanced mechanical stability; long-term performance	[46]
CuBDC, GO	Spatial confinement growth	In situ growth modification	Green synthesis at room temperature	High surface area; precise control of pore structure	Improved separation and catalysis	[47]
GO, UiO-66, chitosan	LbL assembly with UiO-66	Layer-by-layer assembly	Sequential deposition of GO and UiO-66 layers	Expanded channel space (~6.0 Å), molecular sieving, 344% higher water flux (14 L/m^2/h/bar)	Efficient water purification, self-healing, anti-fouling, and separation stability	[48]
GO, HZIF-8, tannic acid (TA), PES	LbL assembly with HZIF-8	Layer-by-layer assembly	Etching ZIF-8 with TA and assembling on PES	High water flux (32.11 kg/m^2/h at 323 K), reduced mass transfer resistance	Enhanced water transport efficiency, layered structure, and surface roughness	[49]

GO, UiO-66 (Zr)-(COOH) 2, PAN	Pressure-assisted self-assembly	Layer-by-layer assembly	Pressure filtration for membrane fabrication	Dense GO layers, superhydrophilic surface, antifouling properties	High water molecule adsorption, uniform dispersion, and selective separation	[50]
GO, Cu-TCPP nanosheets	GO/Cu-TCPP membrane	Physical blending	Filtration-assisted technology for layer assembly	Increased interlayer spacing (0.94–1.00 nm), improved water permeability, and antibacterial properties	Stable heterostructure, enhanced water stability, and synergistic properties of GO and Cu-TCPP	[51,52]
GO, La-MOF-1	La-MOF-1GO membrane	Physical blending	Compression molding	Improved thermal stability, enhanced adsorption capacity, and reduced nanoparticle aggregation	Recyclable, prevents aggregation and detachment of La	[44]
GO, UiO-66, MIL-140A MOFs	UiO-66/MIL-140A blended membranes	Physical blending	Pressure-assisted filtration	Increased water permeability due to wrinkling of GO sheets and expanded interlayer spacing	Enhanced water pathways and structural stability	[53]

Advanced preparation methods for MOF-tuned graphene oxide membranes (GOm)—in-situ growth, Layer-by-Layer (LbL) assembly, and physical blending offer unique strengths for enhancing membrane performance. The in-situ growth method ensures strong interfacial interactions and uniform nanochannel structures, boosting permeability, selectivity, and stability. It also enables antibacterial and antifouling properties, making it ideal for water treatment and molecular separation. The LbL assembly method provides excellent tunability and scalability, using hydrophilic MOFs like UiO-66 and HZIF-8 to improve water flux and sieving performance. Innovations such as pressure-assisted self-assembly also enhance antifouling and mechanical strength while reducing production time. The physical blending method, in addition to being simple and scalable, is also highly cost-effective, enabling the incorporation of various MOFs to improve thermal stability, permeability, and antibacterial properties. Combining this method with LbL assembly helps overcome challenges like uneven dispersion and ensures consistent performance. Together, these techniques showcase the potential for creating efficient, scalable, and sustainable membranes for water treatment and separation applications. Continued development will further enhance their performance and affordability for industrial use.

2.4 Biochar

Biochar is a type of black carbon made from a carbonaceous material through pyrolysis without or under limited oxygen conditions or chemical applications [54]. It combines cost-effectiveness with environmental sustainability, making it attractive for resource-limited settings. Biochar exhibits a porous structure with a high surface area, allowing it to effectively adsorb various contaminants, including heavy metals (e.g., arsenic, lead, and cadmium), organic pollutants, and nutrients like nitrates and phosphates. Its surface chemistry can be further enhanced through chemical activation or functionalization, improving its selectivity and adsorption capacity. In addition to chemical pollutants, biochar can help reduce microbial loads in water. Its adsorption capabilities extend to certain bacteria and pathogens, contributing to safer drinking water.

Biochar is often produced from agricultural and forestry residues, making it a low-cost and environmentally friendly option. Its production process utilizes waste materials and sequesters carbon, contributing to climate change mitigation. Biochar can be used in standalone filtration systems or as a component of composite filters combined with other materials like sand or activated carbon. Its adaptability suits household-level systems and larger-scale community water treatment setups. Despite its benefits, biochar faces challenges such as variability in quality and performance due to differences in feedstock and production conditions. Ongoing research focuses on standardizing production methods and optimizing its functional properties for more effective contaminant removal. With advancements in these areas, biochar holds significant potential as a sustainable solution for global drinking water challenges. Table 11.5 summarizes the modification of biochar by previous research works.

2.5 Metal-oxide biochar

Metal-oxide biochar composites have been developed because of the limitation of pure biochar in environmental remediation and to obtain new materials that exploit the properties of adjuvant and biochar. Table 11.6 gives a general overview of the various methods of preparing metal-oxide biochar composites and their advantages. The synthesis of metal-biochar composite includes impregnation of biomass with metal salts, chemical precipitation, direct pyrolysis, and other emerging methods. Depending on the reagents used and the preparation conditions, nanoscale magnetically responsive biochar composites are obtained. The synthesis of metal-biochar composites involves incorporating metal oxides into the biochar matrix to enhance its physicochemical properties and functional performance. These modifications aim to improve adsorption, catalysis, and degradation of contaminants, offering significant benefits in environmental applications. The key modification methods vary by the type of metal oxide used, the pyrolysis conditions, and the targeted functional properties. For example, biochar modified with CeO_2 (CeO_2-BC) and ZrO_2 (ZrO_2-BC) demonstrates improved sonocatalytic activity. Integrating these metal oxides into the biochar matrix enhances the generation of reactive species during sonocatalysis, making them particularly effective for breaking down organic pollutants. This improvement in activity highlights the critical role of metal oxides in enhancing catalytic reactions facilitated by biochar.

ZnO-modified biochar (ZnO-BC), synthesized at a pyrolysis temperature of 600°C, exhibits an enlarged surface area and increased positive charge. These characteristics improve the composite's adsorption capacity, effectively capturing negatively charged contaminants, such as anions or organic pollutants, from aqueous solutions. This modification showcases

Table 11.5 Modification of biochar.

Materials	Pre-treatment	BET surface area (m^2/g)	Maximum adsorption capacity (mg/g)	Pollutant	Reference
Giand mud crab shell	Pyrolysis (500 °C, 2 h)	60	3139	Dye	[54]
Peanut shell	1. Mix with $Na_2SiO_3 \cdot 9H_2O$ for 48 h 2. Pyrolysis (700 °C, 2 h)	27	240	Heavy metal	[55]
Waste napkin	Pyrolysis (600 °C, 1 h)	2750	1159	Dye	[56]
Litchi peel	Hydrothermal carbonization (180 °C, 12 h)	1006	2468	Dye	[57]
Crayfish shell	Pyrolysis (800 °C, 1.5 h)	665	1079	Dye	[58]

Table 11.6 Various methods for obtaining metal-biochar composites.

Modification methods	Pyrolysis temperature (°C)	Advantage characteristics	References
Biochar modified with CeO_2 (CeO_2-BC)	500	Improving sonocatalytic activity by adding CeO_2 to the biochar matrix	[59]
Biochar modified with ZnO (ZnO-BC)	600	Surface enlargement and reinforcement of the positive charge of the resulting composite.	[60]
MgO-modified biochar (MgO-BC)	600	Production of a highly mesoporous material	[61]
Biochar modified with TiO_2 (TiO_2-BC)	–	Photogeneration of electrons due to interactions between BC and TiO_2	[62]
Biochar modified with ZrO_2 (ZrO_2-BC)	–	Improving sonocatalytic activity by adding ZrO_2 to the biochar matrix	[63]

the advantage of tailoring biochar's surface chemistry to optimize its interaction with specific pollutants. Similarly, MgO-modified biochar (MgO-BC) focuses on creating a highly mesoporous structure. Mesopores enhance the accessibility of adsorption sites, which is especially important for larger molecules or organic contaminants. This structural feature is vital for applications requiring high adsorption capacity and fast pollutant uptake.

TiO_2-modified biochar (TiO_2-BC) introduces a unique photocatalytic property. The interaction between TiO_2 and the biochar matrix allows for the photogeneration of electrons under light irradiation, enabling advanced photocatalytic processes. This modification is highly beneficial for degrading organic pollutants in water, as the reactive species generated can effectively break down contaminants into less harmful by-products. The pyrolysis temperature, ranging between 500 and 600°C, plays a pivotal role in determining the structural and functional outcomes of the composite. Higher temperatures typically enhance carbonization and surface area, which is critical for catalytic and adsorptive performance. However, the choice of metal oxide and the intended application ultimately dictate the specific modification approach.

In conclusion, metal-biochar composites exhibit enhanced capabilities compared to unmodified biochar, with each modification technique offering unique advantages. Integrating metal oxides like CeO_2, ZnO, MgO, TiO_2, and ZrO_2 not only tailors the composite's surface and structural properties but also introduces new functionalities, such as improved sonocatalytic or photocatalytic activity. These advancements position metal-biochar composites as auspicious materials for addressing complex environmental challenges, particularly in water and wastewater treatment applications.

2.6 MXene-polymer composites

The discovery of graphene oxide revolutionized the material world, but the rapidly evolving demands for new materials in the industry prompted researchers to address the deficiencies of graphene-based materials [64]. These efforts led to the discovery of MXenes in 2011, which garnered significant attention in a relatively short period compared to other 2D materials. The first synthesized MXene, derived from Ti3AlC2, resulted in 2D titanium carbide (Ti3C2Tx) [65]. The term "MXene" signifies the loss of A layers after etching the MAX phase and highlights its 2D nature, similar to graphene, the mother of all 2D materials [66]. The thickness of MXene is only a few atoms thick, with a general formula of Mn+1XnTx where n layers of nitrogen or carbon (X) are interleaved with n + 1 layers of transition metal (M). During the etching process, the surface of all units is covered by functional groups such as hydroxyl (OH), oxygen (O), and fluoride (F), where Tx in the formula stands for these functional groups [67]. The oxygen groups are favored in hydrogen evaluation reaction (HER), photo-catalysis, and electrochemical energy [68]. Moreover, the oxygen and hydroxyl groups are appropriate termination groups for higher adsorption capacity [69]. These unique properties of MXenes make them a truly impressive and intriguing area of research in material science and engineering.

The presence of hydroxyl groups on the surface of MXenes enables their use in desalination processes or the efficient removal of heavy metals [70]. The potential for reduction/oxidation reactions in MXenes also makes them suitable for photocatalytic purposes, allowing for the degradation of organic-based pollutants such as oils, dyes, microplastics, and nanoplastics [70]. MXenes possess excellent surface functionality and high electron density, giving them a rare combination of hydrophilicity and metallic conductivity, and thus, they can be employed in water treatment fields such as membrane technology and electrochemical separations. Additionally, MXenes have been used as adsorbents in the water treatment field, as they possess high specific surface area and tunable interlayer spacing. Moreover, it was reported that MXenes possess high antimicrobial properties due to their hydrophilic surface, large surface area, and excellent photothermal absorbance[71,72]. Therefore, MXenes can be valuable in water treatment and reducing microbial activity.

The chemistry of MXene is versatile, and it possesses promising magnetic, electrochemical, electronic, and mechanical properties. However, MXenes have salient shortcomings, and numerous experiments are being conducted to find possible solutions. Composition is among the best solutions because the combination of the 2D structure and flexible layered nature of MXenes makes them intriguing materials for composite fabrication [73]. The potential use of MXenes in various composites means that MXenes can be used in different applications, but MXene-based composites also hold great promise for many applications. Table 11.7 shows the comparison between three methods of MXene-polymer composites.

Preparing MXene-polymer composites relies on achieving a well-dispersed and uniform MXene phase. However, the hydrophilic nature of MXene surfaces creates challenges when mixing with hydrophobic polymers, as physical mixing often leads to poor compatibility and limited long-term stability. In-situ polymerization offers an alternative, relying on electrostatic interactions for monomer adsorption, but it is primarily suited for conductive polymers like PPy, PEDOT, and PANI.

Surface modification is a promising approach to address these challenges, enabling better compatibility with hydrophilic and hydrophobic matrices. It effectively expands the interlayer spacing of MXenes, enhancing mechanical strength, reducing interfacial resistance, improving electron-ion pathways, and preventing layer restacking. Additionally, surface modification allows for synergistic enhancements by grafting materials onto MXene surfaces.

However, this method often requires additional components, such as interlayers or catalysts, and can involve lengthy reaction times. Certain modifications, like silane groups or phosphonic acids, may compromise the MXene structure and reduce electrochemical performance. Ultimately, the

Table 11.7 Different methods to synthesize MXene-polymer composites [74].

Method name	Simplicity	Needed time	Drawbacks	Advantageous
Physical mixing	High	Low	Limited to hydrophilic polymers	Using a large-scale
In-situ polymerization	Medium	Medium	Limited to conductive polymers and lab scale	Direct synthesize on the MXene flakes
Surface modification	Low	Medium to high	Require more time and complexity, and require additional materials	Better physiochemical features, higher electron-ion channels, and lower interfacial resistance

choice of preparation strategy depends on the specific application, balancing simplicity, compatibility, and performance requirements.

2.7 Carbon dots

Carbon dots (CDs) are versatile nanomaterials characterized by low toxicity, biocompatibility, abundant edge defects, and ease of surface functionalization. These properties make them suitable for diverse applications in optoelectronics, catalysis, biomedical imaging, drug delivery, environmental monitoring, energy storage, water purification, and agriculture. CDs can be synthesized through two primary methods: top-down and bottom-up approaches. The top-down approach involves breaking down larger carbon materials like graphite and graphene into nanoscale CDs using laser ablation, ultrasonication, and electrochemical oxidation methods. While this method yields CDs with high crystallinity, well-defined core structures, and functionalized surfaces, it is energy-intensive and not ideal for large-scale production. In contrast, the bottom-up approach assembles CDs from small organic molecules or monomers using hydrothermal, pyrolysis, microwave, and template-assisted synthesis techniques. This method enables the production of tunable CDs with doped elements (e.g., nitrogen, sulfur) and tailored properties, though the resulting CDs are often amorphous.

Recent advancements emphasize using green precursors, such as biomass, agricultural waste (e.g., orange peels, chicken feathers), and microorganisms (e.g., yeast, bacteria), to produce CDs sustainably. These biosources, rich in nitrogen and oxygen, enhance the CDs' ability to adsorb heavy metals and other contaminants, making them ideal for environmental applications like water purification. CDs derived from bioactive molecules have also been shown to retain their biological properties, further expanding their utility in biomedical applications. The focus on green synthesis reduces environmental impact. It supports the development of cost-effective, sustainable materials for industrial and environmental use, showcasing the potential of CDs as innovative solutions for global challenges. Table 11.8 highlights the utilization of green precursors for synthesizing carbon dots.

The utilization of green precursors for synthesizing carbon dots, as summarized in Table 11.8, demonstrates significant progress in sustainable and eco-friendly nanomaterial production. Various biomass-derived materials such as eggshells, spent coffee grounds, Hami melon, crown daisy,

Table 11.8 Utilization of green precursors for synthesizing carbon dots.

Precursor	Synthesis method	Processing conditions	Carbon dot properties	Applications	Advantages	References
Eggshell	Hydrothermal	200°C, 8 h, pH ~7	Nitrogen and sulfur co-doped, good conductivity	Non-invasive detection of metronidazole (aptasensing)	Sustainable, abundant, biodegradable, high performance sensing interface	[75]
Spent coffee grounds	Hydrothermal	200°C, 12 h, neutral pH	Small size, high conductivity, amorphous structure	EDLC (electrochemical double-layer capacitor) electrodes	Waste reutilization, renewable, cost-effective	[76]
Hami melon (fruit biomass)	Hydrothermal	180°C, 6 h, neutral pH	Metal framework-modified, super large Stokes shift, high sensitivity	Detection of thiophanate-methyl residue in leafy vegetables	Renewable, biodegradable, high detection sensitivity	[77]
Crown daisy (plant biomass)	Hydrothermal	180°C, 10 h, neutral pH	Fluorescent, stable photoluminescence, biocompatible	–	Sustainable, renewable, eco-friendly	[78]
Green tea residue	Hydrothermal	200°C, 8 h, neutral pH	Fluorescent, high selectivity for Fe^{3+} ions, biocompatible	Detection of Fe^{3+} ions, imaging of Caco-2 cells	Renewable, cost-effective, biodegradable	[79]

and green tea residue have been effectively employed in hydrothermal methods under moderate conditions (180–200°C) and neutral to slightly acidic pH. These green precursors result in carbon dots with diverse and valuable physicochemical properties, including nitrogen and sulfur doping, high conductivity, fluorescent characteristics, and metal framework modifications. These properties make carbon dots versatile for applications such as non-invasive sensing (e.g., metronidazole detection), electrochemical energy storage (EDLC electrodes), detection of specific ions like Fe^{3+}, residue analysis in leafy vegetables, and bioimaging Caco-2 cells.

The advantages of using these precursors lie in their sustainability, renewability, and cost-effectiveness, as they are abundant waste or biomass materials. Moreover, the synthesized carbon dots exhibit desirable features such as high sensitivity, biocompatibility, stable photoluminescence, and excellent conductivity, making them suitable for both sensing and energy-related applications. Overall, the findings highlight the potential of green precursors to replace conventional materials, contributing to environmentally friendly and economically viable carbon dot production for advanced technological applications.

3. Application of carbon-based materials and composites for drinking water treatment

3.1 Activated carbon

Activated carbon (AC) is one of the most extensively used materials in water treatment due to its high porosity and surface area, which allow for the efficient adsorption of pollutants. In drinking water applications, activated carbon is widely employed to remove organic contaminants, such as pesticides, pharmaceuticals, and natural organic matter (NOM). The adsorption mechanism is driven by hydrophobic interactions and surface binding, which make it highly effective in capturing volatile organic compounds (VOCs) responsible for undesirable taste and odor. Additionally, activated carbon is frequently incorporated into point-of-use filtration systems, providing a practical and accessible method for households to purify water. Its ability to adsorb heavy metals, including lead (Pb) and cadmium (Cd), further underscores its versatility in addressing diverse water quality challenges. Moreover, its reactivation and reuse capabilities make it a sustainable option in industrial-scale water treatment processes, offering long-term benefits and optimism for the future of water treatment. Table 11.9 shows the use of activated carbon for drinking water treatment.

Table 11.9 The use of activated carbon for drinking water treatment.

Precursor	Activator	Pollutant	Maximum adsorption capacity (mg/g)	Reference
Sugarcane bagasse	$ZnCl_2$	Sodium diclofenac	315	[80]
Argania spinasa tree nutshells	H_3PO_4	Carbamazepine	176	[81]
Date press cake	NaOH	Cr (VI)	283	[82]
Rice husk	NH_4Br	Hg	769	[83]
Pistachio wood	NH_4NO_3 and NaOH	Pb^{2+}	190	[84]

3.2 Carbon nanotubes (CNTs)

Carbon nanotubes (CNTs) are renowned for their tubular nanostructure, high surface area, and chemical stability, making them highly effective in water treatment. They exhibit exceptional adsorption capabilities for various pollutants, including heavy metals, hydrophobic organic compounds, and persistent organic pollutants. CNTs can also act as nanocatalysts, accelerating the breakdown of complex organic molecules into more straightforward and less harmful compounds. Their antimicrobial properties allow them to disrupt bacterial membranes, making them suitable for pathogen removal. Moreover, CNTs are increasingly incorporated into filtration membranes to enhance permeability and selectivity, providing an efficient solution for removing organic and inorganic contaminants from water. Recent advancements have focused on functionalizing CNTs with specific chemical groups to improve their adsorption efficiency and reduce fouling during filtration. Table 11.10 shows the use of carbon nanotubes for the disinfection of water.

3.3 Graphene oxides (GO)

Graphene oxides (GO), with their two-dimensional structure and abundant oxygen-containing functional groups, offer excellent performance in water treatment applications. Due to their high binding affinity, they are particularly effective in adsorbing heavy metals, such as arsenic, mercury,

Table 11.10 The use of carbon nanotubes for the disinfection of water.

Type	Microorganism	Parameters	Outcomes	Reference
SWCNTs-OH; SWCNTs-COOH	Paracoccus denitrificans (ATCC19367)	Concentration = 50 mg/L	Modification in the expression of genes related to the free fatty acid/ glycerolipid cycle, with decreased transcription of key genes involved in transcriptional regulation, electron transfer, and substance transport.	[85]
ZnO + MWCNT	E coli	Concentration = 0.5 mg/mL	Complete elimination of bacterial activity	[86]
TiO_2 + magnetic CNT (MCNT)	E coli	MCNT:TiO_2 = 5:1 concentration = 1 g/L cell density = 1–2×10^6 CFU/mL	Complete elimination of bacterial activity	[87]
Magnetic carbon nanotubes/red phosphorus/graphitic carbon nitride	Staphylococcus aureus	MCNTs/RP/CN: CN/RP to MCNTs ratios: 4:1 Concentration: ~10^7 cfu/mL	The disinfection rate was approximately 13.9 times higher than the commercial P25 TiO_2 photocatalyst.	[88]

and lead. Functionalized GO with specific ligands further enhance their removal efficiency for targeted contaminants. In addition, GO can remove organic pollutants, including dyes and pharmaceutical residues, through surface interactions and hydrophilic-hydrophobic balance. The antimicrobial properties of graphene oxides, driven by oxidative stress mechanisms, make them valuable for disinfection applications. GO are also integrated into advanced filtration membranes, enhancing desalination performance and pollutant rejection capabilities. Their ability to form hybrid composites with other materials expands their application potential in multifunctional water treatment systems. Table 11.11 compares the separation performance of different GO-based membranes, heavy metal Ions, and salt rejection.

3.4 Biochar

Biochar, a carbon-rich material from biomass pyrolysis, offers an eco-friendly and cost-effective alternative to activated carbon. Its porous structure and surface chemistry make it highly effective for adsorbing heavy metals such as arsenic (As) and chromium (Cr), as well as nutrients like nitrates and phosphates. Biochar also demonstrates potential in addressing emerging contaminants, such as endocrine-disrupting chemicals and antibiotic residues, by leveraging its tunable surface properties. Adding functional groups or doping with metal ions can enhance its adsorption capacity and selectivity for specific pollutants. Furthermore, biochar is a precursor

Table 11.11 Comparison of separation performance of different GO-based membranes heavy metal Ions and salt rejection.

Membrane category	Permeance (L m^{-2} h^{-1} bar^{-1})	Heavy metal ions/salt	Rejection (%)	Reference
GO/MoS_2-PVA	3.96	NaCl	88.9	[89]
GO/Gr	12.8–20	Na_2SO_4	78.3	[90]
UiO-66-$(COOH)_2$/prGO	20.0 ± 2.5	Cu^{2+}	96.5	[91]
		Cd^{2+}	92.6	
ZIF-67/cGO-HFMs	346.4 ± 11.2	Cu^{2+}	94.5	[92]
		Pb^{2+}	97.8	
SGO@UiO-66TFN	14	Cu^{2+}	99.4	[93]
		Pb^{2+}	99.4	
MOF-801@GO	29.24	NaCl	99.9	[43]

for functionalized materials, enhancing its adsorption and catalytic capabilities when combined with other elements or compounds. Its production process also allows for the utilization of agricultural waste, contributing to circular economy practices. Table 11.12 shows the use of biochar for drinking water treatment.

3.5 Metal-oxide biochar

The integration of metal oxides into biochar significantly enhances its adsorption and catalytic properties. Metal-oxide biochar has shown remarkable efficacy in removing ions such as fluoride and arsenic through specific metal-ion interactions. For example, biochar doped with iron oxides exhibits a strong affinity for arsenic, while titanium oxide-functionalized biochar effectively addresses fluoride contamination. Additionally, it plays a vital role in catalytic degradation processes, particularly advanced oxidation processes (AOPs), where it breaks down organic pollutants into less harmful compounds. The antibacterial properties of metal oxides, such as ZnO and TiO_2, further extend the functionality of this composite material by enabling pathogen disinfection. These properties make metal-oxide biochar a promising material for comprehensive drinking water treatment systems that target multiple contaminants simultaneously. Table 11.13 shows the adsorption capabilities of biochar-metal composites for removing inorganic pollutants from aqueous solutions.

3.6 MXene-polymer composites

MXenes, a family of two-dimensional transition metal carbides and nitrides, exhibit exceptional hydrophilicity, high conductivity, and tunable

Table 11.12 The used of biochar for drinking water treatment.

Precursor	Production temperature (°C)	Pollutant	Maximum adsorption capacity	Reference
Cassava waste residue	500	Norfloxacin (NOR)	38.77 mg/g	[94]
Prosopis juliflora	550 and 850	Microplastics (MP)	>200 mg/g	[51,52]
Spent mushroom	200, 450 and 600	Progesterone	80%	[95]
Wheat straw	700	Tetracycline	268.3 mg/g	[96]
Peanut shell	400	Pb^{2+}	231.9 mg/g	[97]

Table 11.13 Adsorption capabilities of biochar-metal composites for removing inorganic pollutants from aqueous solutions.

Biomass	Modification method	Adsorbates	Maximum adsorption capacity (mg/g)	% removal	Reference
Peanut hull	Zero valent iron	Cr (VI)	223	100.0	[98]
Corn straw	$MnFe_2O_4$	Pb (II)	155	99.0	[99]
Leaf litter	CeO_2, MoS_2 and magnetic	Pb (II)	264	99.6	[100]
Rice husk	Hydroxyapatite	Pb (II)	110	99.5	[101]
Lufa cylindrical fiber	MnO_2	U (VI)	904	100.0	[102]

surface chemistry. When combined with polymers, MXene-polymer composites enhance drinking water treatment performance. These composites are highly effective in adsorbing heavy metals such as chromium lead and organic pollutants. Their sharp edges and oxidative properties contribute to pathogen inactivation, providing an additional disinfection layer. MXene-polymer composites are also being developed for advanced filtration systems, including nanofiltration and desalination, where they demonstrate superior pollutant rejection and durability. Incorporating MXenes into electrochemical systems further enables the degradation of refractory organic pollutants, showcasing their potential in next-generation water treatment technologies. Table 11.14 shows the use of MXene-polymer adsorbents to remove heavy metals in water.

3.7 Carbon dots

Carbon dots (CDs) are nanoscale materials with high surface area, photocatalytic activity, and fluorescence properties. CDs play a crucial role in degrading organic pollutants, such as pesticides and pharmaceutical residues, through photocatalytic reactions under visible light in drinking water treatment. This process generates reactive oxygen species (ROS), breaking contaminants into less toxic by-products. CDs also exhibit strong antimicrobial properties, effectively inactivating bacteria and viruses. Furthermore, their fluorescence properties enable them to function as sensors for detecting pollutants, such as heavy metals and emerging contaminants, while simultaneously facilitating their removal. CDs' low toxicity and environmentally friendly synthesis methods make them highly attractive for sustainable water treatment applications. Emerging research is focused on enhancing their photocatalytic efficiency and exploring their potential in hybrid systems for multifunctional treatment approaches. Table 11.15 summarizes carbon dot-based inorganic composites derived from various precursors and combined with different polymeric materials for the sorption-based removal of heavy metals.

4. Challenges and limitations

Advanced carbon-based materials, such as biochar (BC), activated carbon (AC), graphene oxide membranes (GOm), carbon nanodots (CDs), MXenes, and carbon nanotubes (CNTs), hold significant potential for environmental remediation. However, their widespread adoption is limited by various challenges. Synthesis and production constraints,

Table 11.14 The use of MXene-polymer adsorbents for removal of heavy metals in water.

Adsorbent type	Pollutant	Adsorption mechanism	Maximum adsorption capacity (mg/g)	Reference
MXene/biochar-PDA/PEI	U^{6+}	Chelation, ion exchange, electrostatic interaction	240	[71,72]
Polyethyleneimine@MXene	Cr^{6+}	Hydrogen bonding, electrostatic force, reduction reaction	1240	[103]
Chitin/MXene/poly(L-arginine)	Bilirubin	Both chemisorption monolayer and multilayer adsorption	596	[104]
Polydopamine/MXene	Cr^{6+}	Reduction reaction, electrostatic interactions	862	[105]
PNIPAM/MXene	Cu^{2+}	Thermo-responsive, electrostatic interaction	224	[106]

Table 11.15 Carbon dot-based inorganic composites derived from various precursors and combined with different polymeric materials for the sorption-based removal of heavy metals.

Adsorbent	Carbon dot precursor	Pollutant	Maximum adsorption capacity (mg/g)	Reference
Magnetic adsorbent based on k-carrageenan and N-doped carbon dots	Astragalus gossypinus, ethylenediamine	1. Cu (II) 2. Cd (II) 3. Pb (II) 4. Hg (II)	1. 115 2. 103 3. 120 4. 108	[107]
Magnetic metal-organic frameworks/carbon dots	CMC	U (VI)	606	[108]
Functionalized layered double hydroxide with nitrogen and sulfur co-decorated carbon dots	Sodium citrate, L-cysteine	1. Hg (II) 2. Ag (I)	1. 625 2. 714	[109]
Dual functional nanocomposites of magnetic MnFe2O4 and carbon dot (citric acid and polyethyleneimine CD)	1. Citric acid, glycol 2. Polyethyleneimine, citric acid	U (VI)	194	[110]
Flower-like layered double oxides/carbon dots	Chitosan	U (VI)	354	[111]

including inconsistent properties of BC and AC due to feedstock and pyrolysis variations, high energy demands, and hazardous etching processes for MXenes, present significant obstacles. Similarly, the high fabrication costs and scalability issues of GOm, MXenes, and CNTs hinder their industrial application despite advancements in synthesis technologies reducing costs over time.

Environmental and toxicity concerns further complicate their use. Biochar poses risks of metal leaching and genotoxic effects in soil applications, while MXenes are prone to oxidation and have unassessed environmental impacts. CNTs, despite their versatility, may harm ecosystems and human health due to factors like reactive oxygen species (ROS) generation and aggregation. These concerns necessitate rigorous safety evaluations and environmentally friendly production methods.

Performance limitations are another challenge. Real-world water matrices reduce the adsorption efficiency of BC and AC, while GOm faces issues like poor aqueous stability, swelling, and fouling. MXenes struggle with delamination and limited selectivity, particularly for complex organic contaminants, while CDs require optimization for adsorption efficiency and regeneration methods.

Scalability and economic challenges persist across all materials, particularly for GOm, MXenes, and CNTs, where high production costs and weak interfacial bonding in composites limit widespread application. Additionally, a lack of fundamental understanding of the behaviors of newer composites, coupled with insufficient toxicity and environmental impact assessments, further restricts progress. Lastly, raising awareness about the broader environmental and agricultural benefits of materials like BC and AC and predictive tools like machine learning for optimizing material properties is essential for overcoming these limitations. Addressing these multifaceted challenges is crucial for advancing carbon-based materials toward sustainable and impactful applications.

5. Future directions

Future efforts should focus on addressing the challenges of advanced carbon-based materials by standardizing production methods for biochar (BC), activated carbon (AC), and other materials to ensure consistent properties while developing greener, cost-effective, and scalable synthesis techniques such as fluorine-free etching for MXenes and bio-based precursors for carbon nanodots (CDs). Machine learning and computational tools can optimize material design and predict performance in real-world conditions, enabling site-specific adsorbents and tailored composites with improved stability, selectivity, and efficiency. Comprehensive environmental and toxicity studies are critical to mitigating risks associated with carbon nanotubes (CNTs), MXenes, and related materials, alongside establishing regulatory frameworks for responsible use. Issues like swelling, fouling, and mechanical instability in graphene oxide membranes (GOm) and MXenes can be addressed through cross-linking, polymer reinforcement, and surface modifications to enhance performance. Cost reduction and scalability should be prioritized by advancing decentralized production units for BC and AC and industrial-scale synthesis of CNTs and MXenes, supported by renewable energy-driven applications like solar-assisted photocatalysis. Furthermore, raising awareness about these materials' environmental and agricultural benefits, such as greenhouse gas mitigation and soil improvement, and fostering policy integration will drive their adoption. Fundamental research to elucidate adsorption mechanisms and interfacial behaviors, coupled with energy-efficient production and waste valorization practices, will ensure these materials achieve their full potential in sustainable water and environmental remediation.

6. Conclusion

Carbon-based materials and composites hold immense promise for advancing drinking water treatment due to their unique properties, such as high surface area, tunable functionality, and exceptional adsorption and filtration capabilities. Materials like biochar, activated carbon, carbon nanotubes (CNTs), graphene oxide membranes (GOm), carbon nanodots (CDs), and MXene composites have demonstrated significant potential in removing diverse contaminants, including heavy metals, organic pollutants, and pathogens. However, the widespread application of these materials faces challenges such as high production costs, inconsistent properties, scalability issues, environmental toxicity, and performance limitations under real-world conditions. Future research and innovation should address these limitations by developing cost-effective, environmentally sustainable, and scalable synthesis methods while optimizing material performance through advanced computational tools and machine learning. It will be crucial to integrate these materials with conventional treatment technologies, improve their stability, and minimize potential secondary pollution. Additionally, comprehensive studies on environmental and health impacts, alongside awareness initiatives, are necessary to foster public acceptance and policy integration. With continuous advancements and multidisciplinary collaboration, carbon-based materials and composites have the potential to revolutionize sustainable drinking water treatment, contributing to global efforts to ensure clean and safe water for all.

References

[1] Kalengyo, R.B., Nasr, M., 2024. Water resource management to meet SDG13 on climate change. Water Environ. Sustain. Dev. Goals 157–178. https://doi.org/10.1016/b978-0-443-15354-9.00014-1.

[2] McIntyre, O., 2023. Ensure availability and sustainable management of water and sanitation for all. UN Sustain. Dev. Goals 441–508. https://doi.org/10.1093/law/9780192885173.003.0007.

[3] Laad, M., Ghule, B., 2023. Removal of toxic contaminants from drinking water using biosensors: A systematic review. Groundw. Sustain. Dev. 20, 100888. https://doi.org/10.1016/j.gsd.2022.100888.

[4] Mukhopadhyay, A., Duttagupta, S., Mukherjee, A., 2022. Emerging organic contaminants in global community drinking water sources and supply: A review of occurrence, processes and remediation. J. Environ. Chem. Eng. 10 (3), 107560. https://doi.org/10.1016/j.jece.2022.107560.

[5] Lei, M., Zhou, J., Zhou, Y., Sun, Y., Ji, Y., Zeng, Y., 2022. Spatial distribution, source apportionment and health risk assessment of inorganic pollutants of surface water and groundwater in the southern margin of junggar basin, Xinjiang, China. J. Environ. Manag. 319, 115757. https://doi.org/10.1016/j.jenvman.2022.115757.

[6] Haldar, D., Duarah, P., Purkait, M.K., 2020. MOFs for the treatment of arsenic, fluoride and iron contaminated drinking water: A review. Chemosphere 251, 126388. https://doi.org/10.1016/j.chemosphere.2020.126388.

[7] Senanu, L.D., Kranjac-Berisavljevic, G., Cobbina, S.J., 2023. The use of local materials to remove heavy metals for household-scale drinking water treatment: A review. Environ. Technol. Innov. 29, 103005. https://doi.org/10.1016/j.eti.2023.103005.

[8] Kumar, A., Jeyabalan, J., Priyan V, V., Charan Patra, C., Narayanasamy, S., 2023. Fabrication of a novel bio-polymer adsorbent with high adsorptive capacity towards organic dyes. Ind. Crop. Prod. 203, 117166. https://doi.org/10.1016/j.indcrop.2023.117166.

[9] Kumar, R., Qureshi, M., Vishwakarma, D.K., Al-Ansari, N., Kuriqi, A., Elbeltagi, A., Saraswat, A., 2022. A review on emerging water contaminants and the application of sustainable removal technologies. Case Stud. Chem. Environ. Eng. 6, 100219. https://doi.org/10.1016/j.cscee.2022.100219.

[10] Hawash, H.B., Moneer, A.A., Galhoum, A.A., Elgarahy, A.M., Mohamed, W.A.A., Samy, M., El-Seedi, H.R., Gaballah, M.S., Mubarak, M.F., Attia, N.F., 2023. Occurrence and spatial distribution of pharmaceuticals and personal care products (ppcps) in the aquatic environment, their characteristics, and adopted legislations. J. Water Process Eng. 52, 103490. https://doi.org/10.1016/j.jwpe.2023.103490.

[11] O'Connor, J., Bolan, N.S., Kumar, M., Nitai, A.S., Ahmed, M.B., Bolan, S.S., Vithanage, M., Rinklebe, J., Mukhopadhyay, R., Srivastava, P., Sarkar, B., Bhatnagar, A., Wang, H., Siddique, K.H.M., Kirkham, M.B., 2022. Distribution, transformation and remediation of poly- and per-fluoroalkyl substances (PFAS) in

wastewater sources. Process Saf. Environ. Prot. 164, 91–108. https://doi.org/10.1016/j.psep.2022.06.002.

[12] Zhang, P., Yang, M., Lan, J., Huang, Y., Zhang, J., Huang, S., Yang, Y., Ru, J., 2023. Water quality degradation due to heavy metal contamination: Health impacts and eco-friendly approaches for heavy metal remediation. Toxics 11 (10), 828. https://doi.org/10.3390/toxics11100828.

[13] Rout, S., Yadav, S., Pansare, L.G., Shashidhar, R., Pulhani, V., Vinod Kumar, A., 2025. Simultaneous removal of heavy metals and escherichia coli from water using silver-doped hydroxyapatite-coated activated carbon nano composite-alginate beads. Sustainability 5, 100076. https://doi.org/10.1016/j.nxsust.2024.100076.

[14] World Health Organization, 2024. Guidelines for drinking-water quality: Fourth edition incorporating the first and Second addenda. World Health Organization. https://www.who.int/publications/i/item/9789240045064.

[15] Mane, P.V., Rego, R.M., Yap, P.L., Losic, D., Kurkuri, M.D., 2024. Unveiling cutting-edge advances in high surface area porous materials for the efficient removal of toxic metal ions from water. Prog. Mater. Sci. 146, 101314. https://doi.org/10.1016/j.pmatsci.2024.101314.

[16] Naganathan, K.K., Faizal, A.N., Zaini, M.A., Ali, A., 2021. Adsorptive removal of bisphenol A from aqueous solution using activated carbon from coffee residue. Mater. Today Proc. 47, 1307–1312. https://doi.org/10.1016/j.matpr.2021.02.802.

[17] Liang, J., Li, C., Zhang, S., Wang, S., Hu, X., 2024. Sequential activation of willow wood with ZNCL2 and H3PO4 drastically impacts pore structure of activated carbon. Ind. Crop. Prod. 221, 119387. https://doi.org/10.1016/j.indcrop.2024.119387.

[18] Efimov, M.N., Muratov, D.G., Zhilyaeva, N.A., Vasilev, A.A., Legkov, S.A., Karpacheva, G.P., 2025. Effect of pretreatment temperature and atmosphere on the structure and adsorption properties of Koh-activated porous carbon for methylene blue removal. Inorg. Chem. Commun. 172, 113689. https://doi.org/10.1016/j.inoche.2024.113689.

[19] Wei, S., Qin, Q., Liu, Z., 2024. Thermal behavior analysis and reaction mechanism in the preparation of activated carbon by zncl2 activation of bamboo fibers. J. Anal. Appl. Pyrolysis 179, 106500. https://doi.org/10.1016/j.jaap.2024.106500.

[20] Jaber, L., Backer, S.N., Laoui, T., Abumadi, F., Koujan, M.M., Khalil, K.A., Shanableh, A., Atieh, M.A., 2024. Recent trends in surface impregnation techniques on activated carbon for efficient pollutant removal from wastewater. Desalination Water Treat. 319, 100562. https://doi.org/10.1016/j.dwt.2024.100562.

[21] Thakur, A., Kumar, A., 2024. Unraveling the multifaceted mechanisms and untapped potential of activated carbon in remediation of emerging pollutants: A comprehensive review and critical appraisal of advanced techniques. Chemosphere 346, 140608. https://doi.org/10.1016/j.chemosphere.2023.140608.

[22] Kundu, S., Khandaker, T., Anik, M.A.-A., Hasan, M.K., Dhar, P.K., Dutta, S.K., Latif, M.A., Hossain, M.S., 2024. A comprehensive review of enhanced CO_2 capture using activated carbon derived from biomass feedstock. RSC Adv. 14 (40), 29693–29736. https://doi.org/10.1039/d4ra04537h.

[23] Al-mahbashi, N.M.Y., Kutty, S.R.M., Jagaba, A.H., Rathnayake, U., Hagar, H.S., Al-Dhawi, B.N.S., Al-Shawesh, G.A.M., 2024. Optimization and characterization of paddy stalks-based activated carbon by physical activation: Box-behnken design approach. Nano-Struct. Nano-Objects 38, 101155. https://doi.org/10.1016/j.nanoso.2024.101155.

[24] Vinayagam, M., Suresh Babu, R., Sivasamy, A., Barros, A. L. F. de, 2024. Physical activation assisted porous activated carbon from Strychnos potatorum shells for

high-performance symmetric supercapacitors. Mater. Lett. 371, 136961. https://doi.org/10.1016/j.matlet.2024.136961.
[25] Neema, P., Verma, S.K., Shaz, M.A., 2025. Physically activated resorcinol-formaldehyde derived carbon aerogels for enhanced hydrogen storage. Int. J. Hydrogen Energy 141, 803–810. https://doi.org/10.1016/j.ijhydene.2024.10.015.
[26] Lionetti, V., Poselle Bonaventura, C., Conte, G., De Luca, O., Policicchio, A., Caruso, T., Desiderio, G., Papagno, M., Agostino, R.G., 2024. Production and physical-chemical characterization of walnut shell-derived activated carbons for hydrogen storage application. Int. J. Hydrogen Energy 61, 639–649. https://doi.org/10.1016/j.ijhydene.2024.02.213.
[27] Ali, A.A., Abdul-Hameed, H.M., 2024. Synthesis and characterization of magnetic activated carbon manufactured from palm stones by physical activation to remove lead from aqueous solution. Desalination Water Treat., 100951 https://doi.org/10.1016/j.dwt.2024.100951.
[28] Blanchard, R., Mekonnen, T.H., 2024. Valorization of plastic waste *via* chemical activation and carbonization into activated carbon for functional material applications. RSC Appl. Polym. 2 (4), 557–582. https://doi.org/10.1039/d4lp00016a.
[29] Ren, W., Li, C., Fan, M., Wang, Y., Li, Q., Hu, X., 2025. Production of poplar branch-derived activated carbon with acidic deep eutectic solvent pretreatment coupled with chemical activation. Fuel 381, 133417. https://doi.org/10.1016/j.fuel.2024.133417.
[30] Lamata-Bermejo, I., Alba, M.D., Ramírez-Rico, J., 2024. Influence of the chemical activation with KOH/KNO3 on the CO2 adsorption capacity of activated carbons from pyrolysis of cellulose. J. Environ. Chem. Eng. 12 (6), 114288. https://doi.org/10.1016/j.jece.2024.114288.
[31] Alouiz, I., Benhadj, M., Dahmane, E., Sennoune, M., Amarouch, M.-Y., Mazouzi, D., 2024. Elaboration of fibrous structured activated carbon from olive pomace via chemical activation and low-temperature pyrolysis. Heliyon 10 (20). https://doi.org/10.1016/j.heliyon.2024.e38886.
[32] Hasnain, M.S., Nayak, A.K., 2019. Background: Carbon nanotubes for targeted drug delivery. In: SpringerBriefs in Applied Sciences and Technology, pp. 1–9. https://doi.org/10.1007/978-981-15-0910-0_1.
[33] Song, J., Deng, Q., Huang, M., Kong, Z., 2022. Carbon nanotube enhanced membrane distillation for salty and dyeing wastewater treatment by electrospinning technology. Environ. Res. 204, 111892. https://doi.org/10.1016/j.envres.2021.111892.
[34] Jain, N., Jee Kanu, N., 2021. The potential application of carbon nanotubes in water treatment: A state-of-the-art-review. Mater. Today Proc. 43, 2998–3005. https://doi.org/10.1016/j.matpr.2021.01.331.
[35] Mishra, S., Sundaram, B., 2023. Efficacy and challenges of carbon nanotube in wastewater and water treatment. Environ. Nanotechnol. Monit. Manag. 19, 100764. https://doi.org/10.1016/j.enmm.2022.100764.
[36] Boufades, D., Hammadou Née Mesdour, S., Moussiden, A., Benmebrouka, H., Ghouti, M., Kaddour, O., 2022. Optimization of carbon nanotubes synthesis via pyrolysis over Ni/Al2O3using response surface methodology. Fullerenes, Nanotub. Carbon Nanostruct. 30 (4), 467–475. https://doi.org/10.1080/1536383x.2021.1956475.
[37] Ahmad, S., Liao, Y., Hussain, A., Zhang, Q., Ding, E.-X., Jiang, H., Kauppinen, E. I., 2019. Systematic investigation of the catalyst composition effects on single-walled carbon nanotubes synthesis in floating-catalyst CVD. Carbon 149, 318–327. https://doi.org/10.1016/j.carbon.2019.04.026.
[38] Moo, J.G., Veksha, A., Oh, W.-D., Giannis, A., Udayanga, W.D.C., Lin, S.-X., Ge, L., Lisak, G., 2019. Plastic derived carbon nanotubes for electrocatalytic oxygen

reduction reaction: Effects of plastic feedstock and synthesis temperature. Electrochem. Commun. 101, 11–18. https://doi.org/10.1016/j.elecom.2019.02.014.

[39] Lotfy, V.F., Fathy, N.A., Basta, A.H., 2018. Novel approach for synthesizing different shapes of carbon nanotubes from rice straw residue. J. Environ. Chem. Eng. 6 (5), 6263–6274. https://doi.org/10.1016/j.jece.2018.09.055.

[40] Karami Hoomani, K., Nasouri, K., Mandegari, M., Jalali, S.A., 2025. In situ synthesis of FE3O4/carboxylated multi-walled carbon nanotubes nanostructures on nylon fabric with efficient flame retardant and antibacterial properties. Colloids Surf. A Physicochem. Eng. Asp. 706, 135797. https://doi.org/10.1016/j.colsurfa.2024.135797.

[41] Mandal, M., Alam, I., Sahoo, M.R., Patra, S., Mahapatra, A., Nayak, A.K., Subudhi, S., Mahanandia, P., 2024. Simple pyrolysis synthesis of multi-walled carbon nanotubes as a highly stable electrode material for supercapacitor applications. Chem. Phys. Lett. 853, 141536. https://doi.org/10.1016/j.cplett.2024.141536.

[42] Qian, W., Yang, Q., Zhang, D., Su, J., Ding, Y., Yao, D., 2025. Unlocking the potential of graphene oxide membranes: Microstructural tuning through metal–organic frameworks for water treatment progress. Sep. Purif. Technol. 357, 130140. https://doi.org/10.1016/j.seppur.2024.130140.

[43] Mao, Y., Xu, J., Chen, H., Liu, G., Liu, Z., Cheng, L., Guo, Y., Liu, G., Jin, W., 2023. Hydrophobic metal-organic framework@graphene oxide membrane with enhanced water transport for desalination. J. Membr. Sci. 669, 121324. https://doi.org/10.1016/j.memsci.2022.121324.

[44] Wei, N., Zheng, X., Li, Q., Gong, C., Ou, H., Li, Z., 2020. Construction of lanthanum modified mofs graphene oxide composite membrane for high selective phosphorus recovery and water purification. J. Colloid Interface Sci. 565, 337–344. https://doi.org/10.1016/j.jcis.2020.01.031.

[45] Chen, C., Shen, L., Lin, H., Zhao, D., Li, B., Chen, B., 2024. Hydrogen-bonded organic frameworks for membrane separation. Chem. Soc. Rev. 53 (5), 2738–2760. https://doi.org/10.1039/d3cs00866e.

[46] Zhang, W.-H., Yin, M.-J., Zhao, Q., Jin, C.-G., Wang, N., Ji, S., Ritt, C.L., Elimelech, M., An, Q.-F., 2021. Graphene oxide membranes with stable porous structure for ultrafast water transport. Nat. Nanotechnol. 16 (3), 337–343. https://doi.org/10.1038/s41565-020-00833-9.

[47] Xu, L.-H., Li, Y., Li, S.-H., Lv, M.-Y., Zhao, Z.-P., 2022. Space-confined growth of 2D MOF sheets between go layers at room temperature for superior PDMS membrane-based ester/water separation. J. Membr. Sci. 656, 120605. https://doi.org/10.1016/j.memsci.2022.120605.

[48] Chang, R., Ma, S., Guo, X., Xu, J., Zhong, C., Huang, R., Ma, J., 2019. Hierarchically assembled graphene oxide composite membrane with self-healing and high-efficiency water purification performance. ACS Appl. Mater. Interfaces 11 (49), 46251–46260. https://doi.org/10.1021/acsami.9b18018.

[49] Dai, L., Huang, K., Xiong, Z., Qu, K., Wang, Y., Pang, S., Zhang, D., Xu, F., Lei, L., Guo, X., Xu, Z., 2022. Two-dimensional heterogenous channels incorporated by enhanced-surface hydrophilic hollow zif-8 nanocrystals for ultrafast water permeation. J. Membr. Sci. 661, 120943. https://doi.org/10.1016/j.memsci.2022.120943.

[50] Ying, Y., Liu, D., Zhang, W., Ma, J., Huang, H., Yang, Q., Zhong, C., 2017. High-flux graphene oxide membranes intercalated by metal–organic framework with highly selective separation of aqueous organic solution. ACS Appl. Mater. Interfaces 9 (2), 1710–1718. https://doi.org/10.1021/acsami.6b14371.

[51] Wang, Z., Zhu, J., Xu, S., Zhang, Y., Van der Bruggen, B., 2021a. Graphene-like MOF nanosheets stabilize graphene oxide membranes enabling selective molecular sieving. J. Membr. Sci. 633, 119397. https://doi.org/10.1016/j.memsci.2021.119397.

[52] Wang, J., Sun, C., Huang, Q.-X., Chi, Y., Yan, J.-H., 2021b. Adsorption and thermal degradation of microplastics from aqueous solutions by Mg/Zn modified magnetic biochars. J. Hazard Mater. 419, 126486. https://doi.org/10.1016/j.jhazmat.2021.126486.
[53] Sui, X., Ding, H., Yuan, Z., Leong, C.F., Goh, K., Li, W., Yang, N., D'Alessandro, D.M., Chen, Y., 2019. The roles of metal-organic frameworks in modulating water permeability of graphene oxide-based carbon membranes. Carbon 148, 277–289. https://doi.org/10.1016/j.carbon.2019.03.049.
[54] Mohd Faizal, A.N., Putra, N.R., Abdul Aziz, A.H., Agi, A., Ahmad Zaini, M.A., 2024. Giant mud crab shell biochar: A promising adsorbent for methyl violet removal in wastewater treatment. J. Clean. Prod. 447, 141637. https://doi.org/10.1016/j.jclepro.2024.141637.
[55] Sun, M., Cheng, B., Liu, H., Wang, X., Tian, X., Yu, X., Dai, J., 2024. Silicon-modified peanut shell biochar: Novel passivation to mitigate cadmium bio-accumulation in spinach. J. Environ. Chem. Eng., 114995 https://doi.org/10.1016/j.jece.2024.114995.
[56] Jin, Y., Zhou, Z., Chen, S., Tian, C., Chen, G., 2025. Waste napkin biochar with high-performance designed for antibiotic rapidly removal. Environ. Technol. Innov. 37, 103921. https://doi.org/10.1016/j.eti.2024.103921.
[57] Wu, J., Yang, J., Feng, P., Huang, G., Xu, C., Lin, B., 2020. High-efficiency removal of dyes from wastewater by fully recycling litchi peel biochar. Chemosphere 246, 125734.
[58] Mohd Faizal, A.N., Wen, C.H., Putra, N.R., Zaini, A.S., Agi, A., Nordin, A.H., Zaini, M.A., 2025. Crayfish shell biochar for methyl violet adsorption: Equilibrium and kinetic studies. Sustainability 6, 100093. https://doi.org/10.1016/j.nxsust.2024.100093.
[59] Khataee, A., Gholami, P., Kalderis, D., Pachatouridou, E., Konsolakis, M., 2018. Preparation of novel CEO2-biochar nanocomposite for sonocatalytic degradation of a textile dye. Ultrason. Sonochem. 41, 503–513. https://doi.org/10.1016/j.ultsonch.2017.10.013.
[60] Long, L., Xue, Y., Zeng, Y., Yang, K., Lin, C., 2017. Synthesis, characterization and mechanism analysis of modified crayfish shell biochar possessed zno nanoparticles to remove trichloroacetic acid. J. Clean. Prod. 166, 1244–1252. https://doi.org/10.1016/j.jclepro.2017.08.122.
[61] Xu, K., Lin, F., Dou, X., Zheng, M., Tan, W., Wang, C., 2018. Recovery of ammonium and phosphate from urine as value-added fertilizer using wood waste biochar loaded with magnesium oxides. J. Clean. Prod. 187, 205–214. https://doi.org/10.1016/j.jclepro.2018.03.206.
[62] Lu, L., Shan, R., Shi, Y., Wang, S., Yuan, H., 2019. A novel Tio2/biochar composite catalyst for photocatalytic degradation of methyl orange. Chemosphere 222, 391–398. https://doi.org/10.1016/j.chemosphere.2019.01.132.
[63] Khataee, A., Kayan, B., Gholami, P., Kalderis, D., Akay, S., Dinpazhoh, L., 2017. Sonocatalytic degradation of reactive yellow 39 using synthesized zro2 nanoparticles on biochar. Ultrason. Sonochem. 39, 540–549. https://doi.org/10.1016/j.ultsonch.2017.05.023.
[64] Solangi, N.H., Mazari, S.A., Mubarak, N.M., Karri, R.R., Rajamohan, N., Vo, D.-V.N., 2023. Recent trends in mxene-based material for biomedical applications. Environ. Res. 222, 115337. https://doi.org/10.1016/j.envres.2023.115337.
[65] Anasori, B., Gogotsi, Y., 2022. Mxenes: Trends, growth, and future directions. Graphene 2D Mater. 7 (3–4), 75–79. https://doi.org/10.1007/s41127-022-00053-z.
[66] Dadashi Firouzjaei, M., Karimiziarani, M., Moradkhani, H., Elliott, M., Anasori, B., 2022. Mxenes: The two-dimensional influencers. Mater. Today Adv. 13, 100202. https://doi.org/10.1016/j.mtadv.2021.100202.

[67] Mehta, V., Saini, H.S., Srivastava, S., Kashyap, M.K., Tankeshwar, K., 2021. Assessment of mo2n monolayer as Li-ion battery anodes with high cycling stability. Mater. Today Commun. 26, 102100. https://doi.org/10.1016/j.mtcomm.2021.102100.
[68] He, S., Sun, X., Zhang, H., Yuan, C., Wei, Y., Li, J., 2021. Preparation strategies and applications of mxene-polymer composites: A review. Macromol. Rapid Commun. 42 (19). https://doi.org/10.1002/marc.202100324.
[69] Ghanbari, R., Marandi, A., Zare, E.N., 2023. Development of melamine-based covalent organic framework-MOF pearl-like heterostructure integrated poly (Ether-block-amide) for CO2/CH4 separation. J. Environ. Chem. Eng. 11 (2), 109269. https://doi.org/10.1016/j.jece.2023.109269.
[70] Dixit, F., Zimmermann, K., Dutta, R., Prakash, N.J., Barbeau, B., Mohseni, M., Kandasubramanian, B., 2022. Application of mxenes for water treatment and energy-efficient desalination: A review. J. Hazard Mater. 423, 127050. https://doi.org/10.1016/j.jhazmat.2021.127050.
[71] Liu, F., Wang, S., Zhao, C., Hu, B., 2023a. Constructing coconut shell Biochar/MXenes composites through self-assembly strategy to enhance U(VI) and cs(i) immobilization capability. Biochar 5 (1). https://doi.org/10.1007/s42773-023-00231-z.
[72] Liu, Y., Chen, X., Sun, J., Xu, N., Tang, Q., Ren, J., Chen, C., Lei, W., Zhang, C., Liu, D., 2023b. Large-scale production of MXenes as nanoknives for antibacterial application. Nanoscale Adv. 5 (23), 6572–6581. https://doi.org/10.1039/d3na00744h.
[73] Javaid, A., Latif, S., Imran, M., Hussain, N., Bilal, M., Iqbal, H.M.N., 2022. MXene-based hybrid composites as photocatalyst for the mitigation of pharmaceuticals. Chemosphere 291, 133062. https://doi.org/10.1016/j.chemosphere.2021.133062.
[74] Ghanbari, R., Nazarzadeh Zare, E., 2024. Engineered mxene-polymer composites for water remediation: Promises, challenges and future perspective. Coord. Chem. Rev. 518, 216089. https://doi.org/10.1016/j.ccr.2024.216089.
[75] Ghiasi, M., Ghanbarzadeh, M., Ghaffarinejad, A., Shahdost-Fard, F., 2025. Green nitrogen and sulfur co-doped carbon dots derived from eggshell as a high performance aptasensing interface for non-invasive detection of metronidazole. Talanta 285, 127363. https://doi.org/10.1016/j.talanta.2024.127363.
[76] Arora, G., Sabran, N.S., Liew, C.-W., Ng, C.Y., Low, F.W., Singh, P.K., Jun, H. K., 2024. Characterization of green-synthesized carbon quantum dots from spent coffee grounds for EDLC electrode applications. Chem. Phys. Impact 9, 100767. https://doi.org/10.1016/j.chphi.2024.100767.
[77] Cao, C., Guo, W., 2024. Synthesis of metal framework-modified carbon dots with super large stokes shift using hami melon as a green precursor for detecting thiophanate-methyl residue in leafy vegetables. Food Chem. 460, 140703. https://doi.org/10.1016/j.foodchem.2024.140703.
[78] Hu, J., Wu, S., 2025. Green and sustainable synthesis of fluorescent carbon dots from crown daisy for information encryption application. J. Mol. Struct. 1324, 140794. https://doi.org/10.1016/j.molstruc.2024.140794.
[79] He, Y., Liu, S., Xie, F., Zhou, Y., Yang, X., 2024. Application of green tea residue-derived carbon dots in the detection of fe3+ ions and imaging of Caco-2 cells. J. Food Compos. Anal. 132, 106332. https://doi.org/10.1016/j.jfca.2024.106332.
[80] Abo El Naga, A.O., El Saied, M., Shaban, S.A., El Kady, F.Y., 2019. Fast removal of diclofenac sodium from aqueous solution using sugar cane bagasse-derived activated carbon. J. Mol. Liq. 285, 9–19. https://doi.org/10.1016/j.molliq.2019.04.062.
[81] El Mouchtari, E.M., Daou, C., Rafqah, S., Najjar, F., Anane, H., Piram, A., Hamade, A., Briche, S., Wong-Wah-Chung, P., 2020. TiO_2 and activated carbon of *Argania spinosa* tree nutshells composites for the adsorption photocatalysis removal of

pharmaceuticals from Aqueous Solution. J. Photochem. Photobiol. A: Chem. 388, 112183. https://doi.org/10.1016/j.jphotochem.2019.112183.
[82] Norouzi, S., Heidari, M., Alipour, V., Rahmanian, O., Fazlzadeh, M., Mohammadi-moghadam, F., Nourmoradi, H., Goudarzi, B., Dindarloo, K., 2018. Preparation, characterization and cr(vi) adsorption evaluation of NaOH-activated carbon produced from date press cake; an agro-industrial waste. Bioresour. Technol. 258, 48–56. https://doi.org/10.1016/j.biortech.2018.02.106.
[83] Zeng, Q., Wang, H., Ran, H., Wu, J., Yang, K., 2023. Mercury removal performance and mechanism of biochar co-modified with HNO_3 and NH_4Br under Oxy-combustion atmosphere. Energy Sources, Part A Recovery, Util. Environ. Eff. 45 (3), 8167–8182. https://doi.org/10.1080/15567036.2023.2226100.
[84] Sajjadi, S.-A., Meknati, A., Lima, E.C., Dotto, G.L., Mendoza-Castillo, D.I., Anastopoulos, I., Alakhras, F., Unuabonah, E.I., Singh, P., Hosseini-Bandegharaei, A., 2019. A novel route for preparation of chemically activated carbon from pistachio wood for highly efficient pb(ii) sorption. J. Environ. Manage. 236, 34–44. https://doi.org/10.1016/j.jenvman.2019.01.087.
[85] Zheng, X., Su, Y., Chen, Y., Huang, H., Shen, Q., 2018. Global transcriptional responses of denitrifying bacteria to functionalized single-walled carbon nanotubes revealed by weighted gene-coexpression network analysis. Sci. Total Environ. 613–614, 1240–1249. https://doi.org/10.1016/j.scitotenv.2017.09.193.
[86] Haider, A., AL- Anbari, R., Mohammed, H., Ahmed, D., 2018. Synthesis of multi-walled carbon nanotubes decorated with zinc oxide nanoparticles for removal of pathogenic bacterial. Eng. Technol. J. 36 (10A), 1075–1080. https://doi.org/10.30684/etj.36.10a.8.
[87] Shimizu, Y., Ateia, M., Wang, M., Awfa, D., Yoshimura, C., 2019. Disinfection mechanism of E. coli by CNT-tio2 composites: Photocatalytic inactivation vs. physical separation. Chemosphere 235, 1041–1049. https://doi.org/10.1016/j.chemosphere.2019.07.006.
[88] Yang, H., He, D., Zhang, T., Liu, C., Cheng, F., Zhou, Y., Zhang, Y., Qu, J., 2023. Magnetic carbon nanotubes/red phosphorus/graphitic carbon nitride heterojunction for highly efficient visible-light photocatalytic water disinfection. Chem. Eng. J. 466, 143309. https://doi.org/10.1016/j.cej.2023.143309.
[89] Yadav, S., Ibrar, I., Altaee, A., Samal, A.K., Ghobadi, R., Zhou, J., 2020. Feasibility of brackish water and landfill leachate treatment by GO/MOS2-PVA composite membranes. Sci. Total Environ. 745, 141088. https://doi.org/10.1016/j.scitotenv.2020.141088.
[90] Zhang, P., Wang, Y., Li, P., Luo, X., Feng, J., Kong, H., Li, T., Wang, W., Duan, X., Liu, Y., Li, M., 2022. Improving stability and separation performance of graphene oxide/graphene nanofiltration membranes by adjusting the laminated regularity of stacking-sheets. Sci. Total Environ. 827, 154175. https://doi.org/10.1016/j.scitotenv.2022.154175.
[91] Zhang, P., Gong, J.-L., Zeng, G.-M., Song, B., Liu, H.-Y., Huan, S.-Y., Li, J., 2018. Ultrathin reduced graphene oxide/MOF nanofiltration membrane with improved purification performance at low pressure. Chemosphere 204, 378–389. https://doi.org/10.1016/j.chemosphere.2018.04.064.
[92] Modi, A., Bellare, J., 2020. Zeolitic imidazolate framework-67/carboxylated graphene oxide nanosheets incorporated polyethersulfone hollow fiber membranes for removal of toxic heavy metals from contaminated water. Sep. Purif. Technol. 249, 117160. https://doi.org/10.1016/j.seppur.2020.117160.
[93] He, M., Wang, L., Zhang, Z., Zhang, Y., Zhu, J., Wang, X., Lv, Y., Miao, R., 2020. Stable forward osmosis nanocomposite membrane doped with sulfonated

graphene oxide@metal–organic frameworks for heavy metal removal. ACS Appl. Mater. Interfaces 12 (51), 57102–57116. https://doi.org/10.1021/acsami.0c17405.
[94] Liang, H., Zhu, C., Ji, S., Kannan, P., Chen, F., 2022. Magnetic FE2O3/biochar composite prepared in a molten salt medium for antibiotic removal in water. Biochar 4 (1). https://doi.org/10.1007/s42773-021-00130-1.
[95] Vieira, R.A., Pickler, T.B., Segato, T.C., Jozala, A.F., Grotto, D., 2022. Biochar from fungiculture waste for adsorption of endocrine disruptors in water. Sci. Rep. 12 (1). https://doi.org/10.1038/s41598-022-10165-4.
[96] Li, R., Zhang, Y., Deng, H., Zhang, Z., Wang, J.J., Shaheen, S.M., Xiao, R., Rinklebe, J., Xi, B., He, X., Du, J., 2020. Removing tetracycline and hg(ii) with ball-milled magnetic nanobiochar and its potential on polluted irrigation water reclamation. J. Hazard Mater. 384, 121095. https://doi.org/10.1016/j.jhazmat.2019.121095.
[97] Shan, R., Shi, Y., Gu, J., Wang, Y., Yuan, H., 2020. Single and competitive adsorption affinity of heavy metals toward peanut shell-derived biochar and its mechanisms in aqueous systems. Chin. J. Chem. Eng. 28 (5), 1375–1383. https://doi.org/10.1016/j.cjche.2020.02.012.
[98] Liu, Y., Sohi, S.P., Liu, S., Guan, J., Zhou, J., Chen, J., 2019. Adsorption and reductive degradation of cr(vi) and TCE by a simply synthesized zero valent iron magnetic biochar. J. Environ. Manag. 235, 276–281. https://doi.org/10.1016/j.jenvman.2019.01.045.
[99] Zhang, K., Mao, J., Chen, B., 2019. Reconsideration of heterostructures of biochars: Morphology, particle size, elemental composition, reactivity and toxicity. Environ. Pollut. 254, 113017. https://doi.org/10.1016/j.envpol.2019.113017.
[100] Li, J., Zhang, K., Zhang, H., 2018. Adsorption of antibiotics on microplastics. Environ. Pollut. 237, 460–467. https://doi.org/10.1016/j.envpol.2018.02.050.
[101] Wu, W., Liu, Z., Azeem, M., Guo, Z., Li, R., Li, Y., Peng, Y., Ali, E.F., Wang, H., Wang, S., Rinklebe, J., Shaheen, S.M., Zhang, Z., 2022. Hydroxyapatite tailored hierarchical porous biochar composite immobilized CD(II) and pb(ii) and mitigated their hazardous effects in contaminated water and soil. J. Hazard Mater. 437, 129330. https://doi.org/10.1016/j.jhazmat.2022.129330.
[102] Ioannou, K., Hadjiyiannis, P., Liatsou, I., Pashalidis, I., 2019. U(vi) adsorption by biochar Fiber–MnO2 composites. J. Radioanal. Nucl. Chem. 320 (2), 425–432. https://doi.org/10.1007/s10967-019-06479-9.
[103] Li, L., Shi, X., Huang, T., Zhang, N., Wang, Y., 2023. Synchronously enhanced storage stability and adsorption ability of mxene achieved by grafting polyethylenimine. J. Mater. Chem. A 11 (43), 23438–23451. https://doi.org/10.1039/d3ta04869a.
[104] Yao, M., Zhang, G., Shao, D., Ding, S., Li, L., Li, H., Zhou, C., Luo, B., Lu, L., 2023. Preparation of chitin/mxene/poly(l-arginine) composite aerogel spheres for specific adsorption of bilirubin. Int. J. Biol. Macromol. 243, 125140. https://doi.org/10.1016/j.ijbiomac.2023.125140.
[105] Qing, Q., Shi, X., Hu, S., Li, L., Huang, T., Zhang, N., Wang, Y., 2023. Synchronously enhanced removal ability and stability of mxene through biomimetic modification. Langmuir 39 (27), 9453–9467. https://doi.org/10.1021/acs.langmuir.3c00987.
[106] Peng, C., Li, X., Jiang, P., Peng, W., Tang, J., Li, L., Ye, L., Pan, S., Chen, S., 2022. Thermoresponsive mxene composite system with high adsorption capacity for quick and simple removal of toxic metal ions from aqueous environment. J. Hazard Mater. 440, 129740. https://doi.org/10.1016/j.jhazmat.2022.129740.
[107] Rahmani, Z., Ghaemy, M., Olad, A., 2022. Removal of heavy metals from polluted water using magnetic adsorbent based on κ-carrageenan and n-doped carbon dots. Hydrometallurgy 213, 105915. https://doi.org/10.1016/j.hydromet.2022.105915.

[108] Guo, X., Liu, Q., Liu, J., Zhang, H., Yu, J., Chen, R., Song, D., Li, R., Wang, J., 2019. Magnetic metal-organic frameworks/carbon dots as a multifunctional platform for detection and removal of uranium. Appl. Surf. Sci. 491, 640–649. https://doi.org/10.1016/j.apsusc.2019.06.108.

[109] Asiabi, H., Yamini, Y., Shamsayei, M., Molaei, K., Shamsipur, M., 2018. Functionalized layered double hydroxide with nitrogen and sulfur co-decorated carbon-dots for highly selective and efficient removal of soft HG2+ and Ag+ ions. J. Hazard Mater. 357, 217–225. https://doi.org/10.1016/j.jhazmat.2018.05.055.

[110] Huang, S., Jiang, S., Pang, H., Wen, T., Asiri, A.M., Alamry, K.A., Alsaedi, A., Wang, X., Wang, S., 2019. Dual functional nanocomposites of magnetic $MnFe_2O_4$ and fluorescent carbon dots for efficient U(VI) removal. Chem. Eng. J. 368, 941–950. https://doi.org/10.1016/j.cej.2019.03.015.

[111] Yao, W., Wang, X., Liang, Y., Yu, S., Gu, P., Sun, Y., Xu, C., Chen, J., Hayat, T., Alsaedi, A., Wang, X., 2018. Synthesis of novel flower-like layered double oxides/carbon dots nanocomposites for U(VI) and 241Am(III) efficient removal: Batch and EXAFS studies. Chem. Eng. J. 332, 775–786. https://doi.org/10.1016/j.cej.2017.09.011.

CHAPTER 12

Recycling and reusability of carbon-based materials and composites for water and wastewater treatment

Adikwu Gowon Jacob, Hafeez Adesina Gazali, Ikechukwu Ogadimma Alisi, Abiodun Busuyi Ogbesejana and Uduma Anya Uduma
Department of Industrial Chemistry, Faculty of Physical Sciences, Federal University Dutsin-Ma (FUDMA), Dutsin-Ma, Katsina State, Nigeria

1. Introduction

The increasing global demand for clean water, alongside concerns about water scarcity and the environmental impact of traditional treatment methods, has prompted researchers to seek innovative and sustainable solutions. One significant approach gaining attention is the use of carbon-based materials and composites for water and wastewater treatment. These materials have shown great potential in addressing challenges such as the removal of various contaminants, including heavy metals, organic pollutants, and microorganisms [1].

Carbon-based materials, such as activated carbon, graphene, and carbon nanotubes possess unique physical and chemical properties that make them suitable for these applications. Their high surface area, adsorption capacity, and versatility in functionalization enable efficient removal and separation of contaminants from water. Extensive studies have demonstrated promising results in both laboratory and pilot-scale settings, highlighting the need for further research to scale up and commercialize these technologies [2]. The recycling and reusability of carbon-based materials have emerged as crucial strategies for enhancing the sustainability and cost-effectiveness of water treatment processes. Developing techniques to regenerate and reuse these materials addresses disposal challenges and reduces the environmental

Carbon-Based Materials and Composites for Water and Wastewater Treatment
ISBN 978-0-443-33681-2
https://doi.org/10.1016/B978-0-443-33681-2.00005-3

impact and operational costs associated with spent adsorbents [3]. Effective recycling strategies can contribute to a more circular approach to water management, minimizing waste generation and conserving natural resources.

Researchers have explored various recycling methods, including thermal, chemical, and electrochemical treatments, which have shown the ability to restore adsorption capacity and reactivation of surface properties, allowing for reuse in subsequent treatment cycles [4]. Furthermore, integrating carbon-based materials with other functional components, such as polymers and metal oxides, has led to the development of composites that offer enhanced performance, selectivity, and stability, making them attractive for large-scale applications [5,6]. By utilizing waste streams as feedstock, researchers aim to tackle environmental challenges associated with waste management while reducing reliance on conventional, energy-intensive production methods [7]. Advancements in characterization techniques, such as electron microscopy and spectroscopy, have been critical in understanding the structural and chemical changes during the recycling and reuse of carbon-based materials [8]. This knowledge aids in optimizing regeneration processes and improving the overall performance and durability of these materials [1].

The integration of carbon-based materials into water treatment systems also opens opportunities for hybrid and multifunctional approaches. Combining these materials with technologies such as membrane filtration and photocatalysis can enhance contaminant removal and energy efficiency, addressing a broader range of water quality challenges [2]. As research progresses, the potential integration of these technologies with digital tools, such as machine learning and real-time monitoring systems, can optimize treatment processes and enhance water management systems [3]. Research on recycling and reusability of carbon-based materials and composites is increasingly crucial to meet the growing global demand for clean water. It holds promise for developing sustainable, cost-effective, and resource-efficient water management practices, ultimately ensuring clean water availability for communities worldwide [4].

2. Carbon-based materials

Carbon-based materials have emerged as versatile adsorbents in water treatment, demonstrating unique properties and significant adsorption capabilities. These materials, including activated carbon, graphene, carbon nanotubes, and biochar, have been extensively studied for their effectiveness in removing a wide range of contaminants from water and wastewater [9]. Their remarkable adsorption performance is attributed to their high surface area, tunable pore structures, and the ability to modify surface chemistry to target specific pollutants [10]. Activated carbon is one of the most widely used carbon-based materials, known for its superior adsorption capacity, particularly for organic compounds. It effectively removes various pollutants, such as pesticides, dyes, and pharmaceutical residues [11]. The versatility of activated carbon is enhanced by its customizable pore size distribution and surface functionality through various activation and modification techniques.

Innovative materials such as graphene, carbon nanotubes, and biochar are gaining attention in water treatment. Graphene, noted for its exceptional surface area and unique electronic properties, has shown promise in removing heavy metals, organic pollutants, and pathogens from water [12]. Carbon nanotubes, with their tubular structure and high aspect ratio, exhibit enhanced adsorption performance for heavy metals and organic compounds [13]. Biochar, produced from the pyrolysis of biomass, is recognized for its cost-effectiveness and environmental sustainability. Its physicochemical properties can be tailored to target specific water pollutants, making it effective for adsorbing heavy metals, organic pollutants, and nutrients [14].

The versatility and adsorption capabilities of carbon-based materials have led to extensive research and development in the water treatment industry. As demand for clean water and sustainable solutions grows, optimizing these materials for water purification is increasingly critical [9]. Fig. 12.1 shows the sources of carbon-based materials.

3. Type of carbon-based material

Carbon-based materials have emerged as versatile adsorbents in water treatment, encompassing a range of types, including activated carbon, graphene, carbon nanotubes, and biochar, each with unique properties and applications [9].

3.1 Activated carbon

Activated Carbon is the most widely used carbon-based material in water treatment due to its exceptional adsorption capacity, particularly for organic compounds. It effectively removes pollutants such as pesticides,

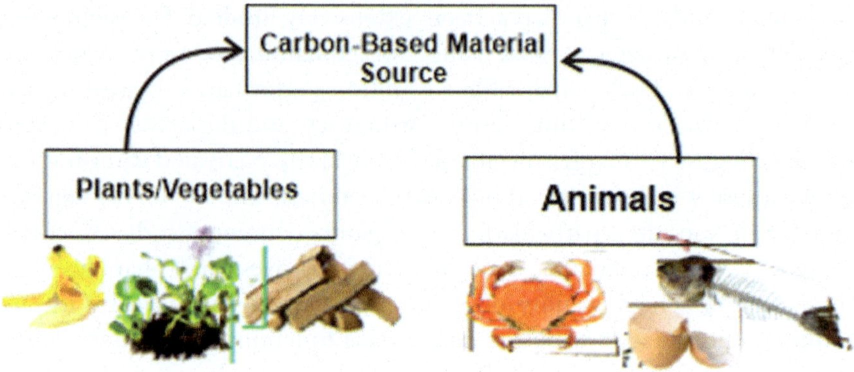

Figure 12.1 Representation of carbon-based sources.

dyes, and pharmaceutical residues [11]. Its versatility arises from the ability to produce it from various carbonaceous precursors, and its porous structure and surface area can be enhanced through activation processes. Additionally, modifications to its pore size and surface functionality expand its applications for targeted contaminant removal [15].

3.2 Graphene

Graphene, is a single layer of carbon atoms in a hexagonal lattice, shows promise in water treatment due to its high surface area and excellent adsorption capacity. It is effective in removing heavy metals, organic pollutants, and pathogens. The development of graphene-based composites, which combine graphene with materials such as metal oxides, enhances adsorption performance and selectivity for specific contaminants [12,16].

3.3 Carbon nanotubes

Carbon Nanotubes is characterized by their cylindrical structure and high aspect ratio, also demonstrate exceptional adsorption capabilities for various contaminants, including heavy metals and organic compounds [13]. Their unique morphology allows for effective pollutant capture. Recent research has focused on synthesizing modified carbon nanotubes to improve their stability, selectivity, and regeneration for enhanced water purification [17].

3.4 Biochar

Biochar is derived from the pyrolysis of biomass, has gained attention for its cost-effectiveness and environmental sustainability. It can be tailored to

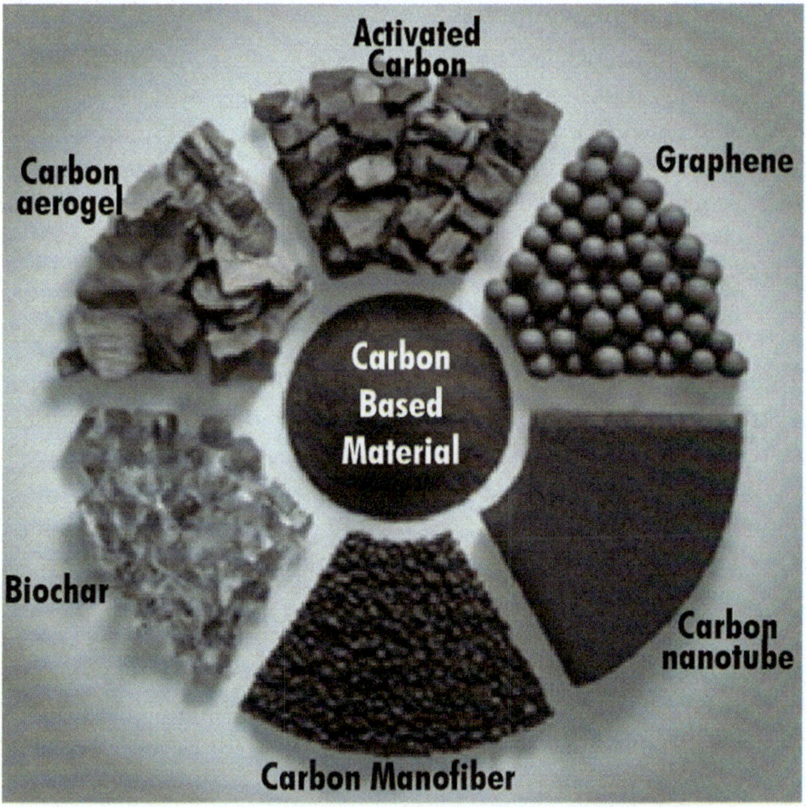

Figure 12.2 Classification of carbon-based materials.

target specific water pollutants, making it a promising alternative to conventional adsorbents. Biochar is versatile in adsorbing a wide range of contaminants, including heavy metals, organic pollutants, and nutrients, positioning it as a sustainable solution for water purification [14,18]. Fig. 12.2 summarizes the various classes of carbon-based materials.

4. Use of carbon-based materials in water treatment applications

Carbon-based materials have gained significant attention in water treatment due to their remarkable adsorption capabilities and versatility in targeting various contaminants.

4.1 Removal of heavy metals

Carbon-based materials are effective in adsorbing heavy metals such as lead, cadmium, and chromium. Activated carbon is widely used for this purpose due to its high surface area and tunable pore structure [19,20]. Graphene-based materials also exhibit excellent adsorption performance for heavy metals, leveraging their unique electronic properties [21]. Additionally, biochar is recognized as a cost-effective and sustainable alternative, with modifications enhancing its ability to target specific metals [22].

4.2 Removal of organic pollutants

These materials excel in removing organic pollutants, including pesticides, dyes, and pharmaceutical residues. Activated carbon is particularly effective due to its high surface area and diverse functional groups that interact with organic contaminants [9]. Recent studies have explored graphene-based composites and carbon nanotubes for organic pollutant adsorption, utilizing their unique properties for targeted modifications [16,17].

4.3 Disinfection and pathogen removal

Carbon-based materials are also investigated for disinfection and pathogen removal. Graphene-based materials have shown antimicrobial properties, effectively inactivating bacteria and viruses [12]. Carbon nanotubes, due to their high aspect ratio, are explored for antibacterial and antiviral applications in water treatment [5].

4.4 Water purification through membrane filtration

The integration of carbon-based materials, such as graphene and carbon nanotubes, into membrane filtration systems enhances performance by improving permeability, selectivity, and antifouling properties [16]. Their unique properties make them promising candidates for advanced water filtration technologies.

4.5 Nutrient removal and recovery

Biochar has been studied for removing and recovering nutrients such as nitrogen and phosphorus from water and wastewater. Its porous structure and surface chemistry can be optimized for selective nutrient adsorption, facilitating recovery for agricultural reuse [23]. This approach promotes sustainability and resource efficiency in water treatment. Fig. 12.3 shows the different uses of carbon-based materials in water and wastewater treatment.

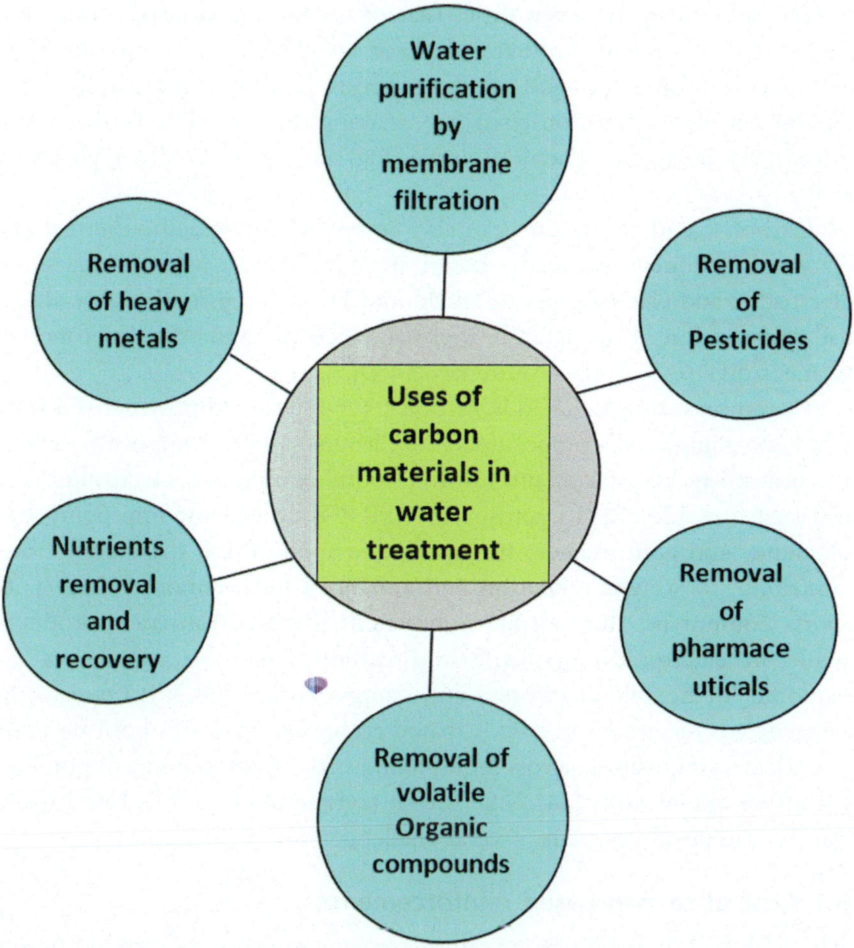

Figure 12.3 A chart depicting various applications of carbon materials in water treatment.

5. Composite material made from carbon-based material

Composite materials made from carbon-based materials are increasingly important across various industries due to their exceptional properties and versatility. These composites typically consist of a polymer matrix reinforced with carbon-based components such as carbon fibers, graphene, or carbon nanotubes [24,25]. The primary advantage of these composites is their outstanding strength-to-weight ratio. Carbon fibers, which have a tensile strength up to seven times that of steel and a density about one-fourth that of steel, make them ideal for weight-sensitive applications in aerospace, automotive, and sporting goods [26,27]. Additionally, they exhibit excellent corrosion resistance, making them suitable for harsh environments in marine, chemical processing, and infrastructure applications [28].

Carbon-based composites can also be tailored for specific thermal and electrical conductivity needs, useful in applications such as heat sinks, electrodes, and electromagnetic shielding [29]. This customizability allows for optimization of mechanical, thermal, electrical, and esthetic properties to meet diverse industry requirements [30].

Common examples include carbon fiber-reinforced polymer (CFRP), graphene-reinforced composites, and carbon–carbon composites, which are utilized in aerospace, automotive, sports equipment, electronics, and infrastructure [25,28]. In aerospace, CFRP is crucial for improving fuel efficiency and performance. In automotive applications, these composites contribute to weight reduction and enhanced fuel economy [26,27]. In sports equipment, they enable lightweight, high-performance products, while in electronics, graphene and carbon nanotube composites are explored for flexible electronics and energy storage [28,30]. Overall, the ongoing advancements in carbon-based composite materials position them as critical to innovation across multiple industries, with significant potential for future applications [24]. The use of conventional carbon-based materials in composites presents several challenges:

5.1 Cost of carbon-based reinforcements

The high cost of carbon-based reinforcements, particularly carbon fibers, is a significant barrier to their adoption. Carbon fibers are more expensive than traditional reinforcements such as glass fibers due to their complex and energy-intensive manufacturing processes [31]. Advanced reinforcements such as graphene and carbon nanotubes also incur high production costs. Research is ongoing to improve cost-effectiveness through alternative manufacturing methods and the use of recycled materials.

5.2 Brittleness and poor impact resistance

Conventional carbon-based composites often exhibit brittleness and poor impact resistance, particularly with high reinforcement content. The rigidity of carbon fibers can lead to sudden failures under dynamic loading conditions, posing safety concerns in applications such as automotive and sports equipment [32]. Researchers are exploring strategies such as incorporating toughening agents and developing hybrid composites to enhance impact resistance.

5.3 Difficulties in processing and manufacturing

The high stiffness and abrasiveness of carbon fibers complicate their processing and manufacturing. These properties can cause tooling wear and part deformation, requiring specialized equipment and techniques [25]. Additionally, achieving uniform dispersion of nanomaterials such as carbon nanotubes or graphene within the matrix can be challenging due to agglomeration.

5.4 Potential health and environmental concerns

The production and handling of certain carbon-based materials, especially carbon nanotubes, raise health and environmental concerns. Some studies suggest potential toxicity, necessitating stringent safety measures during manufacturing and disposal [33]. Efforts are being made to develop safer production methods and explore alternative materials with lower environmental impacts.

5.5 Recycling and waste management challenges

Disposing of and recycling carbon-based composites is complex due to their heterogeneous nature, which complicates the separation of components [34]. This complexity hinders effective recycling solutions. Innovative recycling techniques and comprehensive waste management systems are needed to ensure the sustainability of the carbon-based composite industry.

6. Existing strategies for recycling and reusability of spent carbon-based materials

Certainly, here are overviews of the existing strategies for recycling and reusing spent carbon-based materials from composite applications.

6.1 Recycling of carbon fiber composites

Recycling carbon fiber composites is essential for addressing waste management challenges. Several methods have been developed to recover valuable carbon fibers from end-of-life composites.

6.2 Thermal recycling

This method involves high-temperature treatment (400–700°C) to decompose the polymer matrix, allowing for the recovery of carbon fibers. While effective, thermal recycling may reduce the mechanical properties of the fibers due to potential damage from high temperatures [35,36].

6.3 Chemical recycling

Chemical recycling uses agents such as acids or solvents to dissolve the polymer matrix, facilitating the extraction of carbon fibers with minimal damage. This approach often preserves the original mechanical properties of the fibers, as the process can be optimized to reduce adverse effects on fiber quality [37,38].

6.4 Mechanical recycling

Mechanical recycling involves physically breaking down and separating composite materials through shredding, grinding, or milling. While this method is less complex and energy-intensive, it typically results in shorter fibers with lower mechanical properties due to the physical stress during processing [34,39].

These strategies highlight the potential for recycling and reusing spent carbon-based materials, contributing to more sustainable practices in composite applications.

7. Reusability of carbon-based materials

In addition to recycling, there are also various strategies for the reuse of spent carbon-based materials from composite applications, leveraging their unique properties and characteristics in new applications.

7.1 Filler materials

Recycled carbon fibers, as well as other carbon-based materials such as carbon nanotubes and graphene, can be used as fillers in the production of new composite materials, cement-based products, or thermoplastic compounds. The incorporation of these recycled carbon-based fillers can provide improvements in mechanical, thermal, and electrical properties, making them valuable additives for a wide range of applications [40,41].

7.2 Energy storage applications

Spent carbon-based materials, such as graphite from lithium-ion batteries or carbon fibers from composite materials, can be repurposed and utilized in energy storage devices, including supercapacitors and anodes for lithium-ion batteries. The high surface area, electrical conductivity, and structural characteristics of these carbon-based materials make them suitable for energy storage applications, allowing for the effective reuse of these materials in a new context [42,43].

7.3 Construction and infrastructure

Recycled carbon-based materials, including both fibers and particulate forms, can be incorporated into construction materials, such as concrete and asphalt, to enhance their mechanical properties and durability. The addition of these carbon-based reinforcements can improve the strength, toughness, and overall performance of the construction materials, contributing to more sustainable and resilient infrastructure [39,44].

7.4 Thermal and electrical applications

The inherent thermal and electrical conductivity of carbon-based materials, including recycled forms, can be leveraged in various applications that require efficient heat transfer or electromagnetic shielding. These recycled carbon-based materials can be used as components in heat sinks, thermal management systems, and electromagnetic shielding products, providing a sustainable way to utilize the unique properties of these materials [29,41].

7.5 Challenges and ongoing research

While the recycling and reuse strategies discussed above have shown promising results, the effective management of spent carbon-based composite materials still faces several challenges. The complex nature of these composite structures, the need for specialized equipment and processes, and the potential for property degradation during recycling are some of the key obstacles that need to be addressed. Ongoing research in this field is focused on developing more efficient and cost-effective recycling technologies, exploring novel applications for recycled carbon-based materials, and designing composite materials with improved recyclability and end-of-life considerations [45,46].

8. Synthesis of sustainable carbon-based materials using renewable and waste-derived precursors

The synthesis of sustainable carbon-based materials is increasingly important due to the need to mitigate the environmental impact of traditional materials [47]. Conventional carbon-based materials, such as activated carbons, carbon fibers, and graphene, are commonly used in applications such as energy storage, water treatment, and composite reinforcement [48]. However, their production typically relies on fossil fuel-derived precursors, leading to a significant carbon footprint and depletion of nonrenewable resources [49]. To address these sustainability issues, researchers are exploring the use of renewable and waste-derived precursors for synthesizing carbon-based materials, aiming to create eco-friendlier and resource-efficient production methods [50]. Many renewable precursors are available for carbon-based materials.

8.1 Biomass-derived precursors

Biomass, including a variety of plant-based and organic materials, is a promising renewable source for sustainable carbon-based materials [51]. Agricultural and forestry waste, such as lignocellulosic biomass (e.g., corn stover, sugarcane bagasse) and forestry byproducts (e.g., wood chips, sawdust), can be utilized to produce activated carbons, carbon fibers, and composites [52,53]. Aquatic biomass, including microalgae and macroalgae, is also explored for synthesizing biochar, graphene, and carbon quantum dots due to its high carbon content [54,55]. Additionally, natural polymers such as cellulose, chitin, and lignin can be extracted from biomass for creating various carbon-based materials [56,57].

8.2 Waste-derived precursors

Waste-derived materials are gaining attention as sources for sustainable carbon-based materials [58]. Waste plastics (e.g., polyethylene, polypropylene) can be transformed into carbon nanotubes, graphene, and porous carbons through thermal and chemical treatments [59,60]. This conversion not only addresses plastic waste but also recovers the embodied carbon. End-of-life tires can also serve as a source for producing activated carbons and carbon blacks, contributing to a circular economy [61]. Other waste streams, such as industrial sludge and food waste, can be converted into valuable carbon materials through various treatment processes [62,63].

9. Synthesis techniques for sustainable carbon-based materials

9.1 Pyrolysis and carbonization

Pyrolysis and carbonization are widely used techniques that involve thermal treatment of renewable and waste-derived precursors in inert atmospheres [64]. This process transforms organic matter into carbon-rich solid materials, with specific conditions tailored to optimize properties for applications such as biochar and activated carbons [50]. These methods are scalable and suitable for large-scale production.

9.2 Hydrothermal and solvothermal synthesis

Hydrothermal and solvothermal methods utilize water or solvents under high temperature and pressure to convert biomass or waste-derived precursors into carbon-based materials with unique structures [65]. These techniques allow for the production of carbon quantum dots and graphene, often under milder conditions than pyrolysis [66].

9.3 Chemical and biological activation

Chemical activation uses agents such as phosphoric acid or potassium hydroxide to enhance the porous structure and surface chemistry of carbon-based materials, improving their performance [67,68]. Biological activation methods involving enzymes or microorganisms are also being explored for further property modification [69].

9.4 Hybridization and functionalization

Hybridization and functionalization strategies incorporate other materials, such as metal oxides or polymers, into carbon-based matrices to enhance their functionalities [70]. These approaches expand the properties and applications of sustainable carbon-based materials.

10. Integration of carbon-based materials and composites with water treatment

The quest for sustainable and efficient water treatment solutions has led to the increasing integration of carbon-based materials and composites with various water purification technologies. These carbon-based materials, including activated carbons, carbon nanotubes, graphene, and their derivatives, possess unique physicochemical properties that make them highly versatile and suitable for integration with diverse water treatment processes. By leveraging the synergistic effects of carbon-based materials and their complementary capabilities, researchers and water treatment practitioners have been able to develop advanced, integrated systems that can address a broad spectrum of water contaminants, improve overall treatment efficiency, and enhance the sustainability of water purification efforts [71].

10.1 Carbon–carbon composites

One of the key areas of research has been the development of hybrid carbon–carbon composites, where two or more carbon-based materials are combined to create a synergistic effect. For example, the integration of carbon nanotubes or graphene with activated carbon or biochar can lead to enhanced mechanical properties, increased electrical conductivity, and improved adsorption capabilities [71,72]. The high surface area and porosity of the activated carbon or biochar can provide effective pathways for ion and electron transport, while the carbon nanotubes or graphene can act as conductive reinforcements, improving the overall performance of the hybrid system.

10.2 Carbon–polymer composites

Another promising approach involves the development of hybrid carbon–polymer composites, where carbon-based materials are combined with polymeric matrices to create multifunctional materials [73]. The incorporation of carbon nanotubes, graphene, or carbon fibers into polymer matrices can result in significant improvements in mechanical strength, thermal stability, and electrical conductivity, making these hybrid composites suitable for various applications, such as structural components, energy storage devices, and electromagnetic shielding [74,75].

10.3 Carbon–ceramic hybrids

The integration of carbon-based materials with ceramic components has also led to the emergence of innovative hybrid systems with enhanced performance characteristics. For instance, the combination of carbon materials, such as activated carbon or biochar, with metal oxides or other ceramic phases can result in improved catalytic activity, enhanced energy storage capacity, and superior thermal and mechanical properties [76,77]. These hybrid carbon–ceramic materials have found applications in areas such as energy conversion and storage, water treatment, and high-temperature applications.

11. Performance improvements in hybrid carbon-based systems

11.1 Enhanced adsorption and catalytic activity

The synergistic integration of different carbon-based materials in hybrid systems can lead to significant improvements in adsorption and catalytic performance. The combination of high surface area, porosity, and diverse chemical functionalities provided by the carbon-based components can enhance the overall adsorption capacity and selectivity toward specific target molecules or pollutants [78,79]. Additionally, the integration of carbon materials with catalytically active species, such as metal nanoparticles or metal oxides, can result in enhanced catalytic activity and improved reaction kinetics for various applications, including energy conversion, environmental remediation, and industrial processes.

11.2 Improved energy storage and conversion

Hybrid carbon-based systems have shown immense potential for enhancing the performance of energy storage and conversion devices. The incorporation of carbon nanostructures, such as graphene and carbon nanotubes, into energy storage electrodes can lead to improved charge storage capacity, rate capability, and cycling stability due to their high surface area, electrical conductivity, and structural integrity [80,81]. Similarly, the combination of carbon-based materials with other energy-active components, such as metal oxides or conducting polymers, can result in the development of high-performance supercapacitors, lithium-ion batteries, and fuel cells with enhanced energy density, power density, and efficiency.

11.3 Enhanced mechanical and thermal properties

The synergistic effects observed in hybrid carbon-based systems can also lead to significant improvements in mechanical and thermal properties. The incorporation of carbon nanostructures or carbon fibers into polymer or ceramic matrices can result in enhanced strength, toughness, and thermal stability, making these hybrid materials suitable for structural and high-temperature applications [72,74]. The ability to tailor the composition, morphology, and interfacial interactions within the hybrid system allows for the optimization of mechanical and thermal performance, expanding the potential applications of these sustainable carbon-based materials. Table 12.1 summarized some carbon-based materials and composites used for treatment of water and wastewater.

Table 12.1 Summary of some carbon-based materials and composites used for removal of various pollutants from aqueous solution from 2015 to 2024.

Carbon-based material	Preparation method	Pollutant adsorbed	References
TiO_2-GO sponge	A solution of tetrabutyl titanate dissolved in ethanol was added to a GO dispersion and stirred vigorously, washed with ETOH and deionized H_2O, lyophilized to obtain the TiO_2-GO sponge	Tetracycline	[82]
KOH activated carbon	Activated carbon was produced by hydrothermal treatment of sucrose and was activated with KOH to obtain spherical carbon	Methylene blue	[83]
Activated carbon	Borassus aethiopum shell was carbonized at 500°C for 3 h and separately activated with 0.5M solution of $ZnCl_2$ and H_3PO_4 at 750°C for 1 h, washed with DI water and dried in an oven at a temperature of 1100°C for 3 h	Lead ions	[20]
Fe_3O_4/AC nanocomposite	Activated carbon was dispersed in 400 mL solution of 97.8 g of $FeCl_3.6H_2O$ (28 mM) and 3.8 g of $FeSO_4.7H_2O$ (14 mM) and vigorously stirred at 700 rpm and 70–80°C. NaOH (5.0 M) solution was added drop wise under vigorous stirring at 80°C until the pH raised to 10–11 to precipitate the hydrated iron oxide. It was stirred further at 80°C for 1.0 h, aged at rt for 24h. Fe_3O_4/AC nanocomposite was magnetically separated and washed repeatedly with DI water and ethanol till neutral pH, dried at 80°C for 12 h	Cr(VI), Cu(II) and Cd(II) ions	[84]

C-MOF-74(Zn/Fe)	MOF-74(Zn/Fe) was prepared by dissolving anhydrous zinc acetate (3.13 g) and iron(II) acetate (0.35 g) in DMF (200 mL) and stirred with addition of 2,5-dihydroxyterephthalic acid (0.99 g) dissolved in DMF (200 mL), and 0.55 g of salicylic dissolved in DMF (50 mL)	Methylene blue and methyl orange dyes	[92]
Chitosan/Al_2O_3/Fe_3O_4 nanocomposite	The NC was prepared by past method. For modification by the EDTA, 0.5 g of the NC was added in 200 mL of Na_2EDTA (0.1M)and stirred overnight to adsorb the $EDTA^{2-}$ ions on the surface of magnetic chitosan through electrostatic attraction; next, the particles were washed several times with water and dispersed in phosphate solution (0.003 M and pH = 5), followed by addition of 3-ethyl-1-(3-dimethylaminopropyl) carbodiimide hydrochloride (20 mL, 25.0 g L^{-1}), sonicated for 25 min and stirred for 5.5 h. The NC was washed with disodium hydrogen phosphate (0.1M) till the supernatant was alkalescent and, washed with DI water. Finally, the NC was separated and dried at °C for 12 h	Cd (II), Cu(II) and Zn(II) metal ions	[93]
MWCNTs-KOH@NiNPs nanotube	MWCNTs-KOH@NiNPs were synthesized by deposition/immobilization of NiNPs onto the surface matrix of MWCNTs activated KOH	Lead (II), arsenic (V) and cadmium (II) ions	[94]
Pomegranate peels activated carbon	Pomegranate peels were sorted cleansed with DI water, cut into piece, and sun-dried at 60°C for 96 h. Pomegranate carbon was chemically activated in a single-step using H_3PO_4	Chlorpyrifos pesticide	[95]

12. Conclusion

Carbon-based materials and composites have emerged as highly effective solution for water and wastewater treatment due to their exceptional adsorption capabilities and versatility. The key to the sustainability and cost-effectiveness of these technologies lies in their ability to be recycled and reused. Extensive research has demonstrated the potential of various regeneration techniques, including thermal, chemical and electrochemical methods to restore the adsorption capacity of carbon-based material after they become saturated. This regeneration and reuse process can significantly reduce the operating costs and environmental impact of water treatment systems, making them more economically viable and environmentally friendly.

Despite the promising results, there are still some challenges that need to be addressed to facilitate the widespread adoption of recycled and reusable carbon-based materials and composites for water and wastewater treatment. Scalability of production, cost-effectiveness, and the seamless integration of regeneration and reuse strategies into the overall treatment process are among the key areas that require further research and innovation. Continuous advancement in material design, regeneration techniques, and process optimization will be crucial in overcoming these challenges and unlocking the full potential of these sustainable and eco-friendly technologies.

The recycling and reusability of carbon-materials and composites hold great promise for the water and wastewater treatment industry. By leveraging the unique properties of these materials and addressing the existing challenges, researchers and practitioners can develop more advanced, efficient, and cost-effective treatment solutions that contribute to the preservation of water resources and overall sustainability of water management systems.

Acknowledgment

The authors are grateful to the Department of Industrial Chemistry, Faculty of Physical Sciences, Federal University Dutsin-Ma, Katsina State, Nigeria, for providing the platform to write this chapter.

References

[1] X. Dong, W. Chen, G. Zeng, Emerging carbon-based materials for water and wastewater treatment: Adsorption, catalysis, and beyond, J. Hazard Mater. 452 (2023) 130582.

[2] G. Qi, Q. Liu, Y. Liu, S. Zhang, Recycling and regeneration of carbon-based adsorbents for water treatment: A review, Chem. Eng. J. 429 (2022) 132189.
[3] Y. Yao, F. Xu, M. Chen, Z. Xu, Z. Zhu, Fabrication of carbon-based composites for water and wastewater treatment, Environ. Sci. Pollut. Control Ser. 28 (13) (2021) 16279–16298.
[4] J. Luan, L. Wang, Y. Sun, H. Huang, Recycling and reuse of carbon-based materials for water treatment: Progress, challenges, and perspectives, Environ. Sci. Water Res. Technol. 8 (1) (2022) 20–36.
[5] X. Liu, G. Chen, Y. Zhou, Regeneration and reuse of spent carbon-based adsorbents in water and wastewater treatment: A critical review, Chemosphere 311 (2023) 137681.
[6] A.G. Jacob, R.A. Wahab, S. Chandren, K. Jumbri, W.M.A.W. Mahmood, Physicochemical properties and operational stability of Taguchi design-optimized Candida rugosa lipase supported on biogenic silica/magnetite/graphene oxide for ethyl valerate synthesis, Adv. Powder Technol. 33 (1) (2022) 103374.
[7] N. Zhang, J. Li, H. Huang, Y. Chen, Biochar for water and wastewater treatment: A review of production, characterization, and applications, J. Clean. Prod. 330 (2022) 129874.
[8] A.G. Jacob, R.A. Wahab, M. Misson, Operational stability, regenerability, and thermodynamics studies on biogenic silica/magnetite/graphene oxide nanocomposite-activated Candida rugosa Lipase, Polymers 13 (21) (2021) 3854.
[9] G.Z. Kyzas, M. Kostoglou, Green adsorbents for wastewaters: A critical review, Materials 7 (1) (2014) 333–364.
[10] H. Gao, A.T. Kan, G. Fu, M.B. Tomson, T. Zhang, Adsorption and transport of pharmaceuticals in sediment-water systems, Environ. Sci. Technol. 53 (6) (2019) 3178–3186.
[11] D. Mohan, C.U. Pittman Jr., Activated carbons and low cost adsorbents for remediation of tri-and hexavalent chromium from water, J. Hazard. Mater. 137 (2) (2006) 762–811.
[12] X. Cui, D. Zhao, L. Zhang, Graphene-based composites for water treatment: Adsorption, catalysis, and membrane applications, J. Hazard Mater. 424 (2022) 127585.
[13] Y. Liu, L. Zhao, J. Li, L. Dai, Emerging carbon-based materials for water purification: From fundamentals to applications, Chem. Eng. J. 440 (2023) 136015.
[14] G. Ntalos, A. Grigoriou, A. Zabaniotou, Biochar and activated carbon for water and wastewater treatment: A review, Renew. Sustain. Energy Rev. 143 (2021) 110884.
[15] X. Xu, G. Zeng, D. Huang, C. Lai, L. Qin, X. Liu, C. Zhang, M. Cheng, W. Xiong, Recent advances in tailoring the surface chemistry of activated carbon for enhanced adsorption of organic pollutants, Chem. Eng. J. 430 (2022) 132921.
[16] J. Shen, G. Liu, Y. Ji, Q. Liu, Advances in graphene-based composites for water treatment: Mechanisms, performance, and challenges, Chem. Eng. J. 451 (2023) 139195.
[17] J. Zhu, W. Zhu, L. Xiao, Y. Zhang, Y. Chen, R. Wang, Recent advances in carbon nanotube-based materials for water purification, Environ. Sci. Nano 9 (7) (2022) 1957–1980.
[18] L. Zhang, S. Ye, J. Huang, J. Shen, Integrated carbon-based materials for water and wastewater treatment: Synergistic effects and future prospects, Water Res. 217 (2022) 118399.
[19] D. Mohan, A. Sarswat, Y.S. Ok, C.U. Pittman, Organic and inorganic contaminants removal from water with biochar, a renewable, low cost and sustainable adsorbent - A critical review, Bioresour. Technol. 160 (2014) 191–202.

[20] A.G. Jacob, O.J. Okunola, A.U. Uduma, A. Tijjani, S. Hamisu, Treatment of waste water by activated carbon developed from *Borassus aethiopum*, Niger. J. Mater. Sci. Eng. 6 (1) (2017) 103–107.
[21] M.Z. Alam, S.A. Muyibi, M.F. Mansor, R. Wahid, Graphene for water treatment applications, Separ. Purif. Rev. 50 (1) (2021) 1–25.
[22] H.M. Jang, S. Yoo, Y.K. Choi, S. Park, Y.M. Kim, Biochar for water and wastewater treatment: A review, Chem. Eng. J. 430 (2022) 132921.
[23] J. Zhang, W. Song, G. Zhu, Biochar for nutrient recovery and reuse in water treatment: Mechanisms, performance, and perspectives, J. Clean. Prod. 330 (2022) 129835.
[24] A. Agarwal, V.K. Srivastava, B.P. Singh, Carbon-based composite materials: Advancements, applications, and challenges, Mater. Today Proc. 50 (2022) 2037–2043.
[25] P. Mallick, Fiber-Reinforced Composites: Materials, Manufacturing, and Design, third ed., CRC Press, Boca Raton, FA, USA, 2007.
[26] D.D.L. Chung, A review of self-sensing in carbon fiber structural composite materials, World Sci. Annu. Rev. Funct. Mater. 1 (2023) 2230004.
[27] J.H. Park, S.H. Lee, H.S. Kim, Corrosion resistance of carbon fiber-reinforced polymer composites: A review, Corros. Sci. 208 (2023) 110192.
[28] J. Chen, Y. Li, Z. Xu, Multifunctional carbon-based composite materials for energy storage and conversion, Chem. Eng. J. 449 (2023) 138479.
[29] S. Yasmin, I.M. Daniel, Effect of carbon fiber reinforcement on the thermal conductivity of polymer composites, Compos. Part A Appl. Sci. Manuf. 155 (2022) 106852.
[30] Y. Liu, S. Kumar, I.M. Daniel, Manufacturing and properties of carbon-based composites, Compos. B Eng. 211 (2021) 108604.
[31] T.P. Sathishkumar, J. Naveen, S. Satheeshkumar, Hybrid fiber reinforced polymer composites–A review, J. Reinforc. Plast. Compos. 33 (5) (2014) 454–471.
[32] A. Akinci, S. Sen, C. Ozel, A study on the influence of stacking sequence on the mechanical and thermal properties of carbon fiber/epoxy composite laminates, Compos. B Eng. 58 (2014) 238–246.
[33] K. Aschberger, H.J. Johnston, V. Stone, R.J. Aitken, C.L. Tran, S.M. Hankin, F. M. Christensen, Review of carbon nanotubes toxicity and exposure-appraisal of human health risk assessment based on open literature, Crit. Rev. Toxicol. 40 (8) (2010) 759–790.
[34] N.A. Shuaib, P.T. Mativenga, Effect of mechanical recycling on virgin glass reinforced polypropylene composites, J. Mater. Process. Technol. 229 (2016) 100–109.
[35] J. Palmer, L. Savage, R. Jolly, O.R. Ghita, Sheet moulding compound (SMC) from carbon fibre reinforced sheet thermoplastic material, Compos. Appl. Sci. Manuf. 40 (9) (2009) 1329–1338.
[36] G. Oliveux, L.O. Dandy, G.A. Leeke, Current status of recycling of fibre reinforced polymers: Review of technologies, reuse and resulting properties, Prog. Mater. Sci. 72 (2015) 61–99.
[37] R. Pinero-Hernanz, C. Dodds, J. Hyde, J. Garcia-Serna, M. Poliakoff, E. Lester, M. J. Cocero, Chemical recycling of carbon fibre reinforced composites in near critical and supercritical water, Compos. Appl. Sci. Manuf. 39 (3) (2008) 454–461.
[38] S. Halliwell. In End of Life Options for Composite Waste End of Life Options for Composite Waste Recycle, Reuse or Dispose? National Composites Network Best Practice Guide, 2006.
[39] S.J. Pickering, Recycling technologies for thermoset composite materials-current status, Compos. Appl. Sci. Manuf. 37 (8) (2006) 1206–1215.
[40] S. Yousef, Y.A. El-Shekeil, Utilization of recycled carbon fiber in engineering polymeric composites (Kevlar/Epoxy) for aerospace applications, Am. J. Mater. Sci. 3 (2) (2013) 36–40.

[41] A. Rasheed, S. Renneckar, Consumer waste from carbon fiber reinforced polymer composites and their conversion into activated carbons, Materials 9 (8) (2016) 682.
[42] J. Wang, H.L. Xin, D. Wang, Recent progress on mesoporous carbon materials for advanced energy conversion and storage, Part. Part. Syst. Charact. 31 (5) (2014) 515–539.
[43] G. Ning, C. Xu, J. Cao, Z. Chen, Q. Xue, A. Yu, Z. Qiao, One-step synthesis of graphene-supported ultrafine iron nanoparticles as high-performance anode materials for lithium-ion batteries, Carbon 91 (2015) 224–231.
[44] F.M. Tafti, S.M. Mousavi, J. Reisman, Recycling of carbon fiber reinforced polymer composites for sustainable construction materials: A review, Case Stud. Constr. Mater. 11 (2019) e00287.
[45] O. Fitch-Roy, D. Benson, D. Monciardini, The economics of wind turbine blade recycling: Exploring future solutions, Energy Policy 170 (2022) 113265.
[46] J. Beauson, P. Brøndsted, Wind turbine blades: An end of life perspective, Trends Renew. Energy 2 (1) (2016) 80–101.
[47] K.H. Chua, M.A. Mannan, Sustainability assessment of electricity generation in Malaysia: An overview, Renew. Sustain. Energy Rev. 73 (2017) 685–698.
[48] M. Inagaki, F. Kang, Materials Science and Engineering of Carbon: Fundamentals, second ed., Elsevier Butterworth Heinemann, Oxford, UK, 2014.
[49] F. Deng, H. Zheng, J. Cui, Y. Jiao, Sustainable carbon materials from biomass and waste resources, J. Mater. Chem. A 8 (44) (2020) 23021–23048.
[50] M.M. Titirici, R.J. White, N. Brun, V.L. Budarin, D.S. Su, F. Del Monte, M. J. MacLachlan, Sustainable carbon materials, Chem. Soc. Rev. 44 (1) (2015) 250–290.
[51] A.U. Rajapaksha, M. Vithanage, S.S. Lee, D.C. Seo, D.C. Tsang, Y.S. Ok, Steam activation of biochars facilitates kinetics and pH-resilience of sulfamethazine sorption, J. Soils Sediments 16 (3) (2016) 889–895.
[52] S. Zhai, Q. Liu, Y. Zhao, H. Sun, B. Yang, Y. Weng, A review: research progress in the modification of poly (lactic acid) by lignin and cellulose, Polymers 13 (5) (2021) 776.
[53] M.T. Munir, M. Safdar, M.I. Khan, B.S. Kim, F. Kormin, Biomass-derived carbon materials for energy storage applications: A review, J. Energy Storage 27 (2020) 101020.
[54] V.T. Nguyen, T.B. Nguyen, C.P. Huang, C.W. Chen, X.T. Bui, C.D. Dong, Alkaline modified biochar derived from spent coffee ground for removal of tetracycline from aqueous solutions, J. Water Process Eng. 40 (2021) 101908.
[55] J.S. Cha, S.H. Park, S.C. Jung, C. Ryu, J.K. Jeon, M.C. Shin, Y.K. Park, Production and utilization of biochar: A review, J. Ind. Eng. Chem. 40 (2018) 1–15.
[56] G. Zhang, X. Liu, L. Wang, H. Fu, Recent advances of biomass derived carbon-based materials for efficient electrochemical energy devices, J. Mater. Chem. A 10 (17) (2022) 9277–9307.
[57] H. He, R. Zhang, P. Zhang, P. Wang, N. Chen, B. Qian, B. Dai, Functional carbon from nature: biomass-derived carbon materials and the recent progress of their applications, Adv. Sci. 10 (16) (2023) 2205557.
[58] M. Bernardo, N. Lapa, B. Gonçalves, B. Mendes, F. Pinto, Sustainable gaseous and solid products from the pyrolysis of end-of-life tyres, J. Clean. Prod. 280 (2021) 124489.
[59] Q. Jia, F. Wang, C. Wei, K. Huang, B. Huang, Recycling of waste plastics into graphene and other carbon materials: A review, J. Hazard Mater. 416 (2021) 126232.
[60] A.P. Mouritz, Z. Mathys, Review of methods to recycle end-of-life tyres, Resour. Conserv. Recycl. 76 (2022) 1–17.
[61] U.F.M. Ali, F. Hussin, S.C. Gopinath, M.K. Aroua, M.H. Khamidun, N. Jusoh, S.F. K. Ahmad, Advancement in recycling waste tire activated carbon to potential adsorbent, Environ. Eng. Res. 27 (6) (2022).

[62] J.A. Ávila-Ramírez, D.E. Rivas-Aragón, M.D.R. Rodríguez-Hidalgo, A. Sánchez-Solis, Conversion of industrial and municipal waste into carbon-based materials: A review, Waste Manag. 109 (2020) 15–29.
[63] J. Cao, G. Xiao, X. Xu, D. Shen, R. Li, Study on the pyrolysis of sewage sludge and the characteristics of bio-oil produced at different pyrolysis temperatures, Energy Convers. Manag. 87 (2021) 135–143.
[64] X. Shen, Z. Shi, Z. Zhao, X. Wang, C. Li, J. Huang, G. Liu, Study of the ablation of a carbon/carbon composite at~ 25 MW/m^2 with a nitrogen plasma torch, J. Eur. Ceram. Soc. 40 (15) (2020) 5085–5093.
[65] B. Hu, K. Wang, L. Wu, S.H. Yu, M. Antonietti, M.M. Titirici, Engineering carbon materials from the hydrothermal carbonization process of biomass, Adv. Mater. 22 (7) (2018) 813–828.
[66] C. Zhao, X. Song, Y. Liu, Y. Fu, L. Ye, N. Wang, F. Wang, L. Li, M. Mohammadniaei, M. Zhang, Q. Zhang, J. Liu, Synthesis of graphene quantum dots and their applications in drug delivery, J. Nanobiotechnol. 18 (1) (2020) 142.
[67] Y. Guo, C. Tan, J. Sun, W. Li, J. Zhang, C. Zhao, Porous activated carbons derived from waste sugarcane bagasse for CO_2 adsorption, Chem. Eng. J. 381 (2020) 122736.
[68] H. Marsh, F.R. Reinoso, Activated Carbon, Elsevier, 2006.
[69] K. Mohan, D. Narsimhaswamy, V. Ravi, Graphene/carbon-based materials for advanced energy conversion applications, Adv. Nat. Sci. Nanosci. Nanotechnol. 13 (3) (2022) 033005.
[70] C. Falco, J.P. Marco-Lozar, D. Salinas-Torres, E. Morallon, D. Cazorla-Amorós, M. M. Titirici, D. Lozano-Castello, Tailoring the porosity of chemically activated hydrothermal carbons: influence of the precursor and hydrothermal carbonization temperature, Carbon 62 (2013) 346–355.
[71] J. Ren, S. Tian, L. Ma, Y. Zhang, Z. Jiang, C. Zheng, Synergistic effects of carbon nanotubes and activated carbon in the adsorption of organic pollutants from aqueous solution, Environ. Sci. Pollut. Control Ser. 27 (19) (2020) 24378–24390.
[72] L. Shen, J. Wang, Z. Wang, Q. Xu, H. Wang, S. Dou, Synergistic effects of carbon matrix and carbon nanotubes on the rate capability and cyclability of high-performance supercapacitors, Energy Storage Mater. 17 (2019) 36–44.
[73] N.A. Amran, H.M. Akil, A. Arifin, Recent advances in carbon-based polymer composites: A review, J. Mater. Sci. 56 (1) (2021) 1–35.
[74] S. Rana, R. Alagirusamy, M. Joshi, A review on carbon epoxy nanocomposites, J. Reinforc. Plast. Compos. 38 (13) (2019) archie13–524.
[75] X. Yao, L. Huang, G. Xu, J. Liu, L. Liu, Enhanced mechanical and electrical properties of carbon fiber/polymer composites with carbon nanotube-modified interphase, Compos. Appl. Sci. Manuf. 132 (2020) 105816.
[76] J. Qian, L. Liu, Q. Zhang, W. Zhu, H. Xia, Synergistic effect of porous carbon and metal oxide in hybrid materials for electrochemical energy storage, J. Power Sources 378 (2018) 117–147.
[77] C. Xue, Y. Tu, Y. Lan, J. Wang, S. Wang, Synergistic effects of carbon-based materials and metal oxides in photocatalytic hydrogen generation: A review, Catalysts 9 (2) (2019) 135.
[78] N. Bakhtiari, S.N. Azizi, N. Ghaemi, Graphene oxide/activated carbon hybrid nanomaterial for efficient removal of heavy metal ions from water, Environ. Sci. Pollut. Control Ser. 28 (5) (2021) 5395–5410.
[79] Y. Luo, W. Guo, H.H. Ngo, L.D. Nghiem, F.I. Hai, J. Zhang, X.C. Wang, A review on the occurrence of micropollutants in the aquatic environment and their fate and removal during wastewater treatment, Sci. Total Environ. 473 (2019) 619–641.

[80] L. Zhang, H. Wang, S. Feng, J. Zhang, S. Liu, J. Zhou, Y. Zhao, Hybridization of graphene and carbon nanotube for high-performance supercapacitor and Li-ion battery, Nanoscale 10 (29) (2018) 14184–14193.
[81] X. Li, Y. Zhao, X. Min, J. Xiao, X. Wu, R. Mi, M. Fang, Carbon nanotubes modified graphene hybrid aerogel-based composite phase change materials for efficient thermal storage, Energy Build. 273 (2022) 112384.
[82] L. Zhao, F. Xue, B. Yu, J. Xie, X. Zhang, R. Wu, J. Luo, TiO_2-graphene sponge for the removal of tetracycline, J. Nanoparticle Res. 17 (2015) 1–9.
[83] K.C. Bedin, A.C. Martins, A.L. Cazetta, O. Pezoti, V.C. Almeida, KOH-activated carbon prepared from sucrose spherical carbon: Adsorption equilibrium, kinetic and thermodynamic studies for methylene blue removal, Chem. Eng. J. 286 (2016) 476–484.
[84] M. Jain, M. Yadav, T. Kohout, M. Lahtinen, V.K. Garg, M. Sillanpää, Development of iron oxide/activated carbon nanoparticle composite for the removal of Cr (VI), Cu (II) and Cd (II) ions from aqueous solution, Water Resour. Ind. 20 (2018) 54–74.
[85] X. Wang, C. Jiang, B. Hou, Y. Wang, C. Hao, J. Wu, Carbon composite lignin-based adsorbents for the adsorption of dyes, Chemosphere 206 (2018) 587–596.
[86] W. Xiong, G. Zeng, Z. Yang, Y. Zhou, C. Zhang, M. Cheng, X. Li, Adsorption of tetracycline antibiotics from aqueous solutions on nanocomposite multi-walled carbon nanotube functionalized MIL-53 (Fe) as new adsorbent, Sci. Total Environ. 627 (2018) 235–244.
[87] S. Pu, S. Xue, Z. Yang, Y. Hou, R. Zhu, W. Chu, In situ co-precipitation preparation of a superparamagnetic graphene oxide/Fe 3 O 4 nanocomposite as an adsorbent for wastewater purification: Synthesis, characterization, kinetics, and isotherm studies, Environ. Sci. Pollut. Control Ser. 25 (2018) 17310–17320.
[88] S.K. Sahoo, S. Padhiari, S.K. Biswal, B.B. Panda, G. Hota, Fe_3O_4 nanoparticles functionalized GO/g-C_3N_4 nanocomposite: An efficient magnetic nanoadsorbent for adsorptive removal of organic pollutants, Mater. Chem. Phys. 244 (2020) 122710.
[89] B. D'Cruz, M. Madkour, M.O. Amin, E. Al-Hetlani, Efficient and recoverable magnetic AC-Fe_3O_4 nanocomposite for rapid removal of promazine from wastewater, Mater. Chem. Phys. 240 (2020) 122109.
[90] G. Sharma, M. Naushad, Adsorptive removal of noxious cadmium ions from aqueous medium using activated carbon/zirconium oxide composite: Isotherm and kinetic modelling, J. Mol. Liq. 310 (2020) 113025.
[91] W. Liang, B. Wang, J. Cheng, D. Xiao, Z. Xie, J. Zhao, 3D, eco-friendly metal-organic frameworks@carbon nanotube aerogels composite materials for removal of pesticides in water, J. Hazard Mater. 401 (2021) 123718.
[92] M. del Rio, J.C.G. Escarabajal, G.T. Palomino, C.P. Cabello, Zinc/Iron mixed-metal MOF-74 derived magnetic carbon nanorods for the enhanced removal of organic pollutants from water, Chem. Eng. J. 428 (2022) 131147.
[93] F. Karimi, A. Ayati, B. Tanhaei, A.L. Sanati, S. Afshar, A. Kardan, C. Karaman, Removal of metal ions using a new magnetic chitosan nano-bio-adsorbent; A powerful approach in water treatment, Environ. Res. 203 (2022) 111753.
[94] T.C. Egbosiuba, M.C. Egwunyenga, J.O. Tijani, S. Mustapha, A.S. Abdulkareem, A. S. Kovo, G. Lisak, Activated multi-walled carbon nanotubes decorated with zero valent nickel nanoparticles for arsenic, cadmium and lead adsorption from wastewater in a batch and continuous flow modes, J. Hazard Mater. 423 (2022) 126993.
[95] H. Motaghi, P. Arabkhani, M. Parvinnia, A. Asfaram, Simultaneous adsorption of cobalt ions, azo dye, and imidacloprid pesticide on the magnetic chitosan/activated carbon@ UiO-66 bio-nanocomposite: Optimization, mechanisms, regeneration, and application, Sep. Purif. Technol. 284 (2023) 120258.

Further reading

[1] G. Chen, Y. Wu, Y. Liu, J. Zhang, Z. Wu, X. Liu, Carbon-based nanomaterials for environmental remediation: Progress, challenges, and prospects, Chem. Eng. J. 430 (2022) 133014.
[2] D.L. Chung, Carbon Fiber Composites, Butterworth-Heinemann, Boston, MA, USA, 1994, pp. 3–65.
[3] Z. Dai, S. Sadeghi, A.R. Najafi, A. Mehmanparast, A review on recycling of end-of-life wind turbine blades, Energy Convers. Manag. 238 (2021) 114037.
[4] B. De Vivo, P. Lamberti, G. Spinelli, V. Tucci, G. Clarizia, Challenges and opportunities of nanoparticle-based polymer composites: A review, Appl. Sci. 9 (1) (2019) 66.
[5] O. Fitch-Roy, H. Fairbrass, D. Benson, The politics of recycling: A review of the policy landscape for composite wind turbine blade recycling in the European Union, Renew. Sustain. Energy Rev. 154 (2022) 111845.
[6] T. Hänninen, M. Kanerva, E. Sarlin, M. Hoikkanen, J. Vuorinen, M.L. Räisänen, Carbon fiber production from Finnish lignin, Sustainability 12 (21) (2020) 8960.
[7] J.J. Hernandez, J. Franco Garcia, M.A. Jiménez Benavent, Influence of carbon fiber surface treatment on the fiber-matrix adhesion and mechanical performance of carbon fiber reinforced polymer composites, Materials 13 (11) (2020) 2525.
[8] J. Karger-Kocsis, H. Mahmood, Recent advances in fiber-reinforced asphalt concretes, Compos. B Eng. 78 (2015) 378–387.
[9] M.U. Khan, A. Shahzad, S. Ali, Y. Nawab, M. Sultana, S. Ahmed, R. Ahmed, Recent advances in the recycling of carbon fiber reinforced polymer composites for sustainable future, Polymers 14 (13) (2022) 2682.
[10] H.L. Ornaghi Jr, H.S. Da Silva, A.J. Zattera, Mechanical and dynamic mechanical analysis of hybrid composites made from recycled carbon fibers, Mater. Res. 22 (2) (2019).
[11] G. Wang, L. Zhang, J. Zhang, A review of electrode materials for electrochemical supercapacitors, Chem. Soc. Rev. 41 (2) (2012) 797–828.
[12] M. Mallik, M. Saha, Carbon-based nanocomposites: Processing, electronic properties and applications. In *Carbon nanomaterial electronics: devices and applications*, Springer Singapore, Singapore, 2021, pp. 97–122.
[13] B. Mohan, C.R. Girish, G. Jeppu, P. Patil, Bauhinia monandra-derived mesoporous activated carbon for the efficient adsorptive removal of phenol from wastewater, Sci. Rep. 15 (1) (2025) 31790.

CHAPTER 13

Toxicity and environmental impact of carbon-based materials and composites

Siti Muhamad Nur Husna[1, 2]
[1]Faculty of Pharmacy and Biomedical Sciences, MAHSA University, Jenjarom, Selangor, Malaysia;
[2]Department of Primary Care Medicine, Faculty of Medicine, Universiti Malaya, Wilayah Persekutuan Kuala Lumpur, Malaysia

1. Introduction

With the increase of industrialization, the high demand for materials with superior characteristics is demanded to tailor the specific needs of the key industrial applications. Research interest in finding the best materials has spiked nowadays. The noteworthy contribution that carbon-based materials (CBMs) and their composite have made to materials research has garnered notice. Currently, there are extensive kinds of carbon-based materials including carbon nanotubes (CNTs), carbon nanofiber (CNFs), carbon dots (CDs), carbon nanoparticles (CNPs), fullerenes (C_{60}) and both conventional carbon materials like activated carbon and biochars as well as the novel class of CBMs like graphene and its derivatives [graphene oxide (GO), reduced graphene oxide (rGO), and functionalized graphene nanosheets].

CBMs and their composites have been used widely in the application of biomedical [1,2], wastewater treatments [3,4], electrical industry [5], energy storage and conversion [6], automotive [7], food applications [8] and many others. This is due to carbon-based materials' qualities including low cost, great physical and chemical stability [9], excellent ionic conductivity [10,11], tunable porosity [12], and excellent modification [13,14]. The type of toxicity reaction will be complexly influenced by the physical and chemical properties of the materials, the biological species that they interact with, the form and method of application, and the nature of the medium (soil, soil pore water, and biological surfaces) in which the interaction takes

Carbon-Based Materials and Composites for Water and Wastewater Treatment
ISBN 978-0-443-33681-2
https://doi.org/10.1016/B978-0-443-33681-2.00015-6

place [15]. The direct toxicity of CBMs to humans and animals and environmental contamination by these materials have major negative effects on health.

Notwithstanding the many advantages of CBMs, it is anticipated that organisms have a greater chance of being exposed to harmful materials and possess detrimental effects including DNA damage, skin toxicity, and pulmonary cytotoxicity [16]. It has also been documented that the bioaccumulation and biomagnification phenomena affect members of terrestrial and aquatic/marine ecosystems, including plants, animals and microorganisms [17]. This book chapter will focus mainly on the impact of CBMs and their composite properties on toxicity, mechanism of cytotoxicity, cytotoxicity studies in vitro as well as recent toxicity of the materials toward humans and animals and microorganisms, this is to provide significant guidance toward future research and consideration toward the exploitation of this material.

2. Impact of carbon-based materials and composites properties on toxicity

Understanding how the properties of CBMs and their composite influence their toxicity is essential for ensuring their safe use in various applications. This knowledge helps in designing the best materials with reduced toxicity and in developing guidelines for their safe handling and application. As research progresses, it will be critical to continue examining these properties to mitigate risks and maximize the benefits of CBMs. Physical properties of various CBMs such as length, shape, dispersion, size, surface functionalization and metal impurities influence the toxicity of the materials in organisms [15,16]. Indirect genotoxicity is largely caused by reactive oxygen species (ROS), and it has been shown that tiny size and hence vast surface area enhance the generation of ROS. Hydrogen peroxide (H_2O_2) or superoxide (O-2) anions are produced when the electroactive sites on the CBMs and their composites react with molecular oxygen due to the exponential rise in surface area that occurs as the carbon-based materials and composites shrink in size.

The body's distribution of CBMs and their composites are influenced by their dispersion and agglomeration properties, which also have an impact on the interactions and effects that these materials have on cells. The negative effects of CBMs can be eliminated by a suitable surface coating, which stops the agglomeration and release of harmful ions.

Moreover, surface coatings can alter surface composition and charge, both of which are critical to produce ROS and additional genotoxicity. The toxicity of carbon compounds is mostly caused by residual transition metals catalysts used in their manufacturing. The well-established toxicity of transition metal nanopowders is vastly explained by the variety of their redox chemistry. Similarly, metal-induced toxicity is brought on by metallic catalyst residues in CNTs and CNFs, but only when the metal is present in the carbon matrix in a form that is bioavailable. Reviews on CBMs' toxicity, particularly that of single-walled CNTs (SWCNTs), abound, and they all point to residual metal impurities as one of the primary causes of toxicity. The findings are summarized in Table 13.1.

3. Cytotoxicity mechanism of carbon-based materials and composites

The primary mechanisms of CBMs and their composite cytotoxicity are thought to be the mechanical disruption linked to oxidative stress, mitochondrial damage, and the creation of pro-inflammatory cytokines, depending on the form, length, functionalization, and presence of impurities. Both biotic and abiotic stressors cause oxidative stress, which raises the quantity of ROS due to interference of the carbon-based materials in the organism's system [39,40]. Important biomarkers in detecting oxidative stress are superoxide dismutase (SOD) and glutathione S-transferase (GST) enzyme [41–43]. Fig. 13.1 shows the mechanism of oxidative stress caused by CBMs and their composite.

Secondly, autophagy has been linked to CBM-induced lysosomal damage, even in the absence of inflammasome activation [44]. Autophagy is an intricate process that functions as a survival mechanism by eliminating intracellular infections, damaged organelles, and misfolded or aggregated proteins. Pathogens are engulfed by autophagosomes and directed into lysosomes for destruction and lysosome dysfunction may result in non-apoptotic cell death [45,46]. CD caused the lysosome to enlarge and lose its integrity; CD-induced interleukin (IL)-1β release was lessened when the lysosomal enzyme cathepsin B was inhibited [47]. Cationic CDs also cause pyroptosis in macrophages that is dependent on the inflammasome through lysosomal dysfunction. Significant neutrophil infiltration and fibrosis were not induced by carbon black nanoparticles in pregnant mice [48]. Furthermore, the authors discovered that CNPs might accumulate in lysosomes and reduce cathepsin D, a crucial lysosomal hydrolase, which may

Table 13.1 Impact of CBMs and their composite properties on toxicity.

Properties	Impact of their toxicity	Ref
Size/length	HEK 293T cells demonstrated that both the big and small sizes of graphene and GO considerably decreased cell viability and increased DNA damage	[18]
	The big diameter of mesoporous CNPs significantly reduced the root and shoot lengths in plant	[19]
	Multi-walled CNTs (MWCNTs) with smaller sizes caused ROS and membranes to have multi-pores.	[20]
	Size-dependent cytotoxicity enhancement and DNA polymerase inhibition in nano C_{60} NPs might have more toxicity potency.	[21]
	Small thermally rGO increases toxicity by inducing cell death.	[22]
	Compared to bent MWCNTs, submicron-diameter carbon fibers (SCFs) posed a lower lung risk. Comparing shorter SCFs to smaller SCFs, the latter were more cytotoxic and pulmonary.	[23]
	More lung inflammation is caused by micrometer-sized GO than by nanometer-sized GO.	[24]

Surface properties (functionalization)	Long, short, and carboxylated SWCNTs are most likely to target the nucleus, and mechanical disruption and are most likely to cause DNA damage and more toxicity compared to hydroxylated SWCNT	[25,26]
	Increased oxygen-containing functional groups revealed more toxicity potential to GO.	[22]
	Protein nanocarriers using functionalized graphene oxide and chitosan to prevent enzymatic cleavage and preserve collagenase activity	[27]
	Compared to pristine GO, poly (ethylene glycol)-functionalized GO and poly(acrylic acid)-functionalized GO (GO-PAA) caused reduced toxicity, with GO-PAA being the most biocompatible in both vitro and in vivo settings.	[28]
	Graphene quantum dots (GQDs) were primarily characterized as moderately or highly cytotoxic compared to rGOs	[29].
	In THP-1 cells, cationic carbon nanoparticles with the highest surface charge density caused a considerable amount of oxidative stress, IL-8 production, mitochondrial malfunction, and lysosomal integrity loss.	[30]

Continued

Table 13.1 Impact of CBMs and their composite properties on toxicity.—cont'd

Properties	Impact of their toxicity	Ref
	In comparison to the carboxylfunctionalized group, the amino-functionalized CNTs exhibited greater toxic characteristics.	[31]
	Human myeloperoxidase-driven biodegradation of single-layer and few-layer graphene without chemical functionalization or stabilizing surfactants.	[32]
Metal impurities	The apoptotic pathway regulated by mitochondria is activated when swallowed MWCNTs produce iron overload, which leads to an increase in IL-8 production and an elevation of [Ca2+]i.	[33]
	More severe mitochondrial dysfunction was caused by unpurified MWCNTs because they generated twice as much oxidation as pure CNTs.	[34]
Shape	**Rodlike (e.g., carbon nanotubes)**: High aspect ratios enhance cell membrane penetration and intracellular interactions, causing oxidative stress and potential aggregation, which may reduce bioavailability	[35–38]

	Planar (e.g., graphene): Large surface areas and sharp edges increase membrane disruption and ROS generation due to high surface reactivity **Flake-like CNMs**: Smaller dimensions improve mobility and bioavailability, intensifying cellular interactions and toxicity potential	

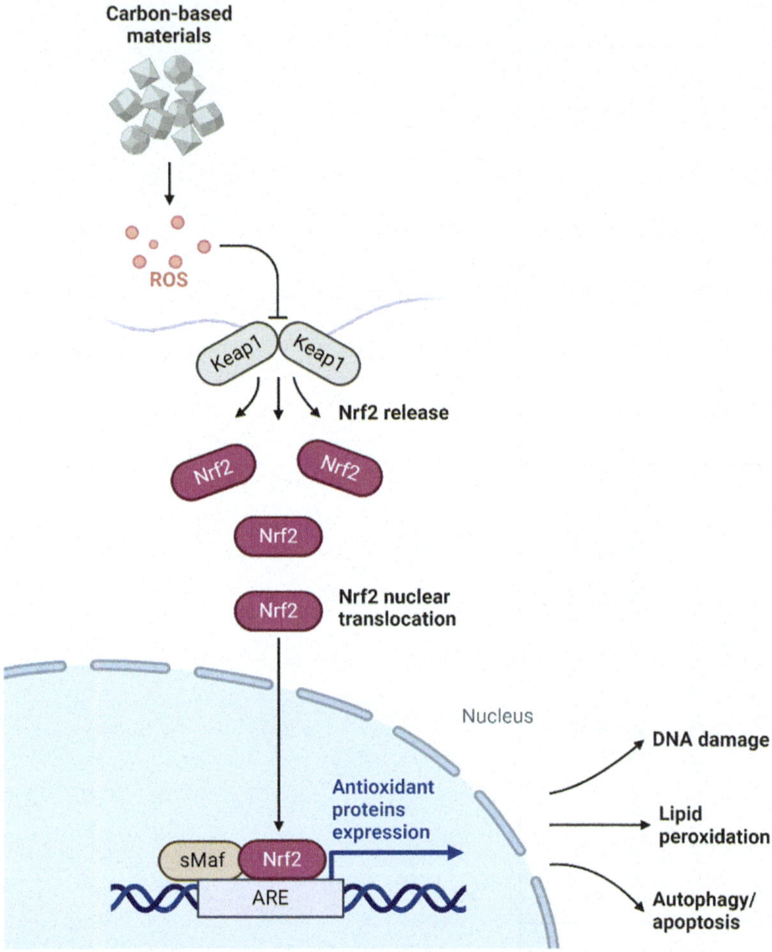

Figure 13.1 Mechanism of oxidative stress caused by CBMs and their composite. CBMs will trigger the production of reactive oxygen species (ROSs) and lead to a subsequent signaling cascade that expresses the antioxidant proteins. Further provides toxicity to the cells in term of DNA damage, lipid peroxidation, autophagy and apoptosis.

be connected to lysosome and autophagy failure and long-term lung injury in dams [48] and in vivo [49]. Macrophages treated with CNPs exhibited necrotic characteristics, including lower intracellular ATP levels, cathepsin B release, lysosome rupture, and the production of reactive oxygen species.

Besides, the mechanism of cytotoxicity was also related to pyroptosis, another type of programmed cell death. Pyroptosis and apoptosis share several characteristics, such as chromatin condensation and DNA damage [50]. It's interesting to note that before the cellular membrane bursts, pyroptotic cells emerge expanding numerous bubble-like protrusions that appear on its surface and pyroptosis can result in inflammation. A dose-dependent neurotoxic effect of MWCNT injection was seen in mice, as demonstrated by increased brain pyroptosis, inflammation, and oxidative stress [51]. MWCNTs caused THP-1 macrophages to undergo pyroptosis [20]. When compared to XFM22 or XFM34, XFM4 more successfully induced pyroptosis, which may serve as a foundation for the thoughtful design of biocompatible MWCNTs.

4. Cytotoxicity studies in vitro

In vitro cytotoxicity studies provide a fundamental finding on the safety of carbon-based materials and composites before their application in real organisms. These studies are reproducible fast and easy to control. The cytotoxicity of carbon fiber-reinforced poly (ether-ether-ketone) (CFR-PEEK) composites was assessed using the CCK-8 assay [52]. The findings showed that the addition of carbon fibers shriveled the cells rapidly decreasing cell viability and cell activity when compared to pure PEEK. Carbon nanomaterials (i.e., MWCNTs and GO) were found to be the most toxic for cells as probed by the MTS assay [53]. GO and MWCNTs significantly increased the fraction of apoptotic cells and decreased cell metabolic activity and/or proliferation, respectively.

N-doped carbon dots (N-CDs) from *Actinidia deliciosa* showed higher cytotoxicity on MCF-7 cells compared to L-929 cells indicating the target selective activity of the prepared N-CDs [54]. CDs produced photo-induced cytotoxicity high significantly harmed cells in high-quantum yield [55] and the degraded molecules were toxic to both normal (HEK-293) and cancerous (HeLa and HepG2) human cells [56]. Cytotoxicity for MWCNTs on the human alveolar epithelial cells is related to dose and time of exposure [the inhibitory concentration of 50% (IC_{50}) was 103.6 as well as 0.65 μg/mL] [57]. Lastly, the toxicity of pluronic (F-127)-dispersed carbon nanomaterials was evaluated in human breast cancer (MCF-7) and epithelial colorectal adenocarcinoma (Caco-2) cell lines [58]. The highest and lowest cell viability was in carbon nanohorns and carbon nanodiamonds, respectively. High hydrophobicity along with the morphology (i.e., sharp edges and conical ends) of carbon nanohorns and carbon nanoparticles attribute to this phenomenon.

5. Toxicity in humans and animals

The toxicity of CBMs and their composite to humans and animals must be studied to establish their ecological safety. These materials can enter the body via trachea drip, inhalation, intravenous injection, ocular, intraperitoneal injection, and oral route [59]. They can cross the mechanical, blood-brain, and blood-placental barriers to collect in the lungs, liver, spleen, and other organs, causing immediate or chronic harm. It leads to neurotoxicity, hepatotoxicity, nephrotoxicity, immunotoxicity, cardiotoxicity, genotoxicity, epigenetic toxicity, dermal toxicity and carcinogenicity that are harmful to human and animal health.

C_{60} has a specific effect on lung tissue, Myojo et al. [60] conducted a toxicity evaluation experiment on rats in which they injected C_{60} into the body by inhalation and intraluminal injection, and the results revealed that the rats' lung tissue would exhibit an inflammatory response. Liver toxicity (i.e., nucleus and vein wall degeneration, and hepatocyte lysis with severe atrophy, impaired sinusoids and inflammation) related to MWCNTs was assessed in 50 mature female Wistar rats, [61]. The same hepatotoxicity (i.e., focal hepatic necrosis and hydropic degeneration of hepatocytes) was observed in 10 rats that received a single intravenous injection of MWCNTs [62]. The appearance of necrotic and apoptotic cells was obvious in high concentrations. Cardiotoxicity observed in mice injected with GO or rGO, histological alterations in heart tissues, together with indications of lipid peroxidation and myocardial enzyme activity [63].

The developmental toxicity of oxidized MWCNTs (O-MWCNTs) was assessed using larvae and capsulated and decapsulated *Artemia salina* cysts [64,65]. Both studies resulted in reduced hatchability with swimming inhibition. Larval mortality rates ranged from 33.8% to 55.7% and this material caused irreparable harm by attaching itself to the body surface, gills, and cysts [65]. Nervous system-related gene expressions were upregulated indicating the neurotoxicity, and immunotoxicity was detected due to increased IL-6, IL-8, tumor necrosis factor-α (TNFα), and interferon-γ (IFN-γ) in RAW264.7 cell line and zebrafish upon the toxicity of GO [66]. A study evaluated whether male rats' lung function and modulatory characteristics of the respiratory epithelium were impacted by inhaled MWCNT [67]. A study comparing the pulmonary toxicity of intratracheal instillation versus inhalation while administering MWCNTs [68]. The toxicity profiles of intratracheal instillation and inhalation are comparable where NM-401 administered intratracheally caused fibrosis and elevated amounts of DNA damage were noted.

As shown by a study on MWCNT-exposed ARPE-19 cells, FA complementation group C (FANCC) was downregulated and hypermethylated in comparison to controls, IL-6 and the apoptosis biomarker caspase three were upregulated, this observation potentially signifying the onset of MWCNT-induced retinal toxicity. MWCNT exposure in ARPE-19 cells leads to hypermethylation and downregulation of FA complementation group C (FANCC), alongside upregulation of IL-6 and caspase 3, indicating potential retinal toxicity [69].

In utero, FANCC caused DNA damage accumulation, apoptosis, and proliferation arrest [70,71] Lymphoid hyperplasia was observed in the bronchial and mediastinal lymph nodes of exposed male and female rats and mice to L-MWNT-102 [72]. The developmental toxicity of GO and graphdiyne (GDY) in zebrafish larvae was compared [73]. Both materials showed a reduction in hatching rate as well as morphological abnormalities as well as behavioral (i.e., hyperactive).

Besides, oxidative stress assay revealed a rise in peroxidase and MDA activity assessed after GO was administered intraperitoneally [74]. *Caenorhabditis elegans* became smaller and had shorter lives after being exposed to GO continuously for multiple generations, impairments of neurons involved in locomotion, increased germline apoptosis, and abnormal expression of reproductive-related genes [75]. Lastly, exposure to CNPs was most likely to cause distinct gene alterations in the pathways responsible for the manufacture of steroids and hormones, delayed hatching of zebrafish embryos and a higher rate of malformation [76]. The summary of the toxicity effects of CBMs and their composites on humans is illustrated in Fig. 13.2.

6. Toxicity in plants

Exposure to the CBMs and their composites on plants and their interactions result in a variety of physiological changes in the chosen plant species. They often influence seed germination, the growth and development of seedlings, and affect shoots, leaves and roots development and overall plant health (Fig. 13.3). It will be harmful for people who consume these plants contaminated by carbon-based materials and their health may be in jeopardy. According to Ren et al. (2018) [26], sulfonated graphene NPs enhanced the accumulation of polychlorinated biphenyls (PCB8 and

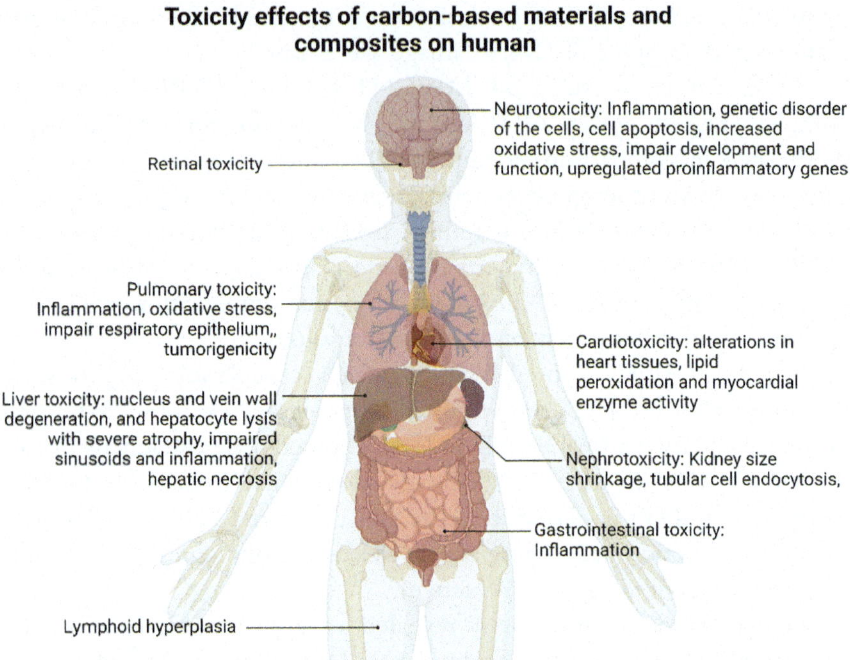

Figure 13.2 Toxicity effects of CBMs and their composites on humans.

PCB18) in the roots of *Z. mays* seedlings. The effect of foliar MWCNT exposure on *Salvia verticillata*'s oxidative stress responses, enzymatic and non-enzymatic antioxidants in the leaves [77]. Effects of rGO and GO on the growth of rice seedlings grown hydroponically for 3 weeks [78]. It was discovered that the primary source of oxidative damage in shoots was the GO treatments (100 and 250 mg/L), which increased the iron (Fe) translocation and produced excessive Fe accumulation in shoots (2.2 and 3.6 times greater than control). When considered collectively, GO affected plant growth even if it decreased the amount of Cd accumulated in rice seedlings [79], it also negatively impacted plant robustness in the presence of co-contamination with Cd.

In this study, GO and graphene oxide quantum dots (GOQDs) were examined for their acute effects on single-celled *Microcystis aeruginosa* [80]. Proteins and chlorophyll *a* were inhibited by GOQDs-induced disruption of antioxidant enzymes and increased levels of ROS and malondialdehyde (MDA) in contrast to GO. Additionally, GO and GOQDs absorbed the algal nutrients, leading to nutrient depletion. The toxicity of GO was examined in green algae (*Chlorella vulgaris*, *Scenedesmus obliquus*, and

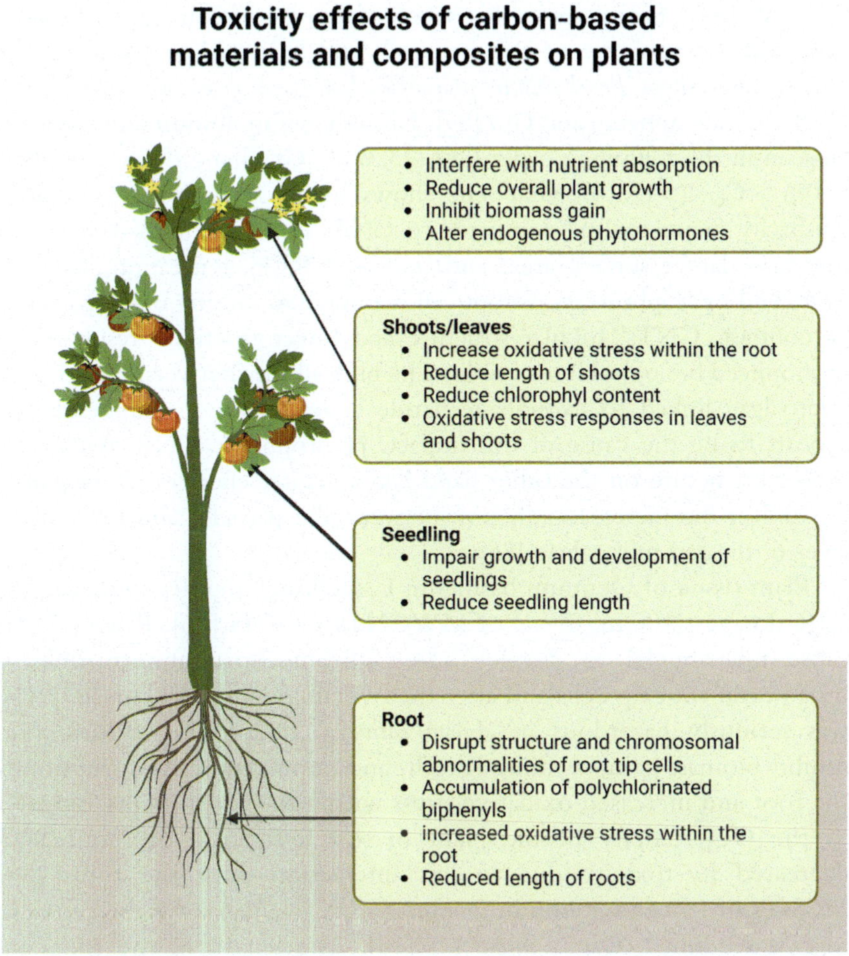

Figure 13.3 Toxicity effects of CBMs and their composites on plants.

Chlamydomonas reinhardtii) [81]. GO toxicity can be seen through decreased development of all examined algal types, and mechanical damage to the cell wall due to oxidative stress and membrane permeability. Silver-reduced GO nanocomposites (MS-Ag-rGO) toxicity revealed changes in nucleoli's location and morphological shrinkage on treated cells *C. vulgaris* cells were confirmed by decreases in growth parameters and changes in antioxidant enzyme activity [82].

Reduced nano C_{60} (nC_{60}) concentrations hindered lindane uptake by the roots, but increased nC_{60} concentrations encouraged lindane uptake by the roots [83]. This suggests that severe pore obstruction caused by nC_{60} disrupted the structure of the root cells and led to plant phytotoxicity. Marine microalgae *Porphyridium purpureum* was used to assess the toxicity of CNTs, C_{60}, graphene, and GO [84]. Growth rate inhibition sensitivity can be seen in the following order GO (23.37) > Graphene (94.88) > CNTs (2.08) > C_{60} (>131.0). Graphene shows higher growth rate inhibition sensitivity in microalgae compared to CNTs due to its two-dimensional structure, larger surface area, and increased ROS generation. Its sharp edges and hydrophobicity disrupt cell membranes, causing oxidative stress. In contrast, CNTs' tubular structure and lower reactivity result in less pronounced biological interactions. The high affinity between particles and microalgae linked to exopolysaccharide coverage on *P. purpureum* cells appears to be the cause of this impact. Hydrophobic CNPs with a less ordered structure on the other hand had a greater effect on *P. purpureum* because of the increased adhesion between the particles and the mucous layer of the red microalgae [85].

Plant tissue of Ocimum basilicum L. seedlings appears to be toxic to large dosages (100 mg L−1) of MWCNTs-COOH [86]. R-CQDs can cause micronuclei to develop cause chromosomal abnormalities in broad bean root tip cells and alter the overall genetic makeup [87]. GO was seriously hazardous in *Avena sativa* L hydroponic culture (i.e., inhibit biomass gain, seedling length, and photosynthesis, disruption to the root and increased oxidative stress within the root) [88]. Compared to the GO toxicity of the plant in soil fertigation, it significantly decreased its toxicity due to less interaction with the roots [88]. MCNPs of 150 and 80 nm in diameter were evaluated for their toxicity to rice seedlings (*Oryza sativa* L.) [19]. The length of the roots and shoots decreased and endogenous phytohormones (i.e., brassinolide, indole propionic acid and dihydrozeatinriboside) in plant shoots were significantly higher.

7. Effect of on microbial communities and toxicity mechanism in microorganisms

Numerous biogeochemical processes on Earth are fueled by microorganisms, which are also vital to each elemental cycle. When bacteria integrate these nutrient elements into biomass, assimilatory coupling of the C, N, and P cycles takes place, and its breakdown releases nutrients. CBMs are used extensively, but they typically have negative impacts because they emit ions into the environment, which kill some beneficial microorganisms [89,90]. Through a variety of metabolic routes, nanomaterials demonstrate bactericidal, bacteriostatic, fungicidal, and fungistatic actions. CBMs mostly affect the composition, development, and variety of microbial communities.

CNTs can alter the conformation of microbial enzymes and have an impact on the enzymatic catalytic process of microorganisms, which would further disrupt microbial metabolism [91,92]. C_{60}, rGO, and MWCNTs markedly changed the bacterial community in the soil in the rhizosphere [93]. While, the effects of C_{60}, SWCNT, and graphene on soil bacterial populations throughout long exposure periods [94]. The alpha and beta diversity of soil bacterial communities was shown to be greatly increased by brief and long-term exposure, respectively. A community of microorganisms in a constructed wetland was observed upon exposure to C_{60}, SWCNTs and MWCNTs [95]. It is reported that nitrogen removal capacity was decreased by 24.1%–42.7% due to an interruption in the enzyme production required for nitrogen removal and ROS production.

Prokaryotic community during the reproductive stage, when the pods were filling, including. MWCNTs have the potential to dramatically change the phylogenetic and functional makeup of prokaryotic communities in the soybean rhizosphere during the reproductive stage and alter functional groups connected to C, N, and S cycling [96]. It is revealed that fullerene significantly affected microorganisms in soils with a modest amount of organic matter and clay [97]. This suggests that the presence of clay and organic matter in the soil can mitigate the negative impacts of fullerene on microbial populations. Furthermore, fullerenes cause microbial cells' regular enzyme secretion to be disrupted, which puts off the balance of their metabolism [98]. Researchers discovered that as GO concentration increased (from 50 to 500 mg/kg), the number of microorganisms in the soil dropped in comparison to the control (which lacked GO) [93]. Graphene causes an imbalance between the oxidation and antioxidation of microbial cells, which causes aberrant enzyme and other substance release and ultimately causes cell destruction [99,100]. The summary of the effect of CBMs and their composites on microorganisms is illustrated in Fig. 13.4.

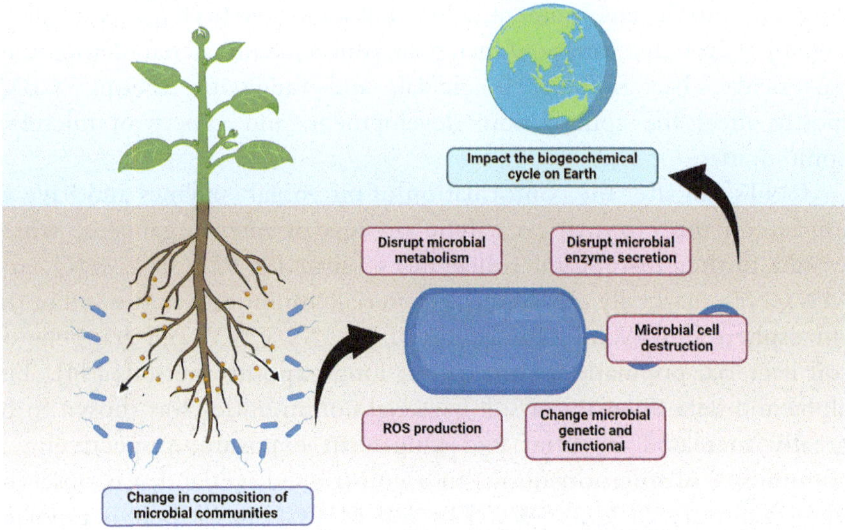

Figure 13.4 Application of CBMs and their composites in agriculture interfere with the microbial communities and provide toxicity to the microbes such as ROS production, disrupt microbial metabolism and enzyme secretion, cell destruction and changes in microbes' genetics, this will further impact the biogeochemical cycles on the Earth that attribute to environmental disturbances.

8. Conclusion

In summary, the exploration of carbon-based materials (CBMs) and their composites highlights both their extensive applications and the significant health and environmental risks associated with their use. These materials, including carbon nanotubes (CNTs), graphene, and carbon dots, are valued for their exceptional physical and chemical properties, enabling their incorporation into diverse industrial sectors ranging from energy storage to biomedical applications. However, the increasing application of CBMs underscores the need to understand their potential toxicity thoroughly. The properties of CBMs—such as size, surface functionalization, and metal impurities—greatly influence their toxicological profiles. Research indicates that the toxicity of these materials can be attributed to mechanisms including oxidative stress, mitochondrial damage, and inflammatory responses. Both in vitro studies and toxicity assessments in humans, animals, and plants have documented adverse effects, such as DNA damage, respiratory issues, and growth impairments. Furthermore, the impact on microbial communities underscores the broader ecological consequences of CBM contamination. As we advance in using CBMs, it is crucial to balance their benefits with a comprehensive understanding of their safety. Future research should focus on developing safer materials, emphasizing the green, sustainable materials' synthesis and functionalization techniques, implementing robust safety guidelines, and mitigating the environmental and health risks associated with CBMs. By doing so, we can harness the potential of these innovative materials while safeguarding both human health and ecological integrity.

References

[1] Joshi, P., Mishra, R., Narayan, R.J., 2021. Biosensing applications of carbon-based materials. Curr. Opin. Biomed. Eng. 18, 100274.
[2] Ye, X., Fan, Q., Shang, L., Ye, F., 2022. Adsorptive carbon-based materials for biomedical applications. Eng. Regen. 3, 352–364.
[3] Sheoran, K., Kaur, H., Siwal, S.S., Saini, A.K., Vo, D.-V.N., Thakur, V.K., 2022. Recent advances of carbon-based nanomaterials (CBNMs) for wastewater treatment: Synthesis and application. Chemosphere 299, 134364.
[4] Priya, A.K., Muruganandam, M., Suresh, S., 2024. Bio-derived carbon-based materials for sustainable environmental remediation and wastewater treatment. Chemosphere, 142731.
[5] Yuan, S., Lai, Q., Duan, X., Wang, Q., 2023. Carbon-based materials as anode materials for lithium-ion batteries and lithium-ion capacitors: A review. J. Energy Storage 61, 106716.
[6] Wu, Q., Yang, L., Wang, X., Hu, Z., 2020. Carbon-based nanocages: A new platform for advanced energy storage and conversion. Adv. Mater. 32, 1904177.
[7] Selvaraj, S.K., Ramesh, R., Narendhra, T.M.V., Agarwal, I.N., Chadha, U., Paramasivam, V., Palanisamy, P., 2021. New developments in carbon-based nanomaterials for automotive brake pad applications and future challenges. J. Nanomater. 2021, 6787435.
[8] Mitura, K., Kornacka, J., Kopczyńska, E., Kalisz, J., Czerwińska, E., Affeltowicz, M., Kaczorowski, W., Kolesińska, B., Frączyk, J., Bakalova, T., 2021. Active carbon-based nanomaterials in food packaging. Coatings 11, 161.
[9] Gopinath, K.P., Vo, D.-V.N., Gnana Prakash, D., Adithya Joseph, A., Viswanathan, S., Arun, J., 2021. Environmental applications of carbon-based materials: A review. Environ. Chem. Lett. 19, 557–582.
[10] Soundaraj, P.V., Dashjav, E., Grüner, D., Prünte, S., Dellen, C., Tietz, F., 2024. Influence of carbon content on the ionic and electronic conductivities of dense Na3V2(PO4)3/C composites. J. Power Sources Adv. 26, 100144. https://doi.org/10.1016/j.powera.2024.100144.
[11] Rajakumar, G., Zhang, X.-H., Gomathi, T., Wang, S.-F., Azam Ansari, M., Mydhili, G., Nirmala, G., Alzohairy, M.A., Chung, I.-M., 2020. Current use of carbon-based materials for biomedical applications—A prospective and review. Processes 8, 355.
[12] Sevilla, M., Alvarez, S., Fuertes, A.B., 2004. Synthesis and characterisation of mesoporous carbons of large textural porosity and tunable pore size by templating mesostructured HMS silica materials. Microporous Mesoporous Mater. 74, 49–58. https://doi.org/10.1016/j.micromeso.2004.06.006.

[13] Wang, Y., Zhang, Z., Yuan, F., Wang, B., 2024. Design and modification of carbon-based materials for high energy density non-aqueous aluminum ion batteries: A review. J. Power Sources 597, 234110. https://doi.org/10.1016/j.jpowsour.2024.234110.
[14] Zuo, Y., Feng, J., Soyol-Erdene, T.-O., Wei, Z., Hu, T., Zhang, Y., Tang, W., 2023. Recent advances in wood-derived monolithic carbon materials: Synthesis approaches, modification methods and environmental applications. Chem. Eng. J. 463, 142332. https://doi.org/10.1016/j.cej.2023.142332.
[15] Samadian, H., Salami, M.S., Jaymand, M., Azarnezhad, A., Najafi, M., Barabadi, H., Ahmadi, A., 2020. Genotoxicity assessment of carbon-based nanomaterials; have their unique physicochemical properties made them double-edged swords? Mutat. Res./Rev. Mutat. Res. 783, 108296. https://doi.org/10.1016/j.mrrev.2020.108296.
[16] Makhado, B.P., Oladipo, A.O., Gumbi, N.N., De Kock, L.A., Andraos, C., Gulumian, M., Nxumalo, E.N., 2024. Unravelling the toxicity of carbon nanomaterials – From cellular interactions to mechanistic understanding. Toxicol. Vitro 100, 105898. https://doi.org/10.1016/j.tiv.2024.105898.
[17] Patil, S.S., Lekhak, U.M., 2020. 12 - Toxic effects of engineered carbon nanoparticles on environment. In: Abd-Elsalam, K.A. (Ed.), Carbon nanomaterials for agri-food and environmental applications. Elsevier, pp. 237–260.
[18] Jia, P.-P., Sun, T., Junaid, M., Yang, L., Ma, Y.-B., Cui, Z.-S., Wei, D.-P., Shi, H.-F., Pei, D.-S., 2019. Nanotoxicity of different sizes of graphene (G) and graphene oxide (GO) in vitro and in vivo. Environ. Pollut. 247, 595–606. https://doi.org/10.1016/j.envpol.2019.01.072.
[19] Hao, Y., Xu, B., Ma, C., Shang, J., Gu, W., Li, W., Hou, T., Xiang, Y., Cao, W., Xing, B., et al., 2019. Synthesis of novel mesoporous carbon nanoparticles and their phytotoxicity to rice (Oryza sativa L.). J. Saudi Chem. Soc. 23, 75–82. https://doi.org/10.1016/j.jscs.2018.05.003.
[20] Wang, S., Ma, J., Guo, S., Huang, Y., Cao, Y., 2020. Transcriptomic analysis revealed that multi-walled carbon nanotubes diameter-dependently induced pyroptosis in THP-1 macrophages. NanoImpact 20, 100270. https://doi.org/10.1016/j.impact.2020.100270.
[21] Song, M., Yuan, S., Yin, J., Wang, X., Meng, Z., Wang, H., Jiang, G., 2012. Size-dependent toxicity of nano-C60 aggregates: More sensitive indication by apoptosis-related bax translocation in cultured human cells. Environ. Sci. Technol. 46, 3457–3464.
[22] Mittal, S., Kumar, V., Dhiman, N., Chauhan, L.K.S., Pasricha, R., Pandey, A.K., 2016. Physico-chemical properties based differential toxicity of graphene oxide/reduced graphene oxide in human lung cells mediated through oxidative stress. Sci. Rep. 6, 39548.
[23] Fujita, K., Obara, S., Maru, J., 2022. Pulmonary toxicity, cytotoxicity, and genotoxicity of submicron-diameter carbon fibers with different diameters and lengths. Toxicology 466, 153063. https://doi.org/10.1016/j.tox.2021.153063.
[24] Rodrigues, A.F., Newman, L., Jasim, D., Mukherjee, S.P., Wang, J., Vacchi, I.A., Ménard-Moyon, C., Bianco, A., Fadeel, B., Kostarelos, K., 2020. Size-dependent pulmonary impact of thin graphene oxide sheets in mice: Toward safe-by-design. Adv. Sci. 7, 1903200.
[25] Jiang, T., Amadei, C.A., Gou, N., Lin, Y., Lan, J., Vecitis, C.D., Gu, A.Z., 2020. Toxicity of single-walled carbon nanotubes (SWCNTs): Effect of lengths, functional groups and electronic structures revealed by a quantitative toxicogenomics assay. Environ. Sci. Nano 7, 1348–1364.

[26] Ren, W., Chang, H., Wang, Y., Teng, Y., Ma, W., Luo, Y., 2018. Effect of sulfonated graphene on uptake, translocation, and metabolism of 2, 4, 4′-trichlorobiphenyl in maize seedlings. Environ. Sci. Pollut. Control Ser. 25, 20084–20096.
[27] Emadi, F., Amini, A., Gholami, A., Ghasemi, Y., 2017. Functionalized graphene oxide with chitosan for protein nanocarriers to protect against enzymatic cleavage and retain collagenase activity. Sci. Rep. 7, 42258.
[28] Xu, M., Zhu, J., Wang, F., Xiong, Y., Wu, Y., Wang, Q., Weng, J., Zhang, Z., Chen, W., Liu, S., 2016. Improved in vitro and in vivo biocompatibility of graphene oxide through surface modification: Poly (acrylic acid)-functionalization is superior to PEGylation. ACS Nano 10, 3267–3281.
[29] Achawi, S., Pourchez, J., Feneon, B., Forest, V., 2021. Graphene-based materials in vitro toxicity and their structure–activity relationships: A systematic literature review. Chem. Res. Toxicol. 34, 2003–2018.
[30] Weiss, M., Fan, J., Claudel, M., Sonntag, T., Didier, P., Ronzani, C., Lebeau, L., Pons, F., 2021. Density of surface charge is a more predictive factor of the toxicity of cationic carbon nanoparticles than zeta potential. J. Nanobiotechnol. 19, 1–19.
[31] Mohammadi, E., Zeinali, M., Mohammadi-Sardoo, M., Iranpour, M., Behnam, B., Mandegary, A., 2020. The effects of functionalization of carbon nanotubes on toxicological parameters in mice. Hum. Exp. Toxicol. 39, 1147–1167.
[32] Kurapati, R., Mukherjee, S.P., Martín, C., Bepete, G., Vázquez, E., Pénicaud, A., Fadeel, B., Bianco, A., 2018. Degradation of single-layer and few-layer graphene by neutrophil myeloperoxidase. Angew. Chem. Int. Ed. 57, 11722–11727.
[33] Tabei, Y., Fukui, H., Nishioka, A., Hagiwara, Y., Sato, K., Yoneda, T., Koyama, T., Horie, M., 2019. Effect of iron overload from multi walled carbon nanotubes on neutrophil-like differentiated HL-60 cells. Sci. Rep. 9, 2224.
[34] Vitkina, T.I., Yankova, V.I., Gvozdeňko, T.A., Kuznetsov, V.L., Krasnikov, D.V., Nazarenko, A.V., Chaika, V.V., Smagin, S.V., Tsatsakis, A.M., Engin, A.B., et al., 2016. The impact of multi-walled carbon nanotubes with different amount of metallic impurities on immunometabolic parameters in healthy volunteers. Food Chem. Toxicol. 87, 138–147. https://doi.org/10.1016/j.fct.2015.11.023.
[35] Murjani, B.O., Kadu, P.S., Bansod, M., Vaidya, S.S., Yadav, M.D., 2022. Carbon nanotubes in biomedical applications: Current status, promises, and challenges. Carbon Lett. 32, 1207–1226.
[36] Xie, C., Zhang, P., Guo, Z., Li, X., Pang, Q., Zheng, K., He, X., Ma, Y., Zhang, Z., Lynch, I., 2020. Elucidating the origin of the surface functionalization-dependent bacterial toxicity of graphene nanomaterials: Oxidative damage, physical disruption, and cell autolysis. Sci. Total Environ. 747, 141546.
[37] Abaszade, R.G., 2022. Synthesis and analysis of flakes graphene oxide. J. Optoelectron. Biomed. Mater. 14, 107–114.
[38] Gies, V., Zou, S., 2018. Systematic toxicity investigation of graphene oxide: Evaluation of assay selection, cell type, exposure period and flake size. Toxicol. Res. 7, 93–101.
[39] Choudhury, F.K., Rivero, R.M., Blumwald, E., Mittler, R., 2017. Reactive oxygen species, abiotic stress and stress combination. Plant J. 90, 856–867.
[40] Tortella, G., Rubilar, O., Pieretti, J.C., Fincheira, P., de Melo Santana, B., Fernández-Baldo, M.A., Benavides-Mendoza, A., Seabra, A.B., 2023. Nanoparticles as a promising strategy to mitigate biotic stress in agriculture. Antibiotics 12, 338.
[41] Weijie, M., Chongnv, W., Xuming, P., Weixin, J., Yuhang, W., Benhui, S., 2020. TiO2 nanoparticles and multi-walled carbon nanotubes monitoring and bioremediation potential using ciliates Pseudocohnilembus persalinus. Ecotoxicol. Environ. Saf. 187, 109825. https://doi.org/10.1016/j.ecoenv.2019.109825.

[42] Deng, X.-Y., Cheng, J., Hu, X.-L., Wang, L., Li, D., Gao, K., 2017. Biological effects of TiO2 and CeO2 nanoparticles on the growth, photosynthetic activity, and cellular components of a marine diatom Phaeodactylum tricornutum. Sci. Total Environ. 575, 87–96.
[43] Cimbaluk, G.V., Ramsdorf, W.A., Perussolo, M.C., Santos, H.K.F., Da Silva De Assis, H.C., Schnitzler, M.C., Schnitzler, D.C., Carneiro, P.G., Cestari, M.M., 2018. Evaluation of multiwalled carbon nanotubes toxicity in two fish species. Ecotoxicol. Environ. Saf. 150, 215–223. https://doi.org/10.1016/j.ecoenv.2017.12.034.
[44] Svadlakova, T., Hubatka, F., Turanek Knotigova, P., Kulich, P., Masek, J., Kotoucek, J., Macak, J., Motola, M., Kalbac, M., Kolackova, M., 2020. Proinflammatory effect of carbon-based nanomaterials: In vitro study on stimulation of inflammasome NLRP3 via destabilisation of lysosomes. Nanomaterials 10, 418.
[45] Maltese, W.A., Overmeyer, J.H., 2015. Non-apoptotic cell death associated with perturbations of macropinocytosis. Front. Physiol. 6, 38.
[46] Mondragon, A.A., Yalonetskaya, A., Ortega, A.J., Zhang, Y., Naranjo, O., Elguero, J., Chung, W.-S., McCall, K., 2019. Lysosomal machinery drives extracellular acidification to direct non-apoptotic cell death. Cell Rep. 27, 11–19.
[47] Arezki, Y., Rapp, M., Lebeau, L., Ronzani, C., Pons, F., 2022. Cationic carbon nanoparticles induce inflammasome-dependent pyroptosis in macrophages via lysosomal dysfunction. Front. Toxicol. 4, 925399.
[48] Liu, X., Tu, B., Jiang, X., Xu, G., Bai, L., Zhang, L., Meng, P., Qin, X., Chen, C., Zou, Z., 2019. Lysosomal dysfunction is associated with persistent lung injury in dams caused by pregnancy exposure to carbon black nanoparticles. Life Sci. 233, 116741. https://doi.org/10.1016/j.lfs.2019.116741.
[49] Yuan, X., Nie, W., He, Z., Yang, J., Shao, B., Ma, X., Zhang, X., Bi, Z., Sun, L., Liang, X., 2020. Carbon black nanoparticles induce cell necrosis through lysosomal membrane permeabilization and cause subsequent inflammatory response. Theranostics 10, 4589.
[50] Yu, P., Zhang, X., Liu, N., Tang, L., Peng, C., Chen, X., 2021. Pyroptosis: Mechanisms and diseases. Signal Transduct. Targeted Ther. 6, 128.
[51] Sallam, A.A., El-Magd, M.A., Ahmed, M.M., Ghamry, H.I., Alshahrani, M.Y., Hegazy, R.A., Magdy, A., El-Fotoh, M.F.A., 2022. Quercetin alleviated multi-walled carbon nanotubes-induced neurotoxicity in mice through inhibition of oxidation, inflammation, and pyroptosis. Biomed. Pharmacother. 151, 113160. https://doi.org/10.1016/j.biopha.2022.113160.
[52] Qin, W., Li, Y., Ma, J., Liang, Q., Tang, B., 2019. Mechanical properties and cytotoxicity of hierarchical carbon fiber-reinforced poly (ether-ether-ketone) composites used as implant materials. J. Mech. Behav. Biomed. Mater. 89, 227–233.
[53] Rozhina, E., Batasheva, S., Miftakhova, R., Yan, X., Vikulina, A., Volodkin, D., Fakhrullin, R., 2021. Comparative cytotoxicity of kaolinite, halloysite, multiwalled carbon nanotubes and graphene oxide. Appl. Clay Sci. 205, 106041. https://doi.org/10.1016/j.clay.2021.106041.
[54] Arul, V., Sethuraman, M.G., 2018. Facile green synthesis of fluorescent N-doped carbon dots from Actinidia deliciosa and their catalytic activity and cytotoxicity applications. Opt. Mater. 78, 181–190.
[55] Esfandiari, N., Bagheri, Z., Ehtesabi, H., Fatahi, Z., Tavana, H., Latifi, H., 2019. Effect of carbonization degree of carbon dots on cytotoxicity and photo-induced toxicity to cells. Heliyon 5.
[56] Liu, Y.-Y., Yu, N.-Y., Fang, W.-D., Tan, Q.-G., Ji, R., Yang, L.-Y., Wei, S., Zhang, X.-W., Miao, A.-J., 2021. Photodegradation of carbon dots cause cytotoxicity. Nat. Commun. 12, 812.

[57] Nasirzadeh, N., Rasoulzadeh, Y., Azari, M.R., Mohammadian, Y., 2020. Cellular toxicity of multi-walled carbon nanotubes on human lung cells. J. Chem. Health Risks 10.
[58] Garriga, R., Herrero-Continente, T., Palos, M., Cebolla, V.L., Osada, J., Muñoz, E., Rodríguez-Yoldi, M.J., 2020. Toxicity of carbon nanomaterials and their potential application as drug delivery systems: In vitro studies in Caco-2 and MCF-7 cell lines. Nanomaterials 10, 1617.
[59] Iavicoli, I., Fontana, L., Pingue, P., Todea, A.M., Asbach, C., 2018. Assessment of occupational exposure to engineered nanomaterials in research laboratories using personal monitors. Sci. Total Environ. 627, 689–702. https://doi.org/10.1016/j.scitotenv.2018.01.260.
[60] Myojo, T., Ono-Ogasawara, M., 2018. Review; risk assessment of aerosolized SWCNTs, MWCNTs, fullerenes and carbon black. KONA Powder Part. J. 35, 80–88.
[61] Kooravand Bardpareh, M., Noori, A., 2018. Liver toxicity of multiwall carbon nanotubes in rats. J. Gorgan Univ. Med. Sci. 20, 36–43.
[62] El-Sayed, S.G., Ahmed, N.H., Ahmed, K.A., Issa, A.M., 2018. The toxic effect of multi wall carbon nanotubes on SWISS albino rat. Egypt. J. Zool. 70, 1–12.
[63] Zhang, J., Cao, H.-Y., Wang, J.-Q., Wu, G.-D., Wang, L., 2021. Graphene oxide and reduced graphene oxide exhibit cardiotoxicity through the regulation of lipid peroxidation, oxidative stress, and mitochondrial dysfunction. Front. Cell Dev. Biol. 9, 616888.
[64] Zhu, S., Luo, F., Tu, X., Chen, W.-C., Zhu, B., Wang, G.-X., 2017. Developmental toxicity of oxidized multi-walled carbon nanotubes on Artemia salina cysts and larvae: Uptake, accumulation, excretion and toxic responses. Environ. Pollut. 229, 679–687. https://doi.org/10.1016/j.envpol.2017.07.020.
[65] Zhu, B., Zhu, S., Li, J., Hui, X., Wang, G.-X., 2018. The developmental toxicity, bioaccumulation and distribution of oxidized single walled carbon nanotubes in Artemia salina. Toxicol. Res. 7, 897–906.
[66] Yang, X., Yang, Q., Zheng, G., Han, S., Zhao, F., Hu, Q., Fu, Z., 2019. Developmental neurotoxicity and immunotoxicity induced by graphene oxide in zebrafish embryos. Environ. Toxicol. 34, 415–423.
[67] Russ, K.A., Thompson, J.A., Kashon, M., Porter, D.W., Friend, S.A., McKinney, W., Fedan, J.S., 2019. Comparison of multi-wall carbon nanotube and nitrogen-doped multi-wall carbon nanotube effects on lung function and airway reactivity in rats. Toxicol. Appl. Pharmacol. 364, 153–163. https://doi.org/10.1016/j.taap.2018.11.004.
[68] Gaté, L., Knudsen, K.B., Seidel, C., Berthing, T., Chézeau, L., Jacobsen, N.R., Valentino, S., Wallin, H., Bau, S., Wolff, H., et al., 2019. Pulmonary toxicity of two different multi-walled carbon nanotubes in rat: Comparison between intratracheal instillation and inhalation exposure. Toxicol. Appl. Pharmacol. 375, 17–31. https://doi.org/10.1016/j.taap.2019.05.001.
[69] Liang, Y., Zhou, Y., Xie, D., Yin, F., Luo, X., 2024. Hypermethylation and low expression of FANCC involved in multi-walled carbon nanotube-induced toxicity on ARPE-19 cells. Environ. Res. 241, 117619. https://doi.org/10.1016/j.envres.2023.117619.
[70] Badra Fajardo, N., Taraviras, S., Lygerou, Z., 2022. Fanconi anemia proteins and genome fragility: Unraveling replication defects for cancer therapy. Trends Cancer 8, 467–481. https://doi.org/10.1016/j.trecan.2022.01.015.
[71] Shen, X., Wang, R., Kim, M.J., Hu, Q., Hsu, C.-C., Yao, J., Klages-Mundt, N., Tian, Y., Lynn, E., Brewer, T.F., et al., 2020. A surge of DNA damage links

transcriptional reprogramming and hematopoietic deficit in Fanconi anemia. Mol. Cell 80. https://doi.org/10.1016/j.molcel.2020.11.040.
[72] National Toxicology, P. *NTP Technical report on the toxicity studies of 1020 long multiwalled carbon nanotubes administered by inhalation to sprague dawley (Hsd: Sprague dawley® SD®) rats and B6C3F1/N mice.* 2019.
[73] Wu, W., Zhao, W., Huang, C., Cao, Y., 2024. Comparison of developmental toxicity of graphene oxide and graphdiyne to zebrafish larvae. Comp. Biochem. Physiol. C Toxicol. Pharmacol. 281, 109924. https://doi.org/10.1016/j.cbpc.2024.109924.
[74] Rhazouani, A., Gamrani, H., Ed-Day, S., Lafhal, K., Boulbaroud, S., Gebrati, L., Fdil, N., Aziz, F., 2023. Sub-acute toxicity of graphene oxide (GO) nanoparticles in male mice after intraperitoneal injection: Behavioral study and histopathological evaluation. Food Chem. Toxicol. 171, 113553. https://doi.org/10.1016/j.fct.2022.113553.
[75] Jin, L., Dou, T.-T., Chen, J.-Y., Duan, M.-X., Zhen, Q., Wu, H.-Z., Zhao, Y.-L., 2022. Sublethal toxicity of graphene oxide in Caenorhabditis elegans under multi-generational exposure. Ecotoxicol. Environ. Saf. 229, 113064. https://doi.org/10.1016/j.ecoenv.2021.113064.
[76] Shi, L., Qian, Y., Shen, Q., He, Y., Jia, Y., Wang, F., 2022. The developmental toxicity and transcriptome analyses of zebrafish (Danio rerio) embryos exposed to carbon nanoparticles. Ecotoxicol. Environ. Saf. 234, 113417. https://doi.org/10.1016/j.ecoenv.2022.113417.
[77] Rahmani, N., Radjabian, T., Soltani, B.M., 2020. Impacts of foliar exposure to multi-walled carbon nanotubes on physiological and molecular traits of Salvia verticillata L., as a medicinal plant. Plant Physiol. Biochem. 150, 27–38. https://doi.org/10.1016/j.plaphy.2020.02.022.
[78] Zhang, P., Guo, Z., Luo, W., Monikh, F.A., Xie, C., Valsami-Jones, E., Lynch, I., Zhang, Z., 2020. Graphene oxide-induced pH alteration, iron overload, and subsequent oxidative damage in rice (Oryza sativa L.): A new mechanism of nanomaterial phytotoxicity. Environ. Sci. Technol. 54, 3181–3190.
[79] You, Y., Liu, L., Wang, Y., Li, J., Ying, Z., Hou, Z., Liu, H., Du, S., 2021. Graphene oxide decreases Cd concentration in rice seedlings but intensifies growth restriction. J. Hazard Mater. 417, 125958. https://doi.org/10.1016/j.jhazmat.2021.125958.
[80] Yan, Z., Yang, X., Lynch, I., Cui, F., 2022. Comparative evaluation of the mechanisms of toxicity of graphene oxide and graphene oxide quantum dots to blue-green algae microcystis aeruginosa in the aquatic environment. J. Hazard Mater. 425, 127898. https://doi.org/10.1016/j.jhazmat.2021.127898.
[81] Yin, J., Fan, W., Du, J., Feng, W., Dong, Z., Liu, Y., Zhou, T., 2020. The toxicity of graphene oxide affected by algal physiological characteristics: A comparative study in cyanobacterial, green algae, diatom. Environ. Pollut. 260, 113847. https://doi.org/10.1016/j.envpol.2019.113847.
[82] Nazari, F., Jafarirad, S., Movafeghi, A., Kosari-Nasab, M., Kazemi, E.M., 2020. Toxicity of microwave-synthesized silver-reduced graphene oxide nanocomposites to the microalga chlorella vulgaris: Comparison with the hydrothermal method synthesized counterparts. J. Environ. Sci. Health Part A 55, 639–649.
[83] He, A., Jiang, J., Ding, J., Sheng, G.D., 2021. Blocking effect of fullerene nanoparticles (nC60) on the plant cell structure and its phytotoxicity. Chemosphere 278, 130474. https://doi.org/10.1016/j.chemosphere.2021.130474.
[84] Pikula, K., Johari, S.A., Santos-Oliveira, R., Golokhvast, K., 2023. The comparative toxic impact assessment of carbon nanotubes, fullerene, graphene, and graphene oxide on marine microalgae Porphyridium purpureum. Toxics 11, 491.

[85] Pikula, K., Chaika, V., Zakharenko, A., Markina, Z., Vedyagin, A., Kuznetsov, V., Gusev, A., Park, S., Golokhvast, K., 2020. Comparison of the level and mechanisms of toxicity of carbon nanotubes, carbon nanofibers, and silicon nanotubes in bioassay with four marine microalgae. Nanomaterials 10, 485.

[86] Gohari, G., Safai, F., Panahirad, S., Akbari, A., Rasouli, F., Dadpour, M.R., Fotopoulos, V., 2020. Modified multiwall carbon nanotubes display either phytotoxic or growth promoting and stress protecting activity in Ocimum basilicum L. in a concentration-dependent manner. Chemosphere 249, 126171. https://doi.org/10.1016/j.chemosphere.2020.126171.

[87] Hu, J., Sang, J., Li, P., Wei, X., Wang, Z., Song, K., 2023. Synthesis and biological toxicity evaluation of red light-emitting carbon quantum dots. Sci. Adv. Mater. 15, 1299–1310.

[88] Chen, L., Yang, S., Liu, Y., Mo, M., Guan, X., Huang, L., Sun, C., Yang, S.-T., Chang, X.-L., 2018. Toxicity of graphene oxide to naked oats (Avena sativa L.) in hydroponic and soil cultures. RSC Adv. 8, 15336–15343.

[89] Ameen, F., Alsamhary, K., Alabdullatif, J.A., Alnadhari, S., 2021. A review on metal-based nanoparticles and their toxicity to beneficial soil bacteria and fungi. Ecotoxicol. Environ. Saf. 213, 112027. https://doi.org/10.1016/j.ecoenv.2021.112027.

[90] Shende, S.S., Rajput, V.D., Gorovtsov, A.V., Harish, Saxena, P., Minkina, T.M., Chokheli, V.A., Jatav, H.S., Sushkova, S.N., Kaur, P., 2021. Interaction of nanoparticles with microbes. In: Plant-microbes-engineered nano-particles (PM-ENPs) nexus in agro-ecosystems: Understanding the interaction of plant, microbes and engineered nano-particles (ENPS), pp. 175–188.

[91] Chen, M., Zeng, G., Xu, P., Yan, M., Xiong, W., Zhou, S., 2017. Interaction of carbon nanotubes with microbial enzymes: Conformational transitions and potential toxicity. Environ. Sci. Nano 4, 1954–1960.

[92] Chen, M., Zeng, G., Xu, P., Zhang, Y., Jiang, D., Zhou, S., 2017. Understanding enzymatic degradation of single-walled carbon nanotubes triggered by functionalization using molecular dynamics simulation. Environ. Sci. Nano 4, 720–727.

[93] Hao, Y., Ma, C., Zhang, Z., Song, Y., Cao, W., Guo, J., Zhou, G., Rui, Y., Liu, L., Xing, B., 2018. Carbon nanomaterials alter plant physiology and soil bacterial community composition in a rice-soil-bacterial ecosystem. Environ. Pollut. 232, 123–136.

[94] Wu, F., Jiao, S., Hu, J., Wu, X., Wang, B., Shen, G., Yang, Y., Tao, S., Wang, X., 2021. Stronger impacts of long-term relative to short-term exposure to carbon nanomaterials on soil bacterial communities. J. Hazard Mater. 410, 124550. https://doi.org/10.1016/j.jhazmat.2020.124550.

[95] Yang, X., He, Q., Guo, F., Sun, X., Zhang, J., Chen, Y., 2021. Impacts of carbon-based nanomaterials on nutrient removal in constructed wetlands: Microbial community structure, enzyme activities, and metabolism process. J. Hazard Mater. 401, 123270. https://doi.org/10.1016/j.jhazmat.2020.123270.

[96] Ge, Y., Shen, C., Wang, Y., Sun, Y.-Q., Schimel, J.P., Gardea-Torresdey, J.L., Holden, P.A., 2018. Carbonaceous nanomaterials have higher effects on soybean rhizosphere prokaryotic communities during the reproductive growth phase than during vegetative growth. Environ. Sci. Technol. 52, 6636–6646.

[97] Chen, M., Sun, Y., Liang, J., Zeng, G., Li, Z., Tang, L., Zhu, Y., Jiang, D., Song, B., 2019. Understanding the influence of carbon nanomaterials on microbial communities. Environ. Int. 126, 690–698. https://doi.org/10.1016/j.envint.2019.02.005.

[98] Chen, M., Zeng, G., Xu, P., Lai, C., Tang, L., 2017. How do enzymes 'Meet' nanoparticles and nanomaterials? Trends Biochem. Sci. 42, 914–930. https://doi.org/10.1016/j.tibs.2017.08.008.

[99] Wu, X., Zhou, Z., Li, K., Liu, S., 2024. Nanomaterials-induced redox imbalance: Challenged and opportunities for nanomaterials in cancer therapy. Adv. Sci. 11, 2308632.

[100] Sahar, S., Sun, S., Zeb, A., Jin, L., Gao, Y., Tian, L., Wang, W., Xu, A.-W., Mao, Z., 2023. Multi-enzyme mimics–cracking the code of subcellular cascade reactions and their potential biological applications. Mater. Chem. Front. 7, 3037–3072.

CHAPTER 14

Challenges and future directions of carbon-based materials and composites in water and wastewater treatment

Asif Hafeez and Khubab Shaker
Department of Materials, National Textile University, Faisalabad, Pakistan

1. Introduction

1.1 Importance of water and wastewater treatment

The presence of hazardous substances in polluted water poses a serious threat to public health and water security globally. Industrial discharges, improper waste disposal, and inadequate sanitation infrastructure have resulted in the contamination of water bodies with a diverse array of harmful materials, including heavy metals, organic pollutants, microplastics, pharmaceutical residues, and various other toxic chemicals [1]. Exposure to these hazardous materials through contaminated water can have severe health-related consequences. Furthermore, the contamination of water sources also contributes to scarcity. The growing demand for water due to industrialization and population growth exacerbates the challenge of providing clean and reliable water supplies. Hence, to mitigate the dangers of water pollution to human health, numerous technologies and approaches have been developed to tackle water contamination and remove hazardous substances from aqueous systems. Adsorption is one of the most cost-effective and efficient techniques for eliminating dyes, heavy metals, and emerging contaminants from wastewater. It is favored for its simplicity, high removal efficiency, and potential for regeneration and reuse. The adsorption process involves the accumulation of pollutants on the surface of a solid absorbent. Moreover, the other most commonly adopted technique for wastewater treatment and purification is membrane

Carbon-Based Materials and Composites for Water and Wastewater Treatment
ISBN 978-0-443-33681-2
https://doi.org/10.1016/B978-0-443-33681-2.00009-0

separation. The membrane-based processes are favored due to their low energy demand, higher separation selectivity for the targeted pollutants, and good stability [2]. In addition to adsorption and membrane-based separation, other wastewater treatment techniques are coagulation, flocculation, and precipitation [3]. These techniques are favored for the removal of a wide range of pollutants including suspended solids, heavy metals, and organic compounds. The choice of the most suitable treatment approach is dictated by the presence of a specific contaminant, the scale of operation, and the overall cost-effectiveness of the system [4].

1.2 Role of carbon-based materials and composites

The pursuit of materials with exceptional properties, tailored to meet the specific demands of key industrial applications has sparked a new area of research interest. This focus is a pivot for recent research activities in the fields of medicine, science, and engineering. For these materials to be applicable in these fields, they must possess enhanced structural, electronic, and optical properties that elevate their functionalities for specific applications. Carbon-based materials (CBMs) have gained significant attention due to their crucial role in the advancement of materials science. These traditional carbon materials like activated carbon and biochars have been used for a long time, however, the emergence of new CBMs like graphene, carbon nanotubes (CNTs), carbon nanofibers (CNFs), fullerenes, and carbon dots have significantly broadened the scope, as shown in Fig. 14.1. CBMs demonstrate exceptional electrical conductivity, as well as high optical, thermal and mechanical characteristics. Their chemical stability, high surface area, and well-defined, adjustable porosity contribute to their suitability for a diverse array of applications including wastewater treatment [6], catalysis [7], energy conversion and storage systems, polymer reinforcement [8], biomedical applications and carbon capture [9].

2. Carbon-based materials for wastewater treatment

The diverse structures and properties of carbon materials are closely tied to the unique characteristics of the carbon atom. Carbon materials are classified as zero-dimensional nanomaterials which include fullerenes (spherical, cage-like structures composed of carbon atoms) and carbon dots (quasi-spherical nanoparticles made of carbon), one-dimensional nanomaterials which encompasses carbon nanotubes (cylindrical structure) and carbon nanofibers (isostructural to carbon nanotubes, however, have a

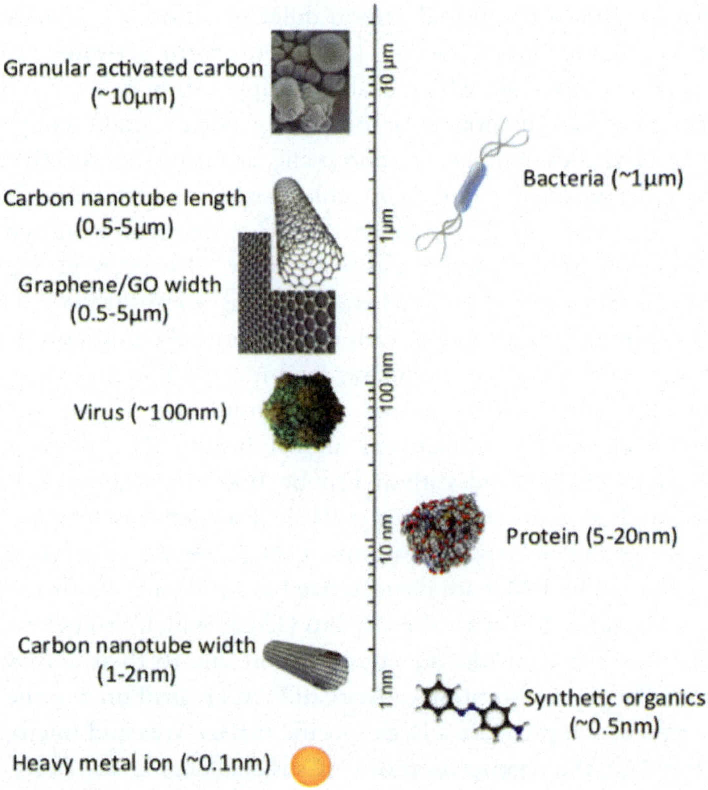

Figure 14.1 Relative size of major classes of carbon-based materials and common water pollutants [5].

distinct graphitic structure along the fiber axis), two-dimensional nanomaterials which include graphene (single atom thick with carbon atoms arranged in a honeycomb lattice) and lastly three-dimensional nanomaterials which includes carbon sponges (porous structure composed of interconnected carbon building blocks) [10].

2.1 Activated carbon

Activated carbon belongs to a broad classification of amorphous carbon-based materials that are manufactured in such a manner that they exhibit a significant level of porosity, high adsorption capacity, and an enhanced surface area which ranges from 500 to 3000 $m^2\ g^{-1}$. Due to these unique properties activated carbon is considered an excellent adsorbent material that is used in the treatment of water and the regulation of air pollution.

The characteristics of activated carbon differ based on the precursor used and the synthesis technique used [11]. In its pure form, activated carbon has a non-polar composition, which enables its application in the removal and remediation of various non-polar or slightly polar organic contaminants including pesticides, phthalates, polycyclic aromatic hydrocarbons, surfactants, polychlorinated biphenyls, chlorinated solvents, phenolic compounds, from water [12]. Activated carbon is frequently utilized in the remediation of drinking water and wastewater. However, its application for treating other types of polluted water such as stormwater is constrained due to the presence of a range of various pollutants like nutrients (nitrogen, phosphorous, etc.), organic pollutants, and metals. The presence of these contaminants induces the risk of bioaccumulation in organisms and detrimental effects on human and animal health. The physiochemical characteristics of activated carbon can be fine-tuned through the pore structure and surface characteristics [13]. Powdered activated carbons (PACs) and granular activated carbons (GACs) are the two types of activated carbons. The PACs are characterized as small particles that settle and are removed more slowly as compared to GACs, which makes PACs more difficult to extract from the aqueous solution due to their excessive dust pollution. The high adsorption capacity of PACs is attributed to the smaller particle size which provides a large specific surface area and microporosity [14]. However, the regeneration of PACs is challenging due to its difficult extraction from the aqueous solution. On the contrary, the GACs that are employed for water treatment can undergo a size reduction to yield PACs and resultantly they can be used to adsorb 2-methyl isobornrneol [15]. The choice between GAC and PAC for water treatment is dictated by factors such as the target pollutant, concentration, flow rate, and adsorption potential. To date, GAC is utilized for the elimination of diverse organic micropollutants, pharmaceuticals, arsenic, carcinogenic chemicals, microplastics, heavy metals, and colorants from wastewater [14].

2.1.1 Synthesis and activation

Activated carbon is prepared through two main activation approaches i.e., thermal (physical) activation and chemical activation. The production of activated carbon involves several key steps: preparing the raw materials, pelletizing, carbonization, and activation. The physical activation involves carbonizing the raw material at 400–500°C to eliminate volatile components, followed by partial gasification with an oxidizing gas like steam, flue gas, or carbon dioxide at 800–1000°C. This method is favored for its convenience, flexibility, and uniform pore formation [11]. This process generates a porous structure with a high surface area and the presence of functional groups [16]. Whereas, chemical activation involves the incorporation of additives during the synthesis process before carbonization. These additives degrade and dehydrate cellulosic materials during carbonization at 250–650°C. Lignin in the raw material is blended with activators like phosphoric acid, and sulfuric acid, followed by carbonization at up to 900 °C [17]. The structural properties, presence of functional groups, and adsorption dynamics are strongly dependent upon the source of the raw material and the activation technique adopted [18].

2.1.2 Modification of activated carbon

The surface modification aim is to enhance the affinity of activated carbon for the removal of specific pollutants such as metals, inorganics, and organics in aqueous media by tailoring the surface properties, pore structure, and the introduction of functional groups [19]. The adsorption efficiency of activated carbon is influenced by its specific surface area, pore size distribution, pore volume, and the type and presence of surface functionality. The adsorption capacity increases with a higher specific surface area. Additionally, pore size distribution is closely related to the raw material composition, degree of activation, and regeneration history. The presence and nature of functionality significantly impact the adsorption of specific pollutants. Surface modification of activated carbon, performed after the activation stage, can be classified into physical, chemical, and biological. Furthermore, oxidative treatments such as acid, base, and impregnation treatment and non-oxidative treatments by ozone, surfactant, and plasma have also been reported to enhance the adsorption capabilities of activated carbon [20,21].

2.1.3 Limitations of activated carbon

Activated carbon, while effective in the removal of various metal, organic, and inorganic pollutants from the aqueous solutions through physical adsorption, has its limitations in wastewater treatment, particularly its selectivity. Furthermore, its low selectivity is associated with its surface and pore structure characteristics which leads to its non-specific adsorption of a broad range of contaminants present in the wastewater. This lack of selectivity results in the removal of specific pollutants along with the non-target compound from the wastewater [22]. Activated carbon can become saturated with contaminants over time, which requires regeneration to maintain its effectiveness. The regeneration process can be complex and costly which impacts the overall feasibility of using activated carbon in long-term applications [23].

2.2 Carbon nanotubes

Carbon nanotubes (CNTs) are nano-architectured carbon allotropes formed from graphene sheets rolled into cylindrical shapes. They exhibit unique strength, along with exceptional electrical, mechanical, and thermal properties. These diverse properties make CNTs highly attractive for a wide range of applications including electronics, energy storage, catalysis, and environmental applications like wastewater treatment. CNTs are categorized into three main types based on the number of tubes present in the structure: single-walled CNTs (SWCNTs), double-walled CNTs (DWCNTs), and multi-walled CNTs (MWCNTs). CNTs tend to have low dispersibility in solvents due to their hydrophobic characteristics and strong van der Waals interactions which leads to aggregation, limited surface accessibility, and reduced adsorption capacity for contaminants.

2.2.1 Synthesis and modification

CNTs could be produced with desirable characteristics needed for specific applications using different techniques like carbon-arc discharge, laser ablation, sonochemical or hydrothermal, electrolysis, and chemical vapor deposition. In general, those techniques that produce CNTs with the least structural and chemical defects are favored. The non-polar nature of graphene layers makes the CNTs inherently hydrophobic. Functionalization of CNTs is achieved through covalent bonding to the π-conjugated network and the attachment of various chemical groups via non-covalent bonds such as hydrogen bonding, π-π interactions, and van der Waals interactions, as shown in Fig. 14.2 [25]. The large surface area of CNTs leads to undesirable agglomeration which reduces its effectiveness. The modification of CNT is possible through physical and chemical processes such as oxidation, impregnation, or grafting, which results in the formation of covalent or non-covalent bonds with the specific functional group on the end or sidewall of the CNTs [26]. The non-covalent functionalization is favored over covalent bonding because it does not disrupt the structure or surface area of the CNTs [27]. Surface functionalization of CNTs through covalent modifications can stabilize the dispersion of CNTs and improve their interaction with the polymeric matrix. The chemical functionalization of MWCNTs is done through oxidative treatments using

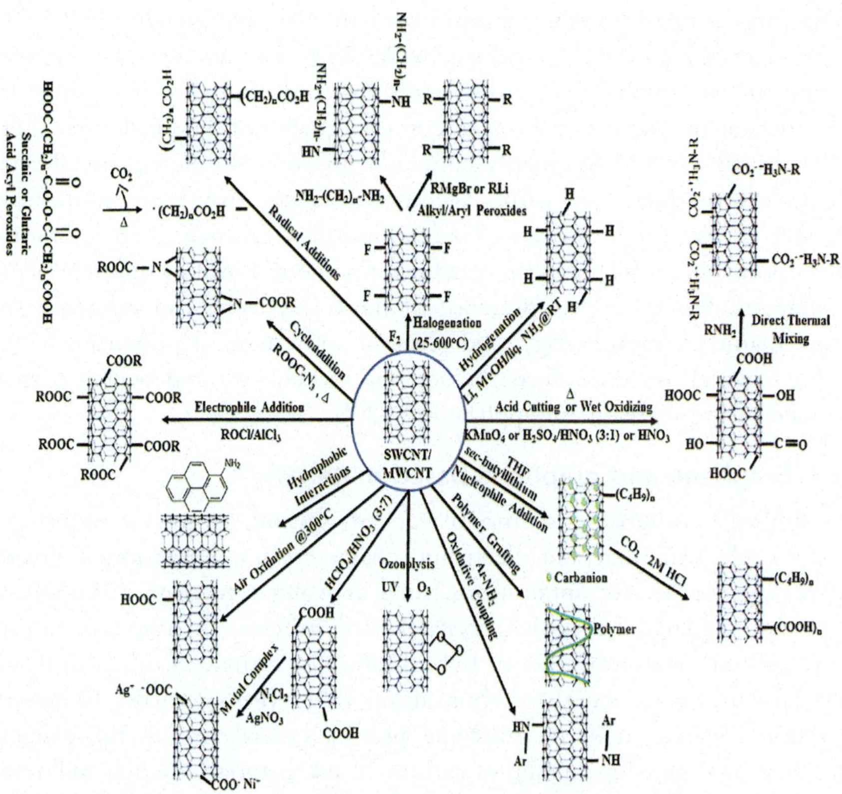

Figure 14.2 Various routes for the surface modification of CNTs [24].

strong acids. The oxidative treatments introduce hydroxyl (-OH) and carboxyl (-COOH) functional groups on the CNT surface, which enhances their dispersibility and adsorption properties. Heavy metals like Pb^{+2} and Zn^{+2} removal from aqueous solution through adsorption with MWCNTs can be enhanced through functionalization of MWCNTs with nitrogen [28].

2.2.2 Limitations

The CNT-based water purification technologies need to be economically viable at an industrial scale, and the production of CNTs needs to be cost-effective without compromising its quality. The raw material costs and the manufacturing process play a vital role in determining the overall viability of the technology. Catalytic chemical vapor deposition is recognized as a viable strategy for facilitating the large-scale manufacturing of MWCNTs.

The process provides better control over the diameter, length, and structural characteristics of the produced MWCNTs. Factors such as the choice of precursors, catalysts, reaction conditions, and post-processing steps are the influencing factors in the final cost and quality of the produced CNTs [24]. Pristine CNTs frequently contain a mixture of different metal catalysts, ash residues, and carbonaceous substances that were introduced during the synthesis process. These impurities can act as an additional adsorption site, which can be beneficial for the removal of multiple water pollutants. However, the presence of these impurities can influence the pore diameter, morphology, and physical and chemical properties of the CNTs, which in turn affects the adsorption behavior and overall performance in water treatment applications [29].

2.3 Graphene and graphene-based materials

Graphene, comprising a monolayer of carbon atoms, demonstrates remarkable structural and electronic characteristics, rendering it greatly appealing for diverse applications, such as water remediation. Graphene possesses a planar hexagonal crystal lattice structure in two dimensions (2D), which arises from the sp^2 hybridization of a single 2s orbital and two 2p orbitals located on the carbon atoms [30]. Moreover, the hexagonal structural arrangement of graphene plays a crucial role as the primary foundational unit for a range of carbon-based materials, such as fullerenes (0D), CNT (1D), and graphite (3D) [31]. The 2D nanostructure and the unique electronic properties of graphene make it a promising material for water purification applications. These 2D nanomaterials exhibit high surface area and the ability to functionalize with various moieties enables its adsorption capacity and selectivity toward various waterborne pollutants. Graphene oxide (GO) and reduced graphene oxide (rGO) represent two fundamental derivatives of graphene, each possessing unique characteristics and potential applications in water purification. GO is an oxidized form of graphene, where the graphene sheets are intercalated with various functional groups such as hydroxyl (-OH), epoxy (-O-), and carboxyl (-COOH) groups. The introduction of these moieties on graphene during the oxidation process increases the hydrophilicity and dispersibility of GO in the aqueous media. Furthermore, the functional groups present on GO provide additional active sites for the adsorption of various water-borne pollutants, such as heavy metals, organic compounds, and water-borne pathogens [32]. rGO is synthesized by the reduction of GO, which involves the elimination or reduction of the oxidized functional groups. The process of reduction can be carried out using chemical, thermal, or electrochemical techniques, which restores the hexagonal graphene-like structure. In addition, the elimination of functional groups improves the electrical conductivity and mechanical properties and reduces hydrophilicity as compared to GO [33].

2.3.1 Synthesis and modification

Graphene is synthesized through two strategies i.e., top-down and bottom-up. Top-down synthesis includes solid-phase and liquid-phase exfoliation and oxidative-reduction techniques. While bottom-up synthesis includes thermal pyrolysis, chemical vapor deposition, epitaxial growth, and chemical synthesis [34]. The functionalized graphene exhibits higher adsorption due to its high surface area and enhanced electrostatic interactions with the adsorbates. Additionally, the π–π interactions between the interlayers of graphene and adsorbates contribute toward the enhanced adsorption capacity. In general, pristine graphene tends to agglomerate, and have poor solvent dispersion and incompatibility due to strong interplanar interactions, which limits the potential application of pristine graphene. To address the agglomeration and compatibility of graphene, surface modification of graphene using various polymers and functional groups is employed [35]. Graphene oxide (GO) is enhanced by the addition of hydrophilic groups, which also enable additional modification with inorganic and organic materials. This versatility makes GO suitable for applications in fuel cells, charge storage, water purification, photocatalysis, and antimicrobial processes. GO can be transformed into rGO by removing the oxygen-containing functional groups through chemical, thermal, or electrochemical methods, resulting in a conjugated sp^2 hybridized structure [31]. The abundant functional groups present in GO can interact with heavy metal ions through ion exchange, surface complexation, and chelation. These interactions and high specific surface area enhance the adsorption properties of GO for the removal and binding of heavy metal ions. Pure Fe_3O_4 nanoparticles have been widely used for wastewater treatment due to their high specific surface area, biocompatibility, and magnetic characteristics. However, Fe_3O_4 nanoparticles tend to aggregate due to their size and susceptibility to oxidation, forming α/γ-Fe_3O_4. To address this issue, Fe_3O_4 has been combined with GO to create a magnetic graphene/iron oxide composite, which has the synergistic effect of both Fe_3O_4 and GO in terms of sorption capacity and separation efficiency [36].

2.3.2 Limitations

Graphene and graphene-based materials have shown promising properties for water treatment applications due to their high surface area, durability, and adsorption capabilities [37]. However, they do face several limitations that hinder their widespread adoption, particularly in wastewater treatment. Graphene and graphene-based materials are susceptible to agglomeration and restacking of the nanosheets, which reduces their effective surface area and adsorption capacity. The synthesis and large-scale production of high-quality graphene and graphene-based materials remains a challenge, especially in terms of maintaining consistent properties and ensuring cost-effectiveness [38]. Mixed matrix membranes fabricated from graphene are susceptible to fouling, especially in complex wastewater systems containing a wide spectrum of pollutants including organic matter and, salts [39]. Overcoming these limitations through innovative approaches is crucial for maximizing the efficiency and sustainability of graphene and graphene-based materials in water remediation.

3. Composite materials for enhanced performance

Carbon is a unique and versatile element that can form a wide range of molecular structures and nanomaterials, each with its own set of distinctive properties. The electronic configuration of carbon allows it to form strong covalent bonds with a variety of other elements, both metal and nonmetals. This ability to form a diverse range of bonding is the foundation for a wide spectrum of carbon-based materials. At the nanoscale, carbon-based materials exhibit exceptional properties in terms of mechanical strength, electrical conductivity, and thermal stability, which makes them highly attractive for a variety of applications, especially in environmental remediation [40]. The main types of carbon nanomaterials include activated carbon, carbon nanotubes, graphite, and graphene. The development of polymer composites incorporating carbon nanomaterials has gained significant interest, particularly in the field of water treatment and desalination. Polymer-carbon composites possess several remarkable properties that make them highly promising for wastewater treatment applications. The incorporation of carbon-based materials such as activated carbon, carbon nanotubes, and graphene into the polymer matrix enhances the aqueous solubility of the composite material. Furthermore, carbon-based materials inherently possess a high surface area-to-volume ratio which is further amplified when they are incorporated into the polymer matrix. The enhanced surface area of these composite materials is beneficial for adsorption, catalysis, and other water treatment processes that rely on surface interactions [41]. The integration of carbon-based materials into the polymer matrix significantly enhances thermal and mechanical properties which is crucial for the longevity and performance of the composite in water treatment applications. Moreover, the large surface area, adsorption capacity, and tailored functionalities of these carbon-based materials allow the removal of both organic and inorganic pollutants from wastewater [42].

3.1 Polymer-based activated carbon composites

Activated carbon is an amorphous, carbonaceous material with a graphitic lattice structure. Activated carbon is an attractive material for wastewater treatment and desalination due to its low cost, facile production from agricultural waste, high surface area, and porous structure. The combination of activated carbon and polymer is a promising approach, as it can address some of the limitations associated with the use of standalone activated carbon. Activated carbon is incorporated into a cellulose triacetate matrix through an evaporation/precipitation technique for the elimination of heavy metals from wastewater. The composite formation immobilizes the activated carbon, making it easier to separate the adsorbent from the treated water. To overcome the aggregation of activated carbon within the cellulose triacetate matrix, a low solvent diffusion process at a low casting temperature is employed. The fabricated activated carbon-cellulose triacetate composite membrane exhibits mechanical stability which ensures its handling and long-term performance. The composite membrane demonstrates exceptional arsenic acid separation from the aqueous solution [43]. In another study polypyrrole-activated carbon composite was fabricated through a carbonization process for the removal of lead (II) ions from the aqueous media. Polypyrrole was combined with activated carbon in the presence of a chemical activator. The adsorption capacity of the polypyrrole-activated carbon composite improved with the increasing contact time and higher lead ion concentrations. The composite material exhibited a combination of monolayer adsorption and multilayer adsorption mechanisms. The study revealed a maximum adsorption capacity of 50 mg/g for Pb^{+2} ions, which was achieved after 4 h of contact time at the pH of 5.5. The adsorption process of polypyrrole-activated carbon is dominated by chemisorption rather than a diffusion-controlled mechanism, which implies the lead ions form a strong covalent bond with the functional groups or active sites present on the composite material [44].

3.2 Polymer-based carbon nanotube composites

CNTs have a one-dimensional cylindrical nanostructure with an extraordinary graphitic nature providing a high surface area which makes them an ideal candidate for water treatment and desalination applications. Despite their excellent properties, pristine CNTs lack surface functional groups and exhibit poor solvent dispersibility. To overcome these limitations of pristine CNTs, surface functionalization is adopted to introduce specific functional groups on the CNT surface. The integration of CNTs into composite membranes is a common approach to address various challenges in water treatment including separation of targeted pollutants and membrane biofouling.

3.2.1 Oil separation

The composite membrane was fabricated using a copolymer of poly(N-isopropyl acrylamide)-co-(acrylamide) and single-walled carbon nanotubes (SWCNTs). The hydrophobic nature of CNTs is employed to selectively interact and capture the oil phase in the water-oil mixture. The fabricated composite membrane demonstrated a 99.99% exclusion of oil from the oil-water mix [45]. The polydopamine functionalized SWCNT also exhibits photo-thermal response which increases the water flux through the membrane upon illumination, as shown in Fig. 14.3.

In another study, MWCNTs are incorporated in a block copolymer membrane of polysulfone/polyether. The oil separation from the water-oil mixture of the membrane improved from 91.4% to 99.79% upon incorporation of MWCNTs, while the permeate flux of the membrane also increased up to three times. The improved oil rejection and permeate flux upon incorporation of MWCNTs is attributed to the high surface area, hydrophobicity, and tortuosity for the selective transport of water [46]. The studies highlighting the application of CNTs in polymeric membranes for oil-water separation highlight the promising prospects of CNTs in enhancing the performance and functionality of these membranes.

3.2.2 Heavy metal separation

Functionalization of CNTs with polymers offers enhanced adsorption properties and overall performance. The functionalization of CNTs (physical or chemical) is an effective approach to enhance the adsorption properties, hydrophilicity, and water solubility. One of the key applications of polymer functionalized CNTs is their application as an efficient adsorbent for the removal of heavy metal ions, such as As, Cr, Pb, Cd, Cu,

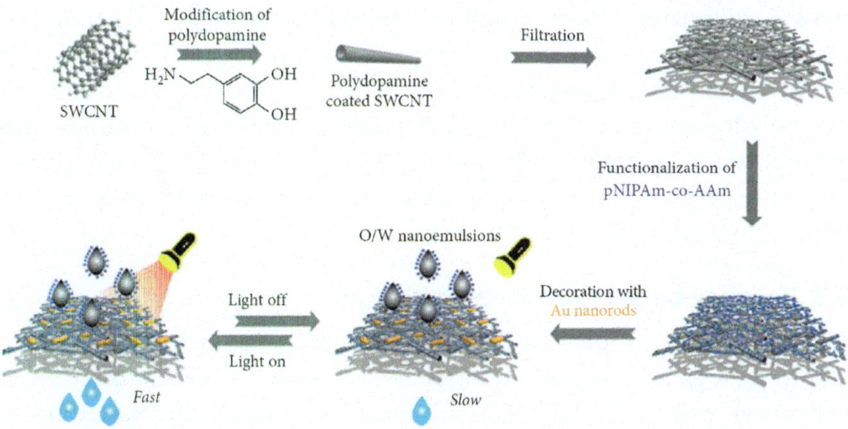

Figure 14.3 Schematic illustration of the synthesis of SWCNTs integration in copolymer membrane of poly(N-isopropyl acrylamide)-co-(acrylamide) for oil separation from water-oil mixture [45].

Ni, Zn, and Fe from wastewater. CNT grafted with polyhydroxybutyrate demonstrated exceptional adsorption of heavy metal ions from industrial wastewater through batch adsorption process. The adsorption mechanism involves ion exchange and electrostatic interactions between the polymer-functionalized CNT and the heavy metal ions [47]. In another study, CNTs coated with polyamidoamine as the polymer functionalization. The polymer-coated CNTs demonstrated an exceptionally high adsorption capacity of 432 mg/g for As, 494 mg/g for Co, and 470 mg/g for Zn ions. The adsorption mechanism involves a combination of ion exchange, chelation, and electrostatic interaction between polyamidoamine-coated CNTs and metal ions [48].

3.2.3 Dye separation

Composite research on CNTs has received significant attention for improving their surface characteristics and enhancing the adsorption of organic pollutants, particularly organic dyes. The polymer coatings or grafting can introduce additional functional groups, increase the surface area, and improve the interactions between the CNTs and the organic dyes. The CNTs coated with dopamine self-polymerize to produce a polydopamine-CNT composite. The polydopamine coating serves as an intermediary layer to facilitate the subsequent grafting of poly(sodium-p-styrene sulfonate). The composite showed remarkable adsorption of methylene blue dye, with an adsorption capacity of 174 mg/g. The improved adsorption performance is attributed to the synergistic effects of the polydopamine and poly(sodium-p-styrene sulfonate) coatings. The polydopamine provides additional binding sites and functional groups, while the poly(sodium-p-styrene sulfonate) enhances the interactions with the organic dye molecules [49].

3.3 Polymer-based graphene composites

Graphene, a 2D carbon-based material consisting of a single layer of carbon atoms, has gained significant attention due to its exceptional physiochemical properties, making it an attractive material for numerous applications, including wastewater treatment. Graphene-based composites can possess a three-dimensional (3D) structure, which provides higher porosity, and a larger specific surface area as compared to individual graphene sheets. These composite materials exhibit excellent mechanical, electrical conductivity, and thermal stability which contribute to the overall performance and robustness of the adsorbent materials. Its unique structural and inherent properties provide an advantage to the composite enabling it to achieve higher adsorption capacity for both organic and inorganic pollutants. Graphene-based composites can be fabricated into membranes for use in ultrafiltration systems. Graphene combined with sodium alginate-based nanocomposite adsorbent possesses a well-aligned structure, with the inclusion of defects and free-volume cavities. These free-volume cavities are crucial for the separation performance and high permeate flux, as shown in Fig. 14.4. The structural defects and free-volume cavities with the graphene-sodium alginate nanocomposite play a vital role in enhancing the selectivity and permeability of the water channels [50].

The membrane was fabricated by grafting reduced graphene oxide with a polydopamine polymer. The membrane exhibited a high separation efficiency of oil from water, reaching up to 99.6%. This remarkable performance is attributed to the hydrophilic nature of the rGO and the superhydrophobic properties of the polydopamine polymer. The reduced graphene-based composite membrane showed improved chemical stability under harsh operating conditions (acidic and high salt concentrations) [51].

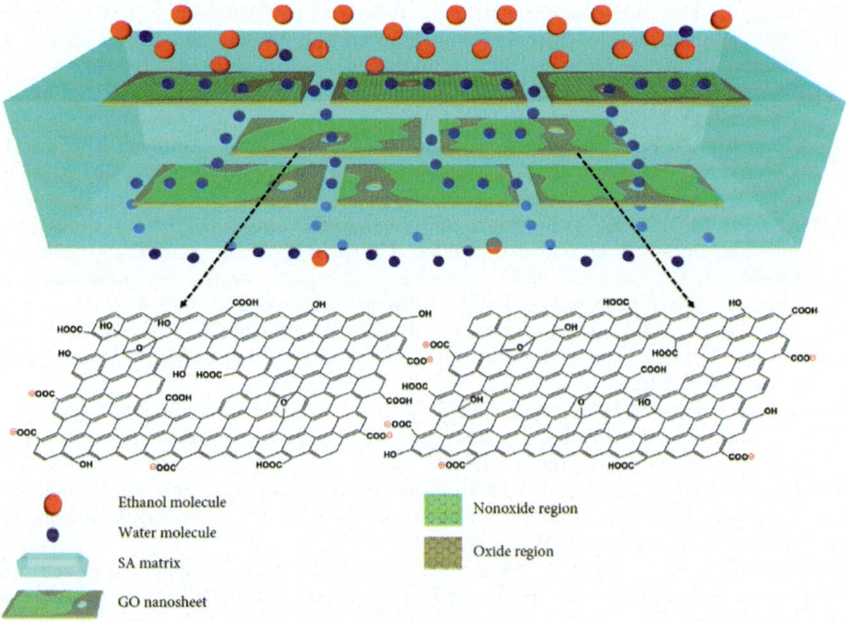

Figure 14.4 Schematic illustration of the separation mechanism of selective water permeation through graphene-sodium alginate [50].

References

[1] Gu, S., Kang, X., Wang, L., Lichtfouse, E., Wang, C., 2019. Clay mineral adsorbents for heavy metal removal from wastewater: A review. Environ. Chem. Lett. 17 (2), 629–654.

[2] Lau, W.-J., Emadzadeh, D., Shahrin, S., Goh, P.S., Ismail, A.F., 2018. Ultrafiltration membranes incorporated with carbon-based nanomaterials for antifouling improvement and heavy metal removal. In: Carbon-based polymer nanocomposites for environmental and energy applications. Elsevier, pp. 217–232.

[3] Bashir, A., Malik, L.A., Ahad, S., Manzoor, T., Bhat, M.A., Dar, G.N., Pandith, A.H., 2019. Removal of heavy metal ions from aqueous system by ion-exchange and biosorption methods. Environ. Chem. Lett. 17 (2), 729–754.

[4] Chowdhury, Z.Z., Sagadevan, S., Johan, R.B., Shah, S.T., Adebesi, A., Md, S.I., Rafique, R.F., 2018. A review on electrochemically modified carbon nanotubes (CNTs) membrane for desalination and purification of water. Mater. Res. Express 5 (10), 102001.

[5] Smith, S.C., Rodrigues, D.F., 2015. Carbon-based nanomaterials for removal of chemical and biological contaminants from water: A review of mechanisms and applications. Carbon N Y 91, 122–143.

[6] Zhou, X., Wang, Y., Gong, C., Liu, B., Wei, G., 2020. Production, structural design, functional control, and broad applications of carbon nanofiber-based nanomaterials: A comprehensive review. Chem. Eng. J. 402, 126189.

[7] Qiu, B., Xing, M., Zhang, J., 2018. Recent advances in three-dimensional graphene based materials for catalysis applications. Chem. Soc. Rev. 47 (6), 2165–2216.

[8] Wang, L., Hou, D., Cao, Y., Ok, Y.S., Tack, F.M.G., Rinklebe, J., O'Connor, D., 2020. Remediation of mercury contaminated soil, water, and air: A review of emerging materials and innovative technologies. Environ. Int. 134, 105281.
[9] Jung, S., Park, Y.-K., Kwon, E.E., 2019. Strategic use of biochar for CO2 capture and sequestration. J. CO2 Util. 32, 128–139.
[10] Gopinath, K.P., Vo, D.-V.N., Gnana Prakash, D., Adithya Joseph, A., Viswanathan, S., Arun, J., 2021. Environmental applications of carbon-based materials: A review. Environ. Chem. Lett. 19 (1), 557–582.
[11] Ahmed, M.J., 2017. Adsorption of non-steroidal anti-inflammatory drugs from aqueous solution using activated carbons: Review. J. Environ. Manag. 190, 274–282.
[12] Björklund, K., Li, L.Y., 2017. Adsorption of organic stormwater pollutants onto activated carbon from sewage sludge. J. Environ. Manag. 197, 490–497.
[13] Borchardt, L., Zhu, Q.-L., Casco, M.E., Berger, R., Zhuang, X., Kaskel, S., Feng, X., Xu, Q., 2017. Toward a molecular design of porous carbon materials. Mater. Today 20 (10), 592–610.
[14] Jjagwe, J., Olupot, P.W., Menya, E., Kalibbala, H.M., 2021. Synthesis and application of granular activated carbon from biomass waste materials for water treatment: A review. J. Bioresour. Bioprod. 6 (4), 292–322.
[15] Pan, L., Takagi, Y., Matsui, Y., Matsushita, T., Shirasaki, N., 2017. Micro-milling of spent granular activated carbon for its possible reuse as an adsorbent: Remaining capacity and characteristics. Water Res. 114, 50–58.
[16] Yamashita, A., Shikada, J., Oshima, T., Baba, Y., 2014. Preparation of activated carbon with high surface area using biomass wastes and its application to methane storage technology. Kagaku Kogaku Ronbunshu 40 (2), 98–103.
[17] Figueiredo, J.L., Pereira, M.F.R., Freitas, M.M.A., Órfão, J.J.M., 1999. Modification of the surface chemistry of activated carbons. Carbon N Y 37 (9), 1379–1389.
[18] Prahas, D., Kartika, Y., Indraswati, N., Ismadji, S., 2008. Activated carbon from jackfruit peel waste by H3PO4 chemical activation: Pore structure and surface chemistry characterization. Chem. Eng. J. 140 (1–3), 32–42.
[19] Ribas, M.C., Adebayo, M.A., Prola, L.D.T., Lima, E.C., Cataluña, R., Feris, L.A., Puchana-Rosero, M.J., Machado, F.M., Pavan, F.A., Calvete, T., 2014. Comparison of a homemade cocoa shell activated carbon with commercial activated carbon for the removal of reactive violet 5 dye from aqueous solutions. Chem. Eng. J. 248, 315–326.
[20] Sato, S., Yoshihara, K., Moriyama, K., Machida, M., Tatsumoto, H., 2007. Influence of activated carbon surface acidity on adsorption of heavy metal ions and aromatics from aqueous solution. Appl. Surf. Sci. 253 (20), 8554–8559.
[21] Santiago, M., Stüber, F., Fortuny, A., Fabregat, A., Font, J., 2005. Modified activated carbons for catalytic wet air oxidation of phenol. Carbon N Y 43 (10), 2134–2145.
[22] Allahkarami, E., Monfared, A.D., 2024. Activated carbon adsorbents for the removal of emerging pollutants and its adsorption mechanisms. In: Sustainable technologies for remediation of emerging pollutants from aqueous environment. Elsevier, pp. 79–109.
[23] Mestre, A.S., Campinas, M., Viegas, R.M.C., Mesquita, E., Carvalho, A.P., Rosa, M. J., 2022. Activated carbons in full-scale advanced wastewater treatment. In: Advanced materials for sustainable environmental remediation. Elsevier, pp. 433–475.
[24] Das, R., Abd Hamid, S.B., Ali, MdE., Ismail, A.F., Annuar, M.S.M., Ramakrishna, S., 2014. Multifunctional carbon nanotubes in water treatment: The present, past and future. Desalination 354, 160–179.
[25] Jun, L.Y., Mubarak, N.M., Yee, M.J., Yon, L.S., Bing, C.H., Khalid, M., Abdullah, E.C., 2018. An overview of functionalised carbon nanomaterial for organic pollutant removal. J. Ind. Eng. Chem. 67, 175–186.

[26] Bahgat, M., Farghali, A.A., El Rouby, W.M.A., Khedr, M.H., 2011. Synthesis and modification of multi-walled carbon nano-tubes (MWCNTs) for water treatment applications. J. Anal. Appl. Pyrolysis 92 (2), 307–313.
[27] Ali, S., Rehman, S.A.U., Luan, H.-Y., Farid, M.U., Huang, H., 2019. Challenges and opportunities in functional carbon nanotubes for membrane-based water treatment and desalination. Sci. Total Environ. 646, 1126–1139.
[28] Oyetade, O.A., Skelton, A.A., Nyamori, V.O., Jonnalagadda, S.B., Martincigh, B.S., 2017. Experimental and DFT studies on the selective adsorption of Pb 2+ and Zn 2+ from aqueous solution by nitrogen-functionalized multiwalled carbon nanotubes. Sep. Purif. Technol. 188, 174–187.
[29] Agnihotri, S., Mota, J.P.B., Rostam-Abadi, M., Rood, M.J., 2006. Adsorption site analysis of impurity embedded single-walled carbon nanotube bundles. Carbon N Y 44 (12), 2376–2383.
[30] Bello, M.M., Raman, A.A.A., 2019. Synergy of adsorption and advanced oxidation processes in recalcitrant wastewater treatment. Environ. Chem. Lett. 17 (2), 1125–1142.
[31] Donga, C., Mishra, S.B., Abd-El-Aziz, A.S., Mishra, A.K., 2021. Advances in graphene-based magnetic and Graphene-Based/TiO2 nanoparticles in the removal of heavy metals and organic pollutants from industrial wastewater. J. Inorg. Organomet. Polym. Mater. 31 (2), 463–480.
[32] Hosseinkhani, O., Hamzehlouy, A., Dan, S., Sanchouli, N., Tavakkoli, M., Hashemipour, H., 2023. Graphene oxide/ZnO nanocomposites for efficient removal of heavy metal and organic contaminants from water. Arab. J. Chem. 16 (10), 105176.
[33] Safian, M.T., Haron, U.S., Mohamad Ibrahim, M.N., 2020. A review on bio-based graphene derived from biomass wastes. Bioresources 15 (4), 9756–9785.
[34] Bhuyan, MdS.A., Uddin, MdN., Islam, MdM., Bipasha, F.A., Hossain, S.S., 2016. Synthesis of graphene. Int. Nano Lett. 6 (2), 65–83.
[35] Kumar, V., Kim, K.-H., Park, J.-W., Hong, J., Kumar, S., 2017. Graphene and its nanocomposites as a platform for environmental applications. Chem. Eng. J. 315, 210–232.
[36] Liu, X., Ma, R., Wang, X., Ma, Y., Yang, Y., Zhuang, L., Zhang, S., Jehan, R., Chen, J., Wang, X., 2019. Graphene oxide-based materials for efficient removal of heavy metal ions from aqueous solution: A review. Environ. Pollut. 252, 62–73.
[37] Chatterjee, D., Sivashanmugam, K., 2024. Utilization of graphene and rGO membranes for water and wastewater treatments. Phys. Sci. Rev. 9 (11), 3369–3386. https://doi.org/10.1515/psr-2023-0046.
[38] Reddy, P.R., Gogoi, A., Reddy, K.A., 2023. Cationic stabilized layered graphene oxide (GO) membrane for shale gas wastewater treatment: An atomistic insight. Desalination 559, 116621.
[39] Mohammed, M.N., Aljibori, H.S.S., Jweeg, M.J., Oqaili, F. Al, Abdullah, T.A., Abdullah, O.I., Meharban, F., Rashed, R.T., Aldulaimi, M., Al-Azawi, K., 2024. A comprehensive review on graphene oxide based nanocomposites for wastewater treatment. Pol. J. Chem. Technol. 26 (1), 64–79.
[40] Salehi, E., Heidary, F., Daraei, P., Keyhani, M., Behjomanesh, M., 2020. Carbon nanostructures for advanced nanocomposite mixed matrix membranes: A comprehensive overview. Rev. Chem. Eng. 36 (6), 723–748.
[41] Gandhi, M.R., Vasudevan, S., Shibayama, A., Yamada, M., 2016. Graphene and graphene-based composites: A rising star in water purification - A comprehensive overview. ChemistrySelect 1 (15), 4358–4385.
[42] Sabzehmeidani, M.M., Mahnaee, S., Ghaedi, M., Heidari, H., Roy, V.A.L., 2021. Carbon based materials: A review of adsorbents for inorganic and organic compounds. Mater. Adv. 2 (2), 598–627.

[43] Terrazas-Bandala, L.P., Gonzalez-Sanchez, G., Garcia-Valls, R., Gumi, T., Beurroies, I., Denoyel, R., Torras, C., Ballinas-Casarrubias, L., 2014. Influence of humidity, temperature, and the addition of activated carbon on the preparation of cellulose acetate membranes and their ability to remove arsenic from water. J. Appl. Polym. Sci. 131 (8).

[44] Alghamdi, A.A., Al-Odayni, A.-B., Saeed, W.S., Al-Kahtani, A., Alharthi, F.A., Aouak, T., 2019. Efficient adsorption of lead (II) from aqueous phase solutions using polypyrrole-based activated carbon. Materials 12 (12), 2020.

[45] Hu, L., Gao, S., Ding, X., Wang, D., Jiang, J., Jin, J., Jiang, L., 2015. Photothermal-responsive single-walled carbon nanotube-based ultrathin membranes for On/Off switchable separation of oil-in-water nanoemulsions. ACS Nano 9 (5), 4835–4842.

[46] Saadati, J., Pakizeh, M., 2017. Separation of oil/water emulsion using a new PSf/pebax/F-MWCNT nanocomposite membrane. J. Taiwan Inst. Chem. Eng. 71, 265–276.

[47] Bankole, M.T., Abdulkareem, A.S., Mohammed, I.A., Ochigbo, S.S., Tijani, J.O., Abubakre, O.K., Roos, W.D., 2019. Selected heavy metals removal from electroplating wastewater by purified and polyhydroxylbutyrate functionalized carbon nanotubes adsorbents. Sci. Rep. 9 (1), 4475.

[48] Hayati, B., Maleki, A., Najafi, F., Gharibi, F., McKay, G., Gupta, V.K., Harikaranahalli Puttaiah, S., Marzban, N., 2018. Heavy metal adsorption using PAMAM/CNT nanocomposite from aqueous solution in batch and continuous fixed bed systems. Chem. Eng. J. 346, 258–270.

[49] Xie, Y., He, C., Liu, L., Mao, L., Wang, K., Huang, Q., Liu, M., Wan, Q., Deng, F., Huang, H., Zhang, X., Wei, Y., 2015. Carbon nanotube based polymer nanocomposites: Biomimic preparation and organic dye adsorption applications. RSC Adv. 5 (100), 82503–82512.

[50] Cao, K., Jiang, Z., Zhao, J., Zhao, C., Gao, C., Pan, F., Wang, B., Cao, X., Yang, J., 2014. Enhanced water permeation through sodium alginate membranes by incorporating graphene oxides. J. Membr. Sci. 469, 272–283.

[51] Liu, N., Zhang, M., Zhang, W., Cao, Y., Chen, Y., Lin, X., Xu, L., Li, C., Feng, L., Wei, Y., 2015. Ultralight free-standing reduced graphene oxide membranes for oil-in-water emulsion separation. J. Mater. Chem. A Mater. 3 (40), 20113–20117.

Index

'*Note:* Page numbers followed by "f" indicate figures and "t" indicate tables.'

B

C

D

I

J

K

L

M

R

S

Printed and bound by CPI Group (UK) Ltd, Croydon, CR0 4YY
14/09/2026
02216408-0005